T0319276

TRANSLATIONAL CARDIOMETABOLIC GENOMIC MEDICINE

TRANSLATIONAL CARDIOMETABOLIC GENOMIC MEDICINE

Edited by

ANNABELLE RODRIGUEZ-OQUENDO

AMSTERDAM • BOSTON • CAMBRIDGE • HEIDELBERG
LONDON • NEW YORK • OXFORD • PARIS • SAN DIEGO
SAN FRANCISCO • SINGAPORE • SYDNEY • TOKYO
Academic Press is an imprint of Elsevier

Academic Press is an imprint of Elsevier
125 London Wall, London EC2Y 5AS, UK
525 B Street, Suite 1800, San Diego, CA 92101-4495, USA
225 Wyman Street, Waltham, MA 02451, USA
The Boulevard, Langford Lane, Kidlington, Oxford OX5 1GB, UK

ISBN: 978-0-12-799961-6

British Library Cataloguing-in-Publication Data
A catalogue record for this book is available from the British Library

Library of Congress Cataloging-in-Publication Data
A catalog record for this book is available from the Library of Congress

For information on all Academic Press publications
visit our website at http://store.elsevier.com/

Typeset by TNQ Books and Journals
www.tnq.co.in

Printed and bound in the United States of America

Contents

7. The Genetics of Obesity
RODRIGO ALONSO, MAGDALENA FARÍAS, VERONICA ALVAREZ, ADA CUEVAS

8. The Epidemiology and Genetics of Vascular Dementia: Current Knowledge and Next Steps
ADAM NAJ

9. Genomic Medicine and Ethnic Differences in Cardiovascular Disease Risk
ALEXIS C. FRAZIER-WOOD, STEPHEN S. RICH

10. Genomics-Guided Immunotherapy of Human Epithelial Ovarian Cancer
SAHAR AL SEESI, FEI DUAN, ION I. MANDOIU, PRAMOD K. SRIVASTAVA, ANGELA KUECK

11. Overview of the Intersection of Genomics of Cholesterol Metabolism and Cardiometabolic Disease with Reproductive Health, Especially in Women

ANTHONY M. DEANGELIS, MEAGHAN ROY-O'REILLY,
ANNABELLE RODRIGUEZ-OQUENDO

12. Overview of Intersection of Genomics of Cardiometabolic Disease and Other Disease States, Such as Eye Health (Macular Degeneration)

GARETH J. MCKAY

13. Translating Genomic Research to the Marketplace
STEVEN MYINT

Contributors

Rodrigo Alonso Obesity and Lipid Units, Department of Nutrition, Clínica Las Condes, Santiago de Chile, Chile

Sahar Al Seesi Department of Computer Science & Engineering, University of Connecticut, Storrs, CT, USA; Department of Immunology, Carole and Ray Neag Comprehensive Cancer Center, University of Connecticut School of Medicine, Farmington, CT, USA

Veronica Alvarez Obesity and Lipid Units, Department of Nutrition, Clínica Las Condes, Santiago de Chile, Chile

Thomas E. Cheatham, III Department of Medicinal Chemistry, L.S. Skaggs Pharmacy Institute, University of Utah, Salt Lake City, UT, USA

Ada Cuevas Obesity and Lipid Units, Department of Nutrition, Clínica Las Condes, Santiago de Chile, Chile

Anthony M. DeAngelis University of Connecticut Health Center, Farmington, CT, USA

Fei Duan Department of Immunology, Carole and Ray Neag Comprehensive Cancer Center, University of Connecticut School of Medicine, Farmington, CT, USA

Charles R. Farber Departments of Public Health Sciences and Biochemistry and Molecular Genetics, Center for Public Health Genomics, University of Virginia, Charlottesville, VA, USA

Magdalena Farías Obesity and Lipid Units, Department of Nutrition, Clínica Las Condes, Santiago de Chile, Chile

Alexis C. Frazier-Wood Department of Pediatrics, USDA/ARS Children's Nutrition Research Center, Baylor College of Medicine, Houston, TX, USA

Lita A. Freeman Lipoprotein Metabolism Section, Cardiopulmonary Branch, National Heart, Lung and Blood Institute, Bethesda, MD, USA

Rodrigo Galindo-Murillo Department of Medicinal Chemistry, L.S. Skaggs Pharmacy Institute, University of Utah, Salt Lake City, UT, USA

David Herrington Department of Internal Medicine, Section of Cardiology, Wake Forest School of Medicine, Winston-Salem, NC, USA

Angela Kueck Carole and Ray Neag Comprehensive Cancer Center, University of Connecticut School of Medicine, Farmington, CT, USA

Ion I. Mandoiu Department of Computer Science & Engineering, University of Connecticut, Storrs, CT, USA

Gareth J. McKay Centre for Public Health, Queen's University Belfast, Belfast, Northern Ireland, UK

Larry D. Mesner Departments of Public Health Sciences and Biochemistry and Molecular Genetics, Center for Public Health Genomics, University of Virginia, Charlottesville, VA, USA

Steven Myint Center for Enterprise and Development, Duke-NUS Medical School, Singapore; Nanyang Business School, Nanyang Technological University, Singapore; Inex Private Ltd, Singapore; Plexpress Oy, Finland

Adam Naj Department of Biostatistics and Epidemiology, Center for Clinical Epidemiology and Biostatistics, Perelman School of Medicine, University of Pennsylvania, Philadelphia, PA, USA

Waqas Qureshi Department of Internal Medicine, Section of Cardiology, Wake Forest School of Medicine, Winston-Salem, NC, USA

Theodore P. Rasmussen Department of Pharmaceutical Sciences, University of Connecticut, Storrs, CT, USA; University of Connecticut Stem Cell Institute, University of Connecticut, Storrs, CT, USA; Department of Molecular and Cell Biology, University of Connecticut, Storrs, CT, USA

Alan T. Remaley Lipoprotein Metabolism Section, Cardiopulmonary Branch, National Heart, Lung and Blood Institute, Bethesda, MD, USA

Stephen S. Rich Center for Public Health Genomics, University of Virginia, Charlottesville, VA, USA

Annabelle Rodriguez-Oquendo University of Connecticut Health Center, Farmington, CT, USA

Meaghan Roy-O'Reilly University of Connecticut Health Center, Farmington, CT, USA

Pramod K. Srivastava Department of Immunology, Carole and Ray Neag Comprehensive Cancer Center, University of Connecticut School of Medicine, Farmington, CT, USA; Carole and Ray Neag Comprehensive Cancer Center, University of Connecticut School of Medicine, Farmington, CT, USA

Kasey C. Vickers Department of Medicine, Vanderbilt University, School of Medicine, Nashville, TN, USA

Metabolomics and Cardiovascular Medicine

Waqas Qureshi, David Herrington

Department of Internal Medicine, Section of Cardiology, Wake Forest School of Medicine, Winston-Salem, NC, USA

1. INTRODUCTION

Cardiovascular disease is a complex, multifactorial disease that currently affects 1.5 billion people worldwide [1]. Many causative pathways through which risk factors act are still unclear. Metabolomics has recently emerged as a promising field that may help elucidate intricate details of the relationships between changes in human biology and complex cardiovascular disease phenotypes.

Metabolites are the end-products of metabolic processes occurring in cellular organelles. These are molecules smaller than 1 kDa (1 Da = 1.66×10^{-27} kg). Historically, studies of metabolism have typically focused on a narrow range of metabolites or specific metabolic pathways. In contrast, metabolomics generally involves a more comprehensive assessment of many metabolites and pathways. The Metabolomics Society defines metabolomics as "comprehensive characterization of the small molecule metabolites in biological systems. It can provide an overview of the metabolic status and global biochemical events associated with a cellular or biological system" [2]. Similarly, Nicholson [3] defined metabolomics as a "quantitative measurement of the dynamic multi-parametric metabolic response of living systems to pathophysiologic stimuli or genetic modification..." [3]. The term "metabolome" was first coined by Oliver et al. [4] and Tweeddale et al. [5] in 1998. Fiehn et al. used the term "metabolomics" for the first time in 2001 [6]. Although the term was initially used in the literature pertaining to plants and agricultural research, with advancement in analytical tools and sample extraction methods, human and animal medical fields have also embraced these terms and approaches.

1

A major advantage of studying the metabolome is that metabolites are generally considered to be molecular phenotypes much closer to clinical traits of interest—reflecting the integrated effects of both upstream molecular signals (e.g., genome, epigenome, transcriptome, and proteome) and environmental factors (e.g., diet, psychological stressors, microbiome, and environmental exposures—collectively referred to as the exposome). This concept was built on central dogma from the 1950s that there is a linear and unidirectional flow of information from genome to phenome. More recently, this concept has evolved into a complex network-based framework to account for multiple interconnected factors at all levels that determine a clinical phenotype. Nevertheless, the metabolome remains at the intersection of many factors that ultimately reflect or determine the health of an individual (Figure 1). Thus, it is essential to understand both the metabolome and its interactions with other -omics to apply metabolomics to practice.

The application of this concept in clinical and population research is still relatively new. However, already many important insights about the relationship between the metabolome and clinical cardiovascular disease have emerged, and several studies have identified key genome—metabolome associations that better characterize the molecular signals contributing to pathogenesis of atherosclerosis, hypertension, and other cardiovascular conditions. In this chapter, we provide an overview

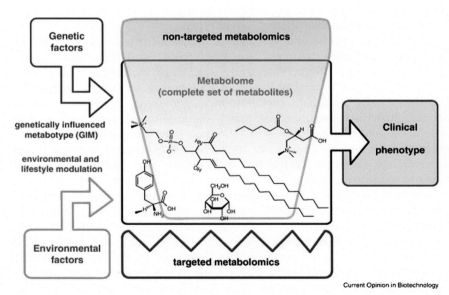

Current Opinion in Biotechnology

FIGURE 1   System biology network. *Adapted from Adamski et al. Curr Opin Biotechnol 2013; 24:39—47.*

of metabolomics trends and techniques, and review current understanding of the relationships among the metabolome, the genome, and the cardiovascular disease.

2. METABOLOMIC METHODS

A metabolomic study is composed of several discrete steps including bio-sampling, separation, metabolite detection, and data analysis (Figure 2). Each step is critically important to ensure valid and interpretable results.

2.1 Bio-specimen Selection and Handling

Particular attention should be given to the type of bio-sample and sampling techniques, as these determine the types of metabolites that need to be studied (Table 1). While collecting bio-samples, care must be taken to avoid degradation and contamination of collected sample. Some necessary considerations for sampling include timing (e.g., 24-h sampling versus collection after a 12- to14-h fasting period), sudden dietary changes (potentially interfere with the metabolome for over a week), bacterial or fungal contamination (especially for urine samples), medication intake (metabolites of medications may change the metabolome), diurnal variation in the hormones (anabolic and catabolic hormones vary during the day and may alter the metabolome), randomization, and transport and storage conditions of samples.

Serum is generally the preferred blood bio-specimen for metabolomic studies. This is because plasma retains fibrinogen and other coagulation cascade proteolytic enzymes that can continue to be active ex vivo and alter the metabolome or interfere with metabolite detection. Furthermore, ethylenediaminetetraacetic acid is typically used as the anticoagulant in plasma samples and can denature chromatograms and proteins in a way that may influence results. Serum has fewer potentially interfering coagulation cascade enzymes and proteins. However, even in serum there remain active enzymes that can continue to modify the metabolome ex vivo. To counteract these enzymatic processes, serum can be collected on ice. It is recommended to use samples stored at $-20\ °C$ within 7 days and samples stored at $-80\ °C$ within 1 month. However, in one report, changes in the metabolome were negligible after storage for 2.5 years [7]. In the case of repeated usage, fewer than three freeze—thaw cycles is advisable [7].

Until now, urine has been the most common sample for human metabolomics studies. It has many advantages, such as a noninvasive method of collection, lack of special preparation for collection, almost no

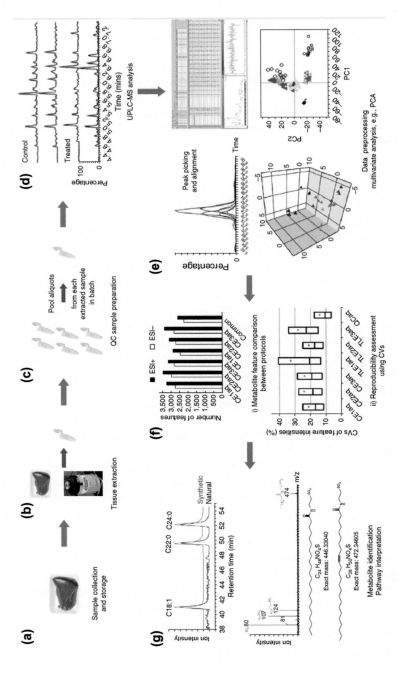

FIGURE 2 Schematic view of steps of a metabolomic study. *Adapted from Want et al. Nature Protoc 2013; 8:17−32.*

TABLE 1   Sample Types Used in Metabolomics Studies

Sample	Advantage	Disadvantage	Approximate volume
Plasma/ serum	Minimally invasive, readily available, provides information of metabolic footprint of many metabolic reactions	May not reflect specific tissue level changes (e.g., cardiac changes) Tissue level metabolic fluxes cannot be measured Requires deproteination of mass spectrometer Anticoagulant used for plasma collection may interfere with the endogenous metabolites	150—550 μL
Urine	Noninvasive, readily available, provides information of metabolic footprint of many metabolic reactions	May not reflect specific tissue level changes (e.g., cardiac changes) Tissue level metabolic fluxes cannot be measured Contains high amount of urea that can damage mass spectrometer Differences in pH of urine leads to difficulty in evaluation of various metabolites especially when nuclear magnetic resonance spectroscopy was used	500—1000 μL
Primary cell culture	Tissue level information can be obtained, metabolic fluxes can be studied	Takes time, expensive, technically demanding, and requires complex procedures, phenotype may change over time	100—600 μL
Tissue	Tissue level information can be obtained, metabolic fluxes can be studied, accurate description of phenotype	Takes time, expensive, technically demanding, and requires complex procedures, invasive for some of the specific tissues	>20 mg

sample pretreatment, and lower protein content, which helps to increase the sensitivity of identifying other metabolites [8]. Urine, like blood or plasma, provides a metabolomic "footprint" of the whole body, not necessarily a single organ. Usually a 24-h urine sample is collected for metabolomics studies; this method requires that participants receive certain instructions and follow specific procedures. Improper procedures may lead to bacterial overgrowth secondary to either contamination or infection that can affect the urine metabolome. Many other types of samples have also been used in metabolomics research (e.g., cell lysates and spinal fluid), although these types of bio-samples are less amenable to large-scale clinical and population research.

Prior to sample separation, typical sample preparation steps are required. Standardization of these sample preparation steps is pivotal to avoid random or, even worse, systematic bias in the results. For example, alcohol-based extraction is a common procedure employed in sample extraction. However, the pyruvate pathway is up-regulated when boiling ethanol or freeze—thaw extraction is used, but down-regulated when cold methanol is used [9]. The ideal method should be highly reproducible, simple, rapid, unselective, and include a metabolism-quenching step. The latter is not frequently used in human studies, but is important because many active enzymes may influence metabolites in the collected sample (especially in plasma or serum samples). If the enzymes continue to remain active, they may change the concentrations of metabolites of interest. Special attention should be paid to this matter when evaluating metabolites that are affected by rate-limiting steps, for example, glucose-6-phosphate or ATP.

2.2 Metabolite Separation Methods

Chromatography is the principal separation technique used to obtain fine resolution of metabolites for metabolomics studies. It plays a key role in obtaining analytical data needed in metabolic profiling. The separation of metabolites is achieved by interactions of analytes with solvent and a variety of stationary phases (e.g., solid or liquid chromatography columns and so on—see below).

The basic theory of chromatography was first described in 1903 when a chromatogram made up of calcium carbonate (stationary phase) was inoculated with a leaf extract (sample) mixed with ethanol (mobile phase). For metabolomics research, a sample (or sample extract) is dissolved in a mobile phase (liquid or gas). This mobile phase is then forced through an immiscible stationary phase. The component of sample that has an affinity for the stationary phase will travel slowly, whereas the component that repels or does not dissolve in the stationary phase is the first one to reach the other end of chromatography column. By making changes in these phases, metabolites can be separated with a high degree of resolution, and a high-quality chromatogram can be obtained.

There are a number of definitions and key concepts that are useful to keep in mind when considering metabolomic studies.

2.2.1 Retention Time

At equilibrium, an analyte in mobile phase and in stationary phase can be given as

$$M_{mobile} \rightleftharpoons M_{stationary}$$

In this equilibrium reaction, the *equilibrium constant K* is defined as the ratio of concentration (in moles) of an analyte in the stationary versus the mobile phase. This is also called the *partition coefficient*. The time needed from injection of sample to formation of peak for a given analyte is known as *retention time* (T_R). The time taken for the mobile phase to pass through column is T_M. A constant called the retention factor (k') is defined as the migration rate of the analyte through the column. This is given as

$$k' = \frac{T_R - T_M}{T_M}$$

The ideal retention factor is from 1 to 5. Too quick a retention factor may obscure the sensitivity of detection, but if the retention factor is slow, chromatography will be too time-consuming [10].

2.2.2 Solvent Strength

Another important factor to consider is the change in solvent (mobile phase) strength during a run. The term *isocratic* is used for solvents that do not change their strength during the entire run (e.g., 60:40 alcohol:water). If the solvent composition changes through the run, it is called a gradient mobile phase. The change may be gradual (linear gradient) or stepwise (step gradient) [11].

2.2.3 Resolution

Resolution (R) is the degree of separation between two analytes. It is given as

$$R = \frac{2(T_{R1} - T_{R2})}{W_1 + W_2}$$

Subscripts 1 and 2 are for analytes 1 and 2. The *selectivity factor* (α) describes the separation between the centers of two peaks of analytes. Resolution is a function of efficiency of the column, selectivity of analytes, and retention time of analytes. These relationships are analyzed using the Purnell equation, which describes the resolution between two peaks on a chromatogram as follows:

$$R = \frac{\sqrt{N_2}}{4} \left(\frac{\alpha - 1}{\alpha} \right) \left(\frac{k'_2}{1 + k'_2} \right)$$

N_2 is the number of theoretical plates in a column of the second peak (a measure of efficiency of the chromatogram), and k'_2 is the retention factor for the second peak.

Baseline resolution occurs when $R > 1.5$ is achieved. Resolution is important for peak identification and helps in accurate integration and quantification of analytes. Improving the resolution of peaks is a primary goal of chromatography. The retention factor can be changed by altering the temperature in gas chromatography (GC) or by changing the composition of the mobile phase in liquid chromatography (LC).

2.2.4 Peak Capacity

Peak capacity (P_c) is the maximum number of peaks that can be separated in a chromatogram with a resolution of 1. Gradient solvents can give higher peak capacities than isocratic solvents. Peak capacity is described in the following equation:

$$P_c = 1 + \left(\frac{T_g}{W_1}\right)$$

where T_g is gradient time and W_1 is width of the peak.

A *Van Deemter plot* depicts relationships between column height and mobile phase velocity (Figure 3). It is usually a J-shaped curve that shows the optimal flow rate of a particular solvent for a given column height.

Below, we discuss the three major forms of separation techniques: GC, LC, and capillary electrophoresis (not exactly chromatography).

The GC is mainly used for volatile organic metabolites. This technique offers high resolution and high sensitivity. It is inexpensive and requires small amount of samples. The technique is limited by the length of time it takes, which increases when derivatization procedures are needed for nonvolatile substances. It is also not suitable for thermally labile samples. This type of chromatography is usually coupled with mass spectrometry

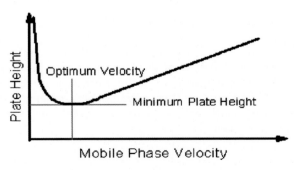

FIGURE 3   Van Deemter plot.

(GC-MS) to provide sensitive and accurate quantification of low-molecular-weight metabolites. Examples of usual known metabolites separated by this technique include cholesterol, fatty acids, hydrophilic vitamins, terpenoids, sugars, and some alkaloids [12].

The LC uses the liquid mobile phase to separate analytes. A smaller particle size (<1 kDa) leads to higher back pressure, and a higher pressure is required to move the particle through the mobile phase. This form of specialized LC is called high-performance liquid chromatography (HPLC). There are many types of HPLC, and the main types used in metabolomics are normal-phase chromatography, reversed-phase chromatography (RPLC), hydrophilic interaction chromatography (HILIC), ion exchange chromatography, size exclusion chromatography, chiral chromatography, and supercritical fluid chromatography.

There are several solvents used in LC that can be distinguished based on their polarity:

$$\text{Polar solvents: Water} > \text{methanol} > \text{acetonitrile} > \text{ethanol} > \text{oxydipropionitrile}$$

$$\text{Non-polar solvents: N} - \text{decane} > \text{N} - \text{hexane} > \text{N} - \text{pentane} > \text{Cyclohexane}$$

HILIC and RPLC are frequently used forms of LC; their relative advantages and disadvantages are listed in Table 2. These techniques are mainly used for separation of secondary neutral metabolites, lipids, flavonoids, and hydrophobic vitamins. Another way to increase resolution of metabolites is utilized in ultra-performance liquid chromatography (UPLC), where the particle size is reduced at the cost of an increase in back pressure.

Capillary electrophoresis is a different type of separation technique that uses an electric field applied to sub-millimeter channels (capillaries) to separate analytes. The separation usually takes about 20 min [13]. This method is useful for the separation of nucleic acids, vitamins (hydrophobic), amino acids, coenzymes, and amino acids.

There are several advantages of linking these separation techniques with analytical tools such as nuclear magnetic resonance (NMR) and MS. Processing samples with these analytical tools will show peaks; however, if the analyte of interest exists in very small concentrations (as is usually true in biological fluids) other compounds with similar-sized peaks may be difficult to differentiate from the analyte of interest. Chromatography uses the physical and chemical properties of the mobile and stationary phases to further increase resolution of this detection process.

TABLE 2   Advantages and Disadvantages of LC Methods

	RPLC	HILIC
Advantages	Well-understood retention mechanisms Widely applicable Faster equilibrium Wide variety of mobile phases available	Retains highly polar analytes not retained by RPC Less interference with matrix components Complementary selectivity to RPC Polar compounds retained more than parent compound High organic mobile phases promotes enhanced ESI-MS response Direct injection of precipitate supernatant without dilution is possible Facilitates use of lower volume sample Increased throughput
Disadvantages	Poor peak shape with basic analytes	Sensitive to sample diluent Mechanism not well understood Longer equilibration time

RPLC; reversed phase liquid chromatography, HILIC; hydrophilic interaction liquid chromatography.

The interface between the separation apparatus and the detectors is also critical. Snyder and Kirkland note desirable characteristics of such an interface [14]:

1. There should be no reduction in chromatographic performance over time.
2. There should be no uncontrolled chemical modification of the analyte during chromatography.
3. The sample transfer to mass spectrometer should be high and efficient.
4. The interface should be reliable, easy to use, and inexpensive.
5. The interface should give a low chemical background.
6. The interface should be capable of operating across a wide variety of chromatography conditions.
7. The interface should not affect the vacuum requirements of the mass spectrometer and should be compatible with all capabilities of the mass spectrometer.
8. The interface should provide quantitative information with reproducibility better than 10%, with low limits of detection. This response should be linear over a wide range of sample sizes.

Flow-injection analysis refers to a technique when the sample is ionized and injected directly into the mass spectrometer without a preceding separation step. This method is mainly used for targeted metabolomic analysis where the unique mass spectra are known beforehand and are easily identified in the mass spectrum. This method takes a few minutes.

2.3 Metabolite Detection

NMR spectroscopy and MS are the primary analytical technologies of metabolomics. Both should be considered as complementary technologies rather than separate methods for a metabolomic study. Almost all of the high abundance molecules can be detected by NMR spectroscopy. For low-abundance molecules, MS is more suitable.

NMR spectroscopy is a technique by which a spectrum is created from resonance of nuclei that, when in a magnetic field, absorb and emit electromagnetic radiation. There are several types of NMR spectroscopy techniques and several different isotopes used for NMR analysis including ^{1}H, ^{13}C, and ^{31}P. Various NMR experiments useful for metabolomics are listed in Table 3.

NMR utilizes the principle that atomic nuclei with nonzero spin when subjected to a strong magnetic field and irradiated with a characteristic radiofrequency wave, emit radiofrequency signals. These signals, which can be detected by an antenna or coil, reflect the number and proximity of the specific nuclear isotopes in question that can be used to infer the quantity and structural features of the compounds in the sample [15]. Because NMR is nondestructive, a sample can be reused many times. Another advantage is that the sample does not require extensive preparation before use.

Limitations of NMR include its lower sensitivity; it can only detect high-abundance metabolites with concentrations >100 nmol/L or $5-10$ µM in size. Also, if one-dimensional NMR is used, many metabolites may have overlapping peaks, especially those in low abundance. This limitation has been resolved by developing sophisticated two-dimensional NMR methods (not discussed in this chapter). Use of LC as a method of separating metabolites improves the sensitivity of analyses (typically <100 metabolites can be resolved at one time).

In contrast, the MS takes advantage of the fact that the velocity of charged particles in an electronic field vary in a precise manner as a function of mass and charge. This permits positive identification of a wide range of molecules including those that are typically part of the metabolome. Since its invention in 1946, time of flight mass spectroscopy (TOF-MS) has grown significantly in its capabilities. The mechanism of ionization has continued to evolve in analyzing biological samples, to address the fact that organic analytes tend to be fragile and often dissociate and break down when hard ionizing techniques are used. Matrix-

TABLE 3   Various Two-dimensional NMR Experiments Used Commonly for Metabolomics

NMR experiments	
Homonuclear experiments	Signals are obtained from usually 1H (same isotope) during both evolution phases of NMR experiment
Correlation spectroscopy (COSY)	Basic and easy form of 2D NMR spectroscopy
Total correlation spectroscopy (TOCSY)	Helpful in identifying impurities
Nuclear Overhauser effect spectroscopy (NOESY)	Provides information about space around the nuclear (e.g., determines protein structure)
Double quantum filtered DQF − COSY	Improves resolution of COSY
Rotating frame Overhauser effect spectroscopy (ROESY)	Provides information about space around the nuclei (especially larger complex molecules)
Heteronuclear experiments	Signals are obtained from two isotopes (usually 1H and 13C) during NMR experiment, helpful for larger molecules (>100 A)
Heteronuclear single quantum correlation (HSQC)	Detects 1H
Heteronuclear multiple bond correlation (HMBC)	Detects 1H
Heteronuclear multiple quantum correlation (HMQC)	Detects 1H, most sensitive out of these three
Other experiments	
J resolved spectroscopy (JRES)	Quick, resolves spectral congestion, and helpful for metabolite quantification
Inadequate	Used for metabolic flux analysis and molecular kinetics

assisted laser desorption ionization (MALDI) [16] is a soft ionization technique that began to be used in TOF-MS in the 1980s and 1990s.

Different ionization techniques now used include electron ionization (formerly known as electron impact) [17,18], electrospray ionization (ESI) [19–21], atmospheric pressure chemical ionization (APCI) [22], desorption electrospray ionization [23], and desorption/ionization on silicon (DIOS) [24,25]. Combinations of two ionization techniques have also been documented (e.g., ESI/APCI, MALDI/DIOS) [26], as has simultaneous use of two ionization techniques (i.e., multimode ionization) [26].

Further modifications of electrodes apply electrostatic force to the ions to improve their resolution before detection. The thermofisher orbitrap and quadrupole ion traps are among the many instruments that use electrostatic or magnetic fields to capture ions in the vacuum system of the mass spectrometer, improving resolution thousands of folds. Due to these improvements, MS has higher sensitivity than NMR and can detect low-abundance metabolites (as low as 1 pmol/L) [27]. However, this sensitivity comes at a cost of low reproducibility. Robust quality control methodology, along with analytical techniques to identify temporal drift or changes in the mass spectrometer, is constantly needed. Internal standards and/or quality control samples are a part of each run to identify errors.

To identify a particular metabolite, its accurate mass, retention time, MS/MS spectrum match, isotope abundance pattern, and fragmentation pattern all should be collected and matched with purified standards under identical conditions [28–30]. This is a time-consuming process. The Metabolomics Standards Initiative has recommended that investigators "putatively annotate" these metabolites without actually knowing their chemical structures unless it is absolutely necessary [28].

3. METABOLOMIC DATA ANALYSIS

There are mainly two forms of metabolomic experiments performed: (a) untargeted metabolomics, a comprehensive agnostic chemometrics approach to evaluate small metabolite molecules that include unknown and known metabolites and (b) targeted metabolomics, a quantitative analysis of a set of chemically known and annotated metabolites. The advantage of targeted metabolomics is that already known pathways are explored by this method and a prior hypothesis is present. On the other hand, the untargeted metabolomics approach can identify novel metabolites and pathways without a prior hypothesis.

Multiple statistical methods have been used to analyze metabolites. For targeted metabolomics, basic statistical tests such as Student's t-test, analysis of variance, and non-parametric tests like the Kruskall–Wallis test may provide adequate statistical means to assess the presence of a signal and its association with a trait of interest. However, many metabolomic signals are highly correlated and thus violate fundamental assumptions of independence for these tests. In such cases, multivariate methods provide an attractive choice. Such analyses are used at various stages of metabolomics—not just to analyze associations in biomarker studies, but also to identify spectral patterns and confirm data robustness of chromatography, MS, and NMR spectra [31].

In unsupervised analyses, no a priori information is utilized to discern various metabolomic patterns. Rather, the correlational structure of the data space is described without regard to its relationship to any external trait or variable. Principal component analysis is frequently used as the first analysis to visualize the data. The elbow method is used to select the principal components that account for the largest proportion of variance in the data. Loading plots and score plots are then used to visualize and identify groups of metabolites and samples that may explain observed clustering in the data. These clustered observations are then checked and screened for various hypothetical characteristics. This approach can also help to discern technical or confounding factors that may have biased the observations. Samples in which medications are present, or from individuals with an inherent metabolic defect, typically stand out on such plots. This method also helps to evaluate the internal controls. Ideally, the controls should not show a specific pattern, since a specific pattern away from the center on a given principal component score plot may identify technical problems. For instance, when a linear orientation of internal controls is evident on a principal component score plot, this may signify a temporal drift in the separation or detection techniques.

On the other hand, supervised analyses attempt to link metabolomic patterns to specific traits or attributes of the sample or the subjects from which the samples were obtained. Thus, unlike unsupervised analyses, supervised analyses depend on a specific prior hypothesis. However, similar to unsupervised analyses, multivariate methods are typically required to address the large number of frequently highly correlated metabolomic signals that are collected. These methods include cluster analysis, partial least squares (PLS) regression, and orthogonal projections to latent structures (OPLS). The main principle of these methods is to organize the data into correlated groups that can explain the covariance between metabolites and the dependent variable. The dependent variable can be a continuous or categorical variable. To estimate risk ratios, metabolite concentrations frequently are changed into standardized units and evaluated in multiple regression models adjusted for confounders. Most studies use a false discovery rate or modified Bonferroni's method to establish a threshold for significance testing [32].

4. METABOLOMIC INFORMATICS RESOURCES

As more and more metabolomic data are acquired, it becomes increasingly important to organize these data and related metadata about chemical structures, biological pathways, bio-sample and study designs, and so on into searchable databases. Fortunately, many such resources already exist and others continue to be developed. These databases provide particular structures, chemical isotopes, molecular weights, mass and charge ratios, and metabolic pathways; however, no one provides all of this information. Table 4 briefly summarizes various

TABLE 4   Metabolomics Databases

Comprehensive metabolic databases	Metabolic pathways databases	Compound specific databases	Spectral databases
The Human Metabolome Database (HMDB) [33–35]	Kyoto Encyclopedia of Genes and Genomes (KEGG) [37–39]	PubChem	The Human Metabolome Database (HMDB)
BiGG [111]	MetaCyc [38,112–116]	Chemical Entities of Biological Interest (ChEBI) [117,118]	The BioMagResBank (BMRB)
SetupX [119]	HumanCyc	ChemSpider [120]	The Madison Metabolomics Consortium Database (MMCD) [121]
BinBase [122]	BioCyc [112,114,123]	Kyoto Encyclopedia of Genes and Genomes Glycan (KEGG) [37–39]	MassBank [124]
SYSTems biology of pseudomonas (SYSTOMONAS) [125]	Reactome [126–128]	In Vivo/In silico Metabolites Database (IIMDB)	The Golm Metabolome Database [129]
Metabolights database [130–133]			The METLIN Metabolite Database [134–137]
LIPIDMAPS			Fiehn GC-MS database
The National Institute of Standards and Technology (NIST) database			The Birmingham Metabolite Library Nuclear Magnetic Resonance database
			mzCloud

metabolomic databases, and some of the commonly used ones are discussed below:

1. **Human Metabolome Database (HMDB)**: The HMDB is the largest available database of known metabolites [33–35]. It includes data on chemical structure, taxonomy, source, accession number, physical and biological properties, among many other characteristics, as well as hypertext links to many other related metabolomic databases.
2. **Metabolic Pathway Databases**: Genome-based studies have shown that almost 30% of metabolites are involved in at least two metabolic reactions, and 12% of metabolites participate in more than 10 reactions. These metabolites are tightly controlled by their respective proteome and transcriptome [36]. For this reason, metabolomic pathway analysis becomes important. The Kyoto Encyclopedia of Genes and Genomes (KEGG) [37–39] is the largest metabolic pathway database. This database includes metabolic pathways for humans and other organisms. Other tools such METATOOL, can be used to identify the structure and stoichiometry of biochemical reaction networks [40]. With advances in DNA microarrays, it is now possible to analyze dynamic events in gene expression. It may be possible in the near future to track the various metabolic pathways and fluxes associated with gene expression.
3. **Compound-Specific Databases**: There are many other dictionaries of molecular structures for thousands and in some cases millions of chemical compounds also available online including one from chemical companies such as Sigma–Aldrich (www.sigmaaldrich. com).

5. CLINICAL AND POPULATION-BASED STUDIES OF METABOLOMICS AND CARDIOMETABOLIC CONDITIONS

There is an extensive body of literature focused on specific metabolic pathways and untargeted metabolomics in cellular intact animal models of cardiometabolic disease. These data have been recently reviewed elsewhere and are not considered here [41,42]. The following section describes results from clinical and population-based metabolomic studies in humans. These studies can be roughly divided into those seeking to better characterize metabolomic pathways contributing to the pathogenesis of various diseases and those seeking to identify novel biomarkers for the purpose of identification of high-risk individuals requiring more aggressive interventions or interventions specifically targeting specific metabolic processes or conditions.

5.1 Atherosclerostic Cardiovascular Disease

There are several examples of targeted metabolomic studies of coronary disease focusing on narrowly defined sets of metabolites in specific metabolic pathways. For instance, in 2009 Wang et al. used HPLC/MS to examine a family of methyl derivatives of amino acids including asymmetrical dimethylarginine (ADMA) and other molecules that thought to play an important role in nitric oxide signaling [43]. In this observational study of 1011 subjects who had undergone coronary angiography, both ADMA and symmetrical dimethylarginine were independently associated with both obstructive coronary disease and 3-year incident major coronary disease events. Similarly, several clinical studies have shown that metabolic profiles consistent with inefficient β-oxidation of fatty acids, including certain acylcarnitines, are independently associated with future coronary events [41]. These studies, along with supportive animal model data, strongly implicate mitochondrial dysfunction as another key contributor to the pathogenesis of atherosclerosis.

In one of the first untargeted metabolomic studies of cardiovascular disease, Brindle et al. investigated the association of full-spectrum ^{1}H-NMR-based metabolomic features with obstructive coronary artery disease (CAD). In this case—control study of 76 subjects, multivariate spectral analysis of serum ^{1}H-NMR data was able to correctly classify angiographic disease status with a specificity of >90% [32]. Although some of the chemical shifts identified in this study could be assigned to lipid moieties such as LDL-, VLDL-, and HDL-cholesterol, it was evident that other information including subtle chemical differences in the lipid particle compositions, the degree of fatty-acid side-chain unsaturation, and lipoprotein—protein molecular interactions also accounted for some of the predictive efficacy of the multivariate models—thereby highlighting the potential additional value of a metabolomic biomarker strategy.

In perhaps the most highly cited cardiovascular disease metabolomics study to date, Wang et al. used LC/MS in a case—control study of 150 subjects to convincingly demonstrate an association between trimethylamine-N-oxide (TMAO), choline, and betaine and acute myocardial infarction, stroke, and mortality [44]. Based on these initial observations, the research team went on to verify a causal link between TMAO and choline by-products and atherosclerosis in an extensive array of histopathological, biochemical, and genetic experiments using atherosclerosis-prone mice. TMAO and choline are by-products of microbial metabolism of phosphatidylcholine, which is predominantly found in eggs, fish, red meat, poultry, milk, and liver. These food groups are energy-rich and have been associated with adverse outcomes. A reduced ratio of arginine to its metabolite (marker of reduced nitric oxide synthase activity) and citrulline are associated with incident cardiovascular disease events [43]. Oxidative stress is a major player in development of coronary heart disease. F2-isoprostane is a marker of lipid peroxidation. Elevated levels of F2-isoprostane were associated with obesity and cardiovascular disease in Framingham Heart Study [45]. Metabolomics has also been used to investigate other cardiovascular conditions including exercise-induced myocardial ischemia [46], in-stent restenosis [47], acute coronary syndromes [48], and cardiogeneic shock [49]. Collectively, these cardiovascular metabolomic studies have identified novel pathways and mechanisms related to the pathogenesis of atherosclerosis and discovered new biomarkers for both acute and chronic coronary disease risk.

5.2 Hypertension and Metabolomics

Hypertension constitutes the most common form of cardiovascular disease, and has been diagnosed in almost 79 million people in the USA [50]. Metabolomic studies in humans have identified a number of metabolites associated with hypertension. Several studies have identified metabolites related to lipid metabolism, particularly metabolites involved in fatty acid synthesis. Prominent metabolites associated with increased risk of hypertension include palmitic acid (a marker of lipogenesis) [51–54], oleic acid (a dietary fatty acid) [53,54], acylcarnitines (transport proteins for fatty acids from the cytoplasm to mitochondria for fatty acid synthesis) [53], and phospholipids (involved in changing the fluidity of cell membrane and hence related to receptor density). Apart from metabolites involved in lipid metabolism, some novel signals have been identified in a few studies with untargeted metabolomics in which alpha-1 acid glycoprotein (an inflammatory marker) [55], and 4-hydroxyhippurate (the end-product of polyphenol metabolism by gut microbes) [56] showed independent associations with hypertension. Very few metabolites have also shown inverse association with hypertension. These metabolites include hippurate (a marker of diet and gut microbial activity) [57], formate (a by-product of fermentation of dietary fiber by gut microbiota) [57], and linoleic acid (an unsaturated dietary fatty acid abundant in vegetable oils). However, these are merely association studies and further intervention studies by modifying the levels of these metabolites are lacking.

5.3 Metabolomics and Heart Failure

Each year, there are 1.1 million new cases of heart failure diagnosed in the USA [50]. Its incidence has still not shown decline as other cardiovascular diseases have in developed countries. A part of this is related to our incomplete understanding of the pathophysiology of heart failure.

Metabolomics have provided important insights into novel pathways affected in heart failure patients. Both targeted and untargeted metabolomics have been used in various studies. The targeted metabolomics demonstrated increased catabolic activity (increased levels of amino acids) and decreased Kreb's cycle metabolites (impairment of oxidative glucose metabolism). In patients with ischemic heart failure, untargeted urinary ^{1}H-NMR-based metabolomics showed decreased excretion of intermediate metabolites of Krebs' cycle (mainly citrate, succinate, and cis-aconitate) and increased excretion of ketone bodies, such as acetone and acetoacetate, suggesting probable activation of anaerobic glycolysis in myocardium in heart failure [58]. Du et al. also confirmed these results using untargeted serum ^{1}H NMR spectroscopy-based metabolomics [59]. Other untargeted metabolomic studies helped in identifying derangements in protein translation and increased RNA degradation by showing increased levels of tyrosine, phenylalanine [60], and pseudouridine in the serum [61]. These initial studies suggested that there is an increased cellular turnover in heart failure patients, which was confirmed by Mirandola et al. who found elevated serum levels of 1-methylnicotinamide in heart failure patients [62]. Even though high levels of branched and unbranched amino acids were observed in some metabolomic studies, it was thought that selective elevation of some of the catabolic products might not be actually due to catabolism but some other unknown alternate pathway [60,63,64]. Ketone bodies such as acetone and acetoacetate have also been found to be elevated in some targeted metabolomic studies that are markers of catabolism as well as cellular anaerobic glycolysis [65–67]. In an attempt to identify biomarkers for prediction of heart failure, pseudouridine (area under the curve = 0.96), 2-oxoglutarate (area under the curve = 0.93), and metabolites involved in lipoprotein metabolism were found to have a sensitivity of 92.31% and specificity of 86.67% [68]. A brief review of metabolomics studies in heart failure patients is given in Table 5.

5.4 Metabolomics and Diabetes

Diabetes affected 12.3% of the US population in 2010, and its prevalence continues to grow as well [69]. Evidence from metabolomic studies has been robust in predicting the risk of diabetes. The Prospective Investigation into Cancer and Nutrition (EPIC)-Potsdam study showed that metabolites obtained from serum of diabetics predicted 7-year risk of diabetes [70]. The authors validated their findings in the Kooperative Gesundheitsforschung in der Region Augsburg (KORA) study. Using flow-injection tandem MS, elevated levels of phenylalanine, tyrosine, isoleucine, and valine were associated with increased risk of diabetes mellitus [70]. This study is consistent with many others that have linked branched chain amino acids (isoleucine, leucine, and valine) and aromatic

TABLE 5 Metabolomics Studies in Heart Failure

Study	Type of metabolomics	Positive association	Negative association
Zheng et al. (ARIC study), 2013 [64]	Untargeted serum GC-MS and LC-MS	Hydroxyleucine or hydroxyisoleucine	Dihydroxy docosatrienoic acid
SADHART-CHF 2010 [63]	Targeted serum GC and LC-MS	Aspartate, glutamate, γ-glutamylleucine, and γ-glutamylglutamine	
Lin et al., 2011 [67]	Targeted serum ^{1}H-NMR metabolomics	Formate, choline, and acetone	
Wang et al., 2013 [68]	Targeted plasma ^{1}H-NMR metabolomics	Lipoproteins	
Marcondes-Braga et al., 2012 [66]	Targeted breath GC-MS	Acetone	
Samara et al., 2013 [65]	Targeted breath selected ion-flow tube mass spectrometry (SIFT-MS)	Acetone and pentane	
Dunn et al., 2007 [61]	Untargeted serum GC-TOF-MS	Pseudouridine, 2-oxaloglutarate	
Du et al., 2014 [59]	Untargeted serum serum ^{1}H NMR-spectroscopy	3-Hydroxybutyrate, acetone, and succinate (AUCs 0.92, 0.90, and 0.86, respectively)	
Tenori et al., 2013 [60]	Untargeted serum and urinary ^{1}H NMR- spectroscopy	Tyrosine, phenylalanine, isoleucine, and creatine	Lysine, L-dopa, lactate, and citrate
Kang et al. [58]	Untargeted urinary ^{1}H NMR-spectroscopy	Acetone, acetoacetate, and acetate	Citrate, succinate, 2-oxoglutarate, and cis-aconitate

ARIC; atherosclerosis in communities study, SADHART-CHF; safety and efficacy of sertraline for depression in congestion heart failure, NMR; nuclear magnetic resonance, AUC; area under the curve.

chain amino acids (tyrosine, phenylalanine, tryptophan, and histidine) with type 2 diabetes mellitus [71—79]. Interestingly, glycine is the only amino acid negatively associated with insulin resistance and risk of diabetes [71,77,80]. Glycine might become depleted in people with diabetes due to divergence of metabolic pathways to gluconeogenesis, or because it helps dispose of incompletely oxidized metabolites such as acylglycines. Diphosphatidylcholines are another group of phospholipids associated with increased risk of diabetes. Conversely, the risk of incident diabetes was inversely associated with sphingomyelin, linoleoyl lyso-phosphatidylcholine, and phosphatidylcholine-containing plasmalogens (C34—C44) [81,82]. In addition to predicting the future risk of diabetes, metabolomic studies have helped in identifying various phenotypes of diabetes [83]. Isolated post-challenge diabetes (i.e., impaired fasting glucose) was also characterized by metabolomic differences compared with normal individuals, including elevated branched-chain amino acids [78,84] lyso-phosphatidylcholines [70,85,86], fatty acids, and acyl carnitines [87]. In KORA F4 study, 1-linoleoyl-glycero-3-phosphocholine was found to be inversely associated with both incident-impaired glucose tolerance and incident diabetes, while glycine was only inversely associated with incident-impaired fasting glucose [88]. In cross-sectional analysis of KORA S4 study, 1-linoleoyl-glycero-3-phosphocholine was able to differentiate between impaired glucose tolerance and type 2 diabetes [88]. UPLC-Q-TOF-MS analysis of plasma metabolomics and skeletal muscle transcriptomics showed concordance of deranged metabolism in several metabolism pathways. Ketogenic and gluconeogenic amino acid, branched-chain amino acids, glycerol, and β-hydroxybutyrate have suggested increased proteolysis, lipolysis, and ketogenesis in type 1 diabetes [89].

5.5 Metabolomics and Obesity

About a third of US population can be classified as obese based on WHO BMI-based criteria. Globally, obesity affects 500 million people, and 1.5 billion are classified as overweight [90]. Multiple genome-wide studies have shown association of obesity with genes [91—93]. However, the functional importance of many of these single-nucleotide polymorphisms (SNPs) is unknown. Diabetes and obesity share many metabolite changes, especially phosphatidylcholines, branched chain amino acids, and lipid metabolism-related metabolites [94]. Targeted serum metabolomics studies have shown positive associations of glycine, glutamate, and glycerophosphotidylcholine (42:0) with obesity, whereas glycerophosphotidylcholine (32:0), glycerophosphotidylcholine (32:1), and glycerophosphotidylcholine (40:5) were found to be inversely associated with obesity [95]. This study was later on confirmed by another study using plasma UPLC-Q-TOF-MS and showed direct associations of lyso-phosphotidylcholine (14:0) and lyso-phosphotidylcholine (18:0) with obesity, whereas lyso-phosphotidylcholine (18:1) was associated with decreased risk of obesity [96]. In another untargeted UPLC-Q-TOF-MS-based study, metabolic derangements in plasma of obese individuals included elevated levels of L-prolyl-L-proline, leucyl-phenylalanine, and decanoylcarnitine in positive ESI mode and N-acetylornithine, 17-hydroxypregnenolone sulfate, 11-beta-hydroxyprogesterone, 5a-dihydrotestosterone sulfate, and glucosylgalactosyl hydroxylysine in negative ESI mode [97]. A more recent study suggested that the metabolome of obese adolescents might differ from older obese individuals. Lower levels of medium- to short-chain acylcarnitines and higher levels of fatty acid oxidation by-products were observed in obese adolescents in a study of 64 individuals [98]. No differences were observed for medium chain (C8 or C10) acylcarnitines, amino acids, and other medium-chain fatty acids in obese versus non-obese adolescents. However, in a targeted MS study, medium- and long-chain fatty acids as well as acylcarnitines were associated with obesity [77]. Other important implicated metabolites include gut microbiome-derived metabolites such as trigonelline, 2-hydroxyisobutyrate, hippuric acid, and xanthine, which were found to be elevated in the urinary metabolome of obese versus non-obese individuals [99].

6. METABOLOMICS AND GENOME-WIDE ASSOCIATION STUDIES

The combination of low-cost genotyping platforms with clinical- and population-scale metabolomic technologies has created remarkable opportunities to map common genetic variants to various aspects of the metabolome. The initial focus of these studies is typically on simple univariate mapping of individual genotypes to individual metabolic features. However, the real power of genome-wide association studies (GWAS) paired with metabolome-wide phenotypes comes from looking in a more integrated way at the entire metabolome. Ratios of metabolites are especially useful to identify genetic variants that influence enzymes or other determinants of specific biochemical reactions responsible for converting one metabolite to another. In some settings the use of metabolite ratios can greatly increase the power, and therefore substantially reduce the sample size required for a successful GWAS analysis [115]. Similarly, network mapping tools and integration with external systems biology information provide mechanisms to describe the influence of genetic variants on entire metabolic pathways, rather than single metabolites.

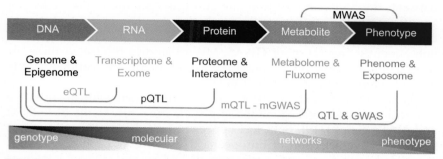

FIGURE 4   Relationship of genome, epigenome, transcriptome, exome, proteome, metabolome, and phenome. *Adapted from Dumas et al. Mol BioSyst 2012; 8:2494–2502.*

Similar concepts can be applied to other "-omics" technologies, including genome-wide evaluation of epigenetic factors that also play pivotal roles in regulating the metabolome (Figure 4). In the following sections, the major clinical and population-based metabolomics-wide or genome-wide (MWAS/GWAS) studies are reviewed.

6.1 Metabolic Phenotypes and GWAS

The first extensive MWAS/GWAS study was performed in a subsample of the KORA F3 cohort of only 284 males using ESI tandem mass spectrometry (MS/MS) and a conventional gene chip array. SNPs with minor homozygote frequency of $\geq 5\%$ were compared across a metabolomic panel of 364 metabolites that included mainly lipidomic species. Additive genetic models showed near genome-wide significant associations between metabolites and SNPs in a linkage disequilibrium block containing the FADS1 gene ($p = 4.5 \times 10^{-8}$). The FADS1 gene encodes for fatty acid delta-5-desaturase that regulates desaturation of fatty acids. The pattern of increased glycerophosphatidylcholines with three or less number of double bonds in their fatty acid side chains and depletion of phosphatidylcholines with four or more double bonds in their fatty acid side chains (including arachidonic acid) strongly suggested that increasing copies of the variant allele for rs174548 lead to greater inefficiency of fatty acid desaturation. This hypothesis was consistent with previous independent GWAS documenting association in the expected directions between this same SNP and both LDL- and HDL-cholesterol concentrations [100,101].

The same study identified a second metabolite-associated SNP (rs4775041, $p = 9.7 \times 10^{-8}$) in proximity to the LIPC gene that encodes triglyceride hydrolase and is a ligand/bridging factor for receptor-mediated lipoprotein uptake, both important components of lipoprotein-related metabolism. In both cases, the strength of association between the SNPs and appropriate substrate/product ratios reflecting specific metabolic reactions were much stronger than individual metabolites themselves or their downstream biochemical products (Figure 5). Subsequently, several other GWAS/MWAS studies replicated these, or related links between FADS1 and LIPC locus variants and various aspects of long-chain fatty acid metabolism [113–116].

In another modest-sized MWAS/GWAS study, metabolite quantitative trait loci (mQTLs) were identified in genetic proximity to the PYROXD2 and NAT8 genes [119]. Both loci appear to have been subject to recent strong positive selection in Europeans. The PYROXD2 locus variant is associated with increased expression of DMAp and TMAu, which is involved in TMAO production, the same metabolite shown to be linked to obstructive CAD by Wang et al. [43]. Also of note in this study is that the identified mQTLs accounted for 40–64% of the population variation in the relevant metabolite—considerably more than is typically seen for GWAS hits associated with more typical serum or plasma analytes such as LDL-cholesterol. Collectively, these studies have helped to identify

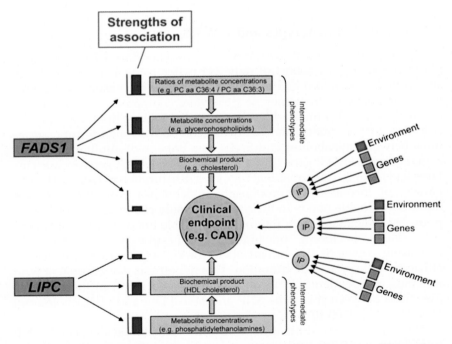

FIGURE 5   Schematic illustration of the role of intermediate phenotypes (IPs), such as metabolic traits, demonstrated at the examples of two genes that code for major enzymes of the long-chain fatty acid metabolism (*FADS1* and *LIPC*). *Adapted from Gieger et al. PLoS Genet November 2008; 4(11): e1000282.*

distinct metabolic intermediate phenotypes under partial regulation by one or more common genetic variants that could contribute to complex clinical traits and disease—including CAD.

More recently, there have been two substantially more comprehensive MWAS/GWAS studies involving thousands of subjects using either NMR spectroscopy or MS in a targeted analysis of 100–400 identified metabolites. The first study, published in 2012 examined 7.7 million SNPs and 216 metabolites identified using ^{1}NMR spectroscopy in 8330 Finnish men [102]. This study identified 33 distinct genetic loci that met strict MWAS/GWAS criteria of significance ($p < 2.31 \times 10^{-10}$). Thirty-one of them were found from GWAS and two from fine-mapping (Figure 6). In some cases these loci accounted for as much as 40% of the heritability of the metabolomic traits. In the second study, 2.1 million SNPs were examined in relation to 529 metabolites quantified using MS in 7824 adult

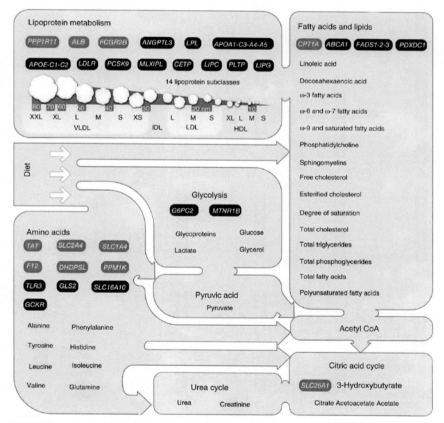

FIGURE 6   Nuclear magnetic resonance spectroscopy measured metabolites and associated genetic loci. *Adapted from Kettunen et al. Nat Genet 2012; 44(3): 269–276.*

individuals from two European population studies [103]. In this study, there were 145 statistically significant associations between SNPs and specific metabolites (Figure 7). The genetic loci identified by these two studies were dominated by enzymes, transporter molecules, or metabolic regulators, and a network view of the gene loci and associated metabolites provides a comprehensive depiction of genome—metabolome interactions (Figure 8).

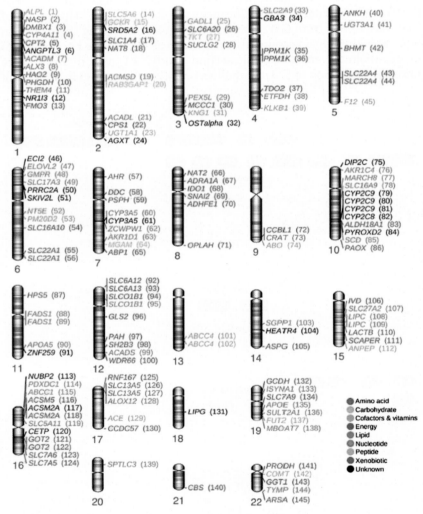

FIGURE 7   Ideogram of metabolomic associations. *Adapted from Shin et al. Nat Genet June 2014; 46(6): 543–550.*

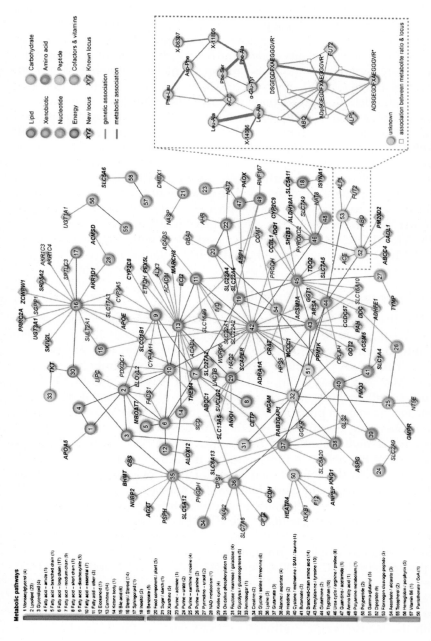

FIGURE 8 A network view of genetic and metabolomic associations. *Adapted from Shin et al. Nat Genet June 2014; 46(6): 543–550.*

6.2 Novel Metabolic Pathway Identification in GWAS

Some of the mQTLs identified in MWAS/GWAS studies recapitulated previously known metabolic pathways. However, in other cases the data suggested entirely new dimensions to previously known pathways or functional genetic variants and/or novel metabolic pathways that were previously unknown and are still incompletely understood. In an mGWAS (meta-GWAS), SNP rs1260326 was found to affect ratios of phosphatidylcholines in diabetic patients [104]. This SNP in glucose kinase receptor gene (GCKR) was associated with low fasting glucose and triglyceride levels in type 2 diabetics [105]. In another study, SNP rs10830963 in melatonin receptor gene (MTNR1B) was found to be associated with fasting glucose levels [106]. This same SNP was found to affect phenylalanine:tryptophan ratio [104]. Since phenylalanine is a precursor of melatonin, these studies suggest a previously unexpected genetic contributor to the melatonin–phenylalanine pathway. Similarly, another SNP rs964184 in the apolipoprotein cluster APOA1–APOC3–APOA4–APOA5 was found to be associated with triglyceride levels [107]. This SNP is also associated with phenylalanine ratios.

6.3 Epigenetics and Metabolomics

The first epigenetic/metabolomic study was also performed in the KORA cohort. In sera of 1814 participants, 649 metabolite traits were examined with respect to the degree of methylation in 457,004 CpG sites [108]. This study identified CpG loci in close proximity to ACADS, PYROXD2, NAT8, ACADM, OPLAH, FADS1, UGT1A, and SULT2A1 as being significantly associated with elements of the metabolome. These gene loci have been previously identified in mGWAS [104,109,110] raising the challenging question of whether the observed association between CpG locus methylation and metabolites could be confounded by the presence of the SNPs. In these cases, it is hard to determine if the SNP interferes with the methylation assay at the CpG site, is functionally related to the degree of methylation at the site or possibly interacts with the SNP effect on the metabolite in question. Interestingly, this study also found seven distinct genetic loci (UGT2B15, TXNIP, DHCR24, MYO5C, ABCG1, SLC25A22, and CPT1A) that did not have evidence of underlying genetic effect associated with metabolic phenotypes. These were related to phospholipids PC ae C42:4, PC ae C42:5, PC ae C44:4, and PC ae C26:0, the lipid traits Chylo-A, VLDL-A, tryptophan, and tyrosine. However, even these methylation–metabolite associations could be subject to confounding by environmental or other external factors that contribute to both the locus-specific methylation and variations in the metabolite in question. It is clear from these data that more functional genomic

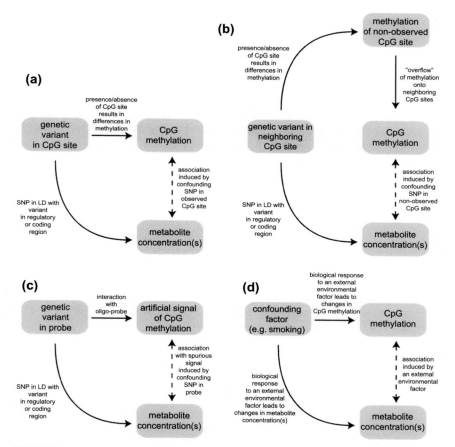

FIGURE 9   Interactions between various -omics and possible scenarios that may result in an observed CpG—metabotype association induced by a confounding genetic variant or by an external environmental factor. *Adapted from Peterson et al. Hum Mol Genet January 15, 2014; 23(2): 534—545.*

investigation on a case-by-case basis will be required to understand the true nature of these relationships (Figure 9). Nevertheless, this study does demonstrate the potential for even more comprehensive mapping of both genetic and epigenetic influences on the metabolome.

7. CONCLUSIONS

This chapter briefly described the complex tools and techniques used in cardiovascular metabolomics and reviewed some of the relevant research in the field. Most of the work to date has focused on targeted metabolomics of known pathways or sets to ~100–500 well-identified metabolites. These studies have produced important insights concerning known and novel metabolic pathways that likely contribute to the pathogenesis of several cardiometabolic traits. Notable among these is the emerging evidence concerning the role of the gut microbiome and its contribution to human metabolism. Combining metabolomics with other "-omics" technologies in the setting of clinical and population research also shows great promise to better define the complex interaction among the genome, the environment, and the metabolome. Ultimately, mapping the topology of these complex relationships may provide new strategies for risk prediction, biomarker detection, and identification of novel and targetable mechanisms to reduce the risk for cardiovascular disease.

References

[1] Sidney S, Rosamond WD, Howard VJ, Luepker RV. The "heart disease and stroke statistics—2013 update" and the need for a national cardiovascular surveillance system. Circulation 2013;127:21–3.
[2] Metabolomics. 2014.
[3] Nicholson JK, Lindon JC. Systems biology: metabonomics. Nature 2008;455:1054–6.
[4] Oliver SG, Winson MK, Kell DB, Baganz F. Systematic functional analysis of the yeast genome. Trends Biotechnol 1998;16:373–8.
[5] Tweeddale H, Notley-McRobb L, Ferenci T. Effect of slow growth on metabolism of escherichia coli, as revealed by global metabolite pool ("metabolome") analysis. J Bacteriol 1998;180:5109–16.
[6] Fiehn O. Combining genomics, metabolome analysis, and biochemical modelling to understand metabolic networks. Comp Funct Genomics 2001;2:155–68.
[7] Pinto J, Domingues MR, Galhano E, Pita C, Almeida MD, Carreira IM, et al. Human plasma stability during handling and storage: impact on NMR metabolomics. Analyst 2014;139(5):1168–77.
[8] Dong H, Zhang A, Sun H, Wang H, Lu X, Wang M, et al. Ingenuity pathways analysis of urine metabolomics phenotypes toxicity of chuanwu in wistar rats by UPLC-Q-TOF-HDMS coupled with pattern recognition methods. Mol Biosyst 2012;8:1206–21.
[9] Vuckovic D. Current trends and challenges in sample preparation for global metabolomics using liquid chromatography-mass spectrometry. Anal Bioanal Chem 2012;403:1523–48.
[10] Wu N, Lippert JA, Lee ML. Practical aspects of ultrahigh pressure capillary liquid chromatography. J Chromatogr A 2001;911:1–12.
[11] Wang A, Carr PW. Comparative study of the linear solvation energy relationship, linear solvent strength theory, and typical-conditions model for retention prediction in reversed-phase liquid chromatography. J Chromatogr A 2002;965:3–23.
[12] Clement RE, Onuska FI, Eiceman GA, Hill HH. Gas chromatography. Anal Chem 1990;62:414R–22R.
[13] Gaspar A, Englmann M, Fekete A, Harir M, Schmitt-Kopplin P. Trends in CE-MS 2005-2006. Electrophoresis 2008;29:66–79.
[14] Köhler J, Chase DB, Farlee RD, Vega AJ, Kirkland JJ. Comprehensive characterization of some silica-based stationary phase for high-performance liquid chromatography. J Chromatogr A 1986;352:275–305.
[15] Bothwell JH, Griffin JL. An introduction to biological nuclear magnetic resonance spectroscopy. Biol Rev Camb Philos Soc 2011;86:493–510.

[16] Vaidyanathan S, Gaskell S, Goodacre R. Matrix-suppressed laser desorption/ionisation mass spectrometry and its suitability for metabolome analyses. Rapid Commun Mass Spectrom RCM 2006;20:1192−8.

[17] Jellum E. Profiling of human body fluids in healthy and diseased states using gas chromatography and mass spectrometry, with special reference to organic acids. J Chromatogr 1977;143:427−62.

[18] Kopka J. Current challenges and developments in GC-MS based metabolite profiling technology. J Biotechnol 2006;124:312−22.

[19] Want EJ, O'Maille G, Smith CA, Brandon TR, Uritboonthai W, Qin C, et al. Solvent-dependent metabolite distribution, clustering, and protein extraction for serum profiling with mass spectrometry. Anal Chem 2006;78:743−52.

[20] Tolstikov VV, Lommen A, Nakanishi K, Tanaka N, Fiehn O. Monolithic silica-based capillary reversed-phase liquid chromatography/electrospray mass spectrometry for plant metabolomics. Anal Chem 2003;75:6737−40.

[21] Nordstrom A, O'Maille G, Qin C, Siuzdak G. Nonlinear data alignment for UPLC-MS and HPLC-MS based metabolomics: quantitative analysis of endogenous and exogenous metabolites in human serum. Anal Chem 2006;78:3289−95.

[22] Aharoni A, Ric de Vos CH, Verhoeven HA, Maliepaard CA, Kruppa G, Bino R, et al. Nontargeted metabolome analysis by use of fourier transform ion cyclotron mass spectrometry. OMICS J Integr Biol 2002;6:217−34.

[23] Chen H, Pan Z, Talaty N, Raftery D, Cooks RG. Combining desorption electrospray ionization mass spectrometry and nuclear magnetic resonance for differential metabolomics without sample preparation. Rapid Commun Mass Spectrom RCM 2006;20:1577−84.

[24] Kim JK, Murray KK. Matrix-assisted laser desorption/ionization with untreated silicon targets. Rapid Commun Mass Spectrom RCM 2009;23:203−5.

[25] Weman H, Zhao QX, Monemar B. Impact ionization of excitons and electron-hole droplets in silicon. Phys Rev B Condens Matter 1987;36:5054−7.

[26] Nordstrom A, Want E, Northen T, Lehtio J, Siuzdak G. Multiple ionization mass spectrometry strategy used to reveal the complexity of metabolomics. Anal Chem 2008;80: 421−9.

[27] Dunn WB, Broadhurst DI, Atherton HJ, Goodacre R, Griffin JL. Systems level studies of mammalian metabolomes: the roles of mass spectrometry and nuclear magnetic resonance spectroscopy. Chem Soc Rev 2011;40:387−426.

[28] Dunn WB, Broadhurst D, Begley P, Zelena E, Francis-McIntyre S, Anderson N, et al. Procedures for large-scale metabolic profiling of serum and plasma using gas chromatography and liquid chromatography coupled to mass spectrometry. Nat Protoc 2011; 6:1060−83.

[29] Xie C, Zhong D, Yu K, Chen X. Recent advances in metabolite identification and quantitative bioanalysis by LC-Q-TOF MS. Bioanalysis 2012;4:937−59.

[30] Wishart DS. Advances in metabolite identification. Bioanalysis 2011;3:1769−82.

[31] Wang X, Sun H, Zhang A, Wang P, Han Y. Ultra-performance liquid chromatography coupled to mass spectrometry as a sensitive and powerful technology for metabolomic studies. J Sep Sci 2011;34:3451−9.

[32] Brindle JT, Antti H, Holmes E, Tranter G, Nicholson JK, Bethell HW, et al. Rapid and noninvasive diagnosis of the presence and severity of coronary heart disease using 1H-NMR-based metabonomics. Nat Med 2002;8:1439−44.

[33] Wishart DS, Jewison T, Guo AC, Wilson M, Knox C, Liu Y, et al. HMDB 3.0—the human metabolome database in 2013. Nucleic Acids Res 2013;41:D801−7.

[34] Wishart DS, Knox C, Guo AC, Eisner R, Young N, Gautam B, et al. HMDB: a knowledgebase for the human metabolome. Nucleic Acids Res 2009;37:D603−10.

[35] Wishart DS, Tzur D, Knox C, Eisner R, Guo AC, Young N, et al. HMDB: the human metabolome database. Nucleic Acids Res 2007;35:D521−6.

[36] Ryan D, Robards K. Metabolomics: the greatest omics of them all? Anal Chem 2006;78: 7954—8.

[37] Posma JM, Robinette SL, Holmes E, Nicholson JK. Metabonetworks, an interactive matlab-based toolbox for creating, customizing and exploring sub-networks from KEGG. Bioinformatics March 15, 2014;30(6):893—5.

[38] Altman T, Travers M, Kothari A, Caspi R, Karp PD. A systematic comparison of the MetaCyc and KEGG pathway databases. BMC bioinformatics 2013;14(1):112.

[39] Tanabe M, Kanehisa M. Using the KEGG database resource. editoral board. In: Baxevanis Andreas D, et al., editors. Current protocols in bioinformatics; 2012. Chapter 1: Unit 1.12.

[40] Liao JC, Hou SY, Chao YP. Pathway analysis, engineering, and physiological considerations for redirecting central metabolism. Biotechnol Bioeng 1996;52:129—40.

[41] Shah SH, Kraus WE, Newgard CB. Metabolomic profiling for the identification of novel biomarkers and mechanisms related to common cardiovascular diseases: form and function. Circulation 2012;126:1110—20.

[42] Barderas MG, Laborde CM, Posada M, de la Cuesta F, Zubiri I, Vivanco F, et al. Metabolomic profiling for identification of novel potential biomarkers in cardiovascular diseases. J Biomed Biotechnol 2011;2011:790132.

[43] Wang ZN, Tang WHW, Cho L, Brennan DM, Hazen SL. Targeted metabolomic evaluation of arginine methylation and cardiovascular risks potential mechanisms beyond nitric oxide synthase inhibition. Arterioscler Thromb Vasc Biol 2009;29: 1383—91.

[44] Wang Z, Klipfell E, Bennett BJ, Koeth R, Levison BS, Dugar B, et al. Gut flora metabolism of phosphatidylcholine promotes cardiovascular disease. Nature 2011;472: 57—63.

[45] Keaney Jr JF, Larson MG, Vasan RS, Wilson PW, Lipinska I, Corey D, et al. Obesity and systemic oxidative stress: clinical correlates of oxidative stress in the Framingham Study. Arterioscler Thromb Vasc Biol 2003;23:434—9.

[46] Sabatine MS, Liu E, Morrow DA, Heller E, McCarroll R, Wiegand R, et al. Metabolomic identification of novel biomarkers of myocardial ischemia. Circulation 2005;112: 3868—75.

[47] Hasokawa M, Shinohara M, Tsugawa H, Bamba T, Fukusaki E, Nishiumi S, et al. Identification of biomarkers of stent restenosis with serum metabolomic profiling using gas chromatography/mass spectrometry. Circ J Off J Jpn Circ Soc 2012;76:1864—73.

[48] Lewis GD, Wei R, Liu E, Yang E, Shi X, Martinovic M, et al. Metabolite profiling of blood from individuals undergoing planned myocardial infarction reveals early markers of myocardial injury. J Clin Invest 2008;118:3503—12.

[49] Nicholls SJ, Wang Z, Koeth R, Levison B, DelFraino B, Dzavik V, et al. Metabolic profiling of arginine and nitric oxide pathways predicts hemodynamic abnormalities and mortality in patients with cardiogenic shock after acute myocardial infarction. Circulation 2007;116:2315—24.

[50] Go AS, Mozaffarian D, Roger VL, Benjamin EJ, Berry JD, Blaha MJ, et al. Heart disease and stroke statistics—2014 update: a report from the American Heart Association. Circulation 2014;129:e28—292.

[51] Wilkinson I, Cockcroft JR. Cholesterol, lipids and arterial stiffness. Adv Cardiol 2007; 44:261—77.

[52] Fagot-Campagna A, Balkau B, Simon D, Warnet JM, Claude JR, Ducimetiere P, et al. High free fatty acid concentration: an independent risk factor for hypertension in the paris prospective study. Int J Epidemiol 1998;27:808—13.

[53] Kenny LC, Broadhurst DI, Dunn W, Brown M, North RA, McCowan L, et al. Robust early pregnancy prediction of later preeclampsia using metabolomic biomarkers. Hypertension 2010;56:741—9.

[54] Odibo AO, Goetzinger KR, Odibo L, Cahill AG, Macones GA, Nelson DM, et al. First-trimester prediction of preeclampsia using metabolomic biomarkers: a discovery phase study. Prenat Diagn 2011;31:990−4.

[55] Boos CJ, Lip GY. Is hypertension an inflammatory process? Curr Pharm Des 2006;12: 1623−35.

[56] Zheng Y, Yu B, Alexander D, Mosley TH, Heiss G, Nettleton JA, et al. Metabolomics and incident hypertension among blacks: the atherosclerosis risk in communities study. Hypertension 2013;62:398−403.

[57] Holmes E, Loo RL, Stamler J, Bictash M, Yap IK, Chan Q, et al. Human metabolic phenotype diversity and its association with diet and blood pressure. Nature 2008; 453:396−400.

[58] Kang SM, Park JC, Shin MJ, Lee H, Oh J, Ryu do H, et al. (1)H nuclear magnetic resonance based metabolic urinary profiling of patients with ischemic heart failure. Clin Biochem 2011;44:293−9.

[59] Du Z, Shen A, Huang Y, Su L, Lai W, Wang P, et al. 1H-NMR-based metabolic analysis of human serum reveals novel markers of myocardial energy expenditure in heart failure patients. PLoS One 2014;9:e88102.

[60] Tenori L, Hu X, Pantaleo P, Alterini B, Castelli G, Olivotto I, et al. Metabolomic finger-print of heart failure in humans: a nuclear magnetic resonance spectroscopy analysis. Int J Cardiol 2013;168:e113−115.

[61] Dunn WB, Broadhurst DI, Deepak SM, Buch MH, McDowell G, Spasic I, et al. Serum metabolomics reveals many novel metabolic markers of heart failure, including pseudouridine and 2-oxoglutarate. Metabolomics Off J Metabolomic Soc 2007;3: 413−26.

[62] Mirandola SR, Melo DR, Schuck PF, Ferreira GC, Wajner M, Castilho RF. Methylmal-onate inhibits succinate-supported oxygen consumption by interfering with mito-chondrial succinate uptake. J Inherit Metabolic Dis 2008;31:44−54.

[63] Steffens DC, Wei J, Krishnan KR, Karoly ED, Mitchell MW, O'Connor CM, et al. Metab-olomic differences in heart failure patients with and without major depression. J Geriatr Psychiatry Neurol 2010;23:138−46.

[64] Zheng Y, Yu B, Alexander D, Manolio TA, Aguilar D, Coresh J, et al. Associations be-tween metabolomic compounds and incident heart failure among African Americans: the ARIC Study. Am J Epidemiol 2013;178:534−42.

[65] Samara MA, Tang WH, Cikach Jr F, Gul Z, Tranchito L, Paschke KM, et al. Single exhaled breath metabolomic analysis identifies unique breathprint in patients with acute decompensated heart failure. J Am Coll Cardiol 2013;61:1463−4.

[66] Marcondes-Braga FG, Gutz IG, Batista GL, Saldiva PH, Ayub-Ferreira SM, Issa VS, et al. Exhaled acetone as a new biomaker of heart failure severity. Chest 2012;142: 457−66.

[67] Lin D, Hollander Z, Meredith A, Stadnick E, Sasaki M, Cohen Freue G, et al. Molecular signatures of end-stage heart failure. J Cardiac Fail 2011;17:867−74.

[68] Wang J, Li Z, Chen J, Zhao H, Luo L, Chen C, et al. Metabolomic identification of diag-nostic plasma biomarkers in humans with chronic heart failure. Mol Biosyst 2013;9: 2618−26.

[69] Selvin E, Parrinello CM, Sacks DB, Coresh J. Trends in prevalence and control of dia-betes in the United States, 1988-1994 and 1999-2010. Ann Intern Med 2014;160:517−25.

[70] Floegel A, Stefan N, Yu Z, Muhlenbruch K, Drogan D, Joost HG, et al. Identification of serum metabolites associated with risk of type 2 diabetes using a targeted metabolo-mic approach. Diabetes 2013;62:639−48.

[71] Fiehn O, Garvey WT, Newman JW, Lok KH, Hoppel CL, Adams SH. Plasma metabo-lomic profiles reflective of glucose homeostasis in non-diabetic and type 2 diabetic obese African-American women. PLoS One 2010;5:e15234.

[72] Wurtz P, Tiainen M, Makinen VP, Kangas AJ, Soininen P, Saltevo J, et al. Circulating metabolite predictors of glycemia in middle-aged men and women. Diabetes Care 2012;35:1749—56.

[73] Wurtz P, Makinen VP, Soininen P, Kangas AJ, Tukiainen T, Kettunen J, et al. Metabolic signatures of insulin resistance in 7098 young adults. Diabetes 2012;61:1372—80.

[74] Laferrere B, Reilly D, Arias S, Swerdlow N, Gorroochurn P, Bawa B, et al. Differential metabolic impact of gastric bypass surgery versus dietary intervention in obese diabetic subjects despite identical weight loss. Sci Transl Med 2011;3:80re82.

[75] Kim DH, Sartor MA, Bain JR, Sandoval D, Stevens RD, Medvedovic M, et al. Rapid and weight-independent improvement of glucose tolerance induced by a peptide designed to elicit apoptosis in adipose tissue endothelium. Diabetes 2012;61: 2299—310.

[76] Shah SH, Crosslin DR, Haynes CS, Nelson S, Turer CB, Stevens RD, et al. Branched-chain amino acid levels are associated with improvement in insulin resistance with weight loss. Diabetologia 2012;55:321—30.

[77] Huffman KM, Shah SH, Stevens RD, Bain JR, Muehlbauer M, Slentz CA, et al. Relationships between circulating metabolic intermediates and insulin action in overweight to obese, inactive men and women. Diabetes Care 2009;32:1678—83.

[78] Wang TJ, Larson MG, Vasan RS, Cheng S, Rhee EP, McCabe E, et al. Metabolite profiles and the risk of developing diabetes. Nat Med 2011;17:448—53.

[79] Cheng S, Rhee EP, Larson MG, Lewis GD, McCabe EL, Shen D, et al. Metabolite profiling identifies pathways associated with metabolic risk in humans. Circulation 2012;125:2222—31.

[80] Escobar-Morreale HF, Samino S, Insenser M, Vinaixa M, Luque-Ramirez M, Lasuncion MA, et al. Metabolic heterogeneity in polycystic ovary syndrome is determined by obesity: plasma metabolomic approach using GC-MS. Clin Chem 2012;58: 999—1009.

[81] Pietilainen KH, Sysi-Aho M, Rissanen A, Seppanen-Laakso T, Yki-Jarvinen H, Kaprio J, et al. Acquired obesity is associated with changes in the serum lipidomic profile independent of genetic effects—a monozygotic twin study. PLoS One 2007; 2:e218.

[82] Barber MN, Risis S, Yang C, Meikle PJ, Staples M, Febbraio MA, et al. Plasma lysophosphatidylcholine levels are reduced in obesity and type 2 diabetes. PLoS One 2012;7:e41456.

[83] Lehmann R. Diabetes subphenotypes and metabolomics: the key to discovering laboratory markers for personalized medicine? Clin Chem 2013;59:1294—6.

[84] Newgard CB. Interplay between lipids and branched-chain amino acids in development of insulin resistance. Cell Metab 2012;15:606—14.

[85] Lehmann R, Franken H, Dammeier S, Rosenbaum L, Kantartzis K, Peter A, et al. Circulating lysophosphatidylcholines are markers of a metabolically benign nonalcoholic fatty liver. Diabetes Care 2013;36:2331—8.

[86] Wang-Sattler R, Yu Z, Herder C, Messias AC, Floegel A, He Y, et al. Novel biomarkers for pre-diabetes identified by metabolomics. Mol Syst Biol 2012;8:615.

[87] McCormick K, Mick GJ, Mattson V, Saile D, Starr D. Carnitine palmitoyltransferase: effects of diabetes, fasting, and ph on the reaction that generates acyl CoA. Metabolism Clin Exp 1988;37:1073—7.

[88] Oresic M. Metabolomics in the studies of islet autoimmunity and type 1 diabetes. Rev Diabet Stud RDS 2012;9:236—47.

[89] Dutta T, Chai HS, Ward LE, Ghosh A, Persson XM, Ford GC, et al. Concordance of changes in metabolic pathways based on plasma metabolomics and skeletal muscle transcriptomics in type 1 diabetes. Diabetes 2012;61:1004—16.

[90] (WHO) WHO. Global strategy on diet and physical activity. 2014.

[91] Dong C, Beecham A, Slifer S, Wang L, McClendon MS, Blanton SH, et al. Genome-wide linkage and peak-wide association study of obesity-related quantitative traits in Caribbean Hispanics. Hum Genet 2011;129:209—19.

[92] Renstrom F, Payne F, Nordstrom A, Brito EC, Rolandsson O, Hallmans G, et al. Replication and extension of genome-wide association study results for obesity in 4923 adults from northern Sweden. Hum Mol Genet 2009;18:1489—96.

[93] Meyre D, Delplanque J, Chevre JC, Lecoeur C, Lobbens S, Gallina S, et al. Genome-wide association study for early-onset and morbid adult obesity identifies three new risk loci in European populations. Nat Genet 2009;41:157—9.

[94] Newgard CB, An J, Bain JR, Muehlbauer MJ, Stevens RD, Lien LF, et al. A branched-chain amino acid-related metabolic signature that differentiates obese and lean humans and contributes to insulin resistance. Cell Metab 2009;9:311—26.

[95] Oberbach A, Bluher M, Wirth H, Till H, Kovacs P, Kullnick Y, et al. Combined proteomic and metabolomic profiling of serum reveals association of the complement system with obesity and identifies novel markers of body fat mass changes. J Proteome Res 2011;10:4769—88.

[96] Kim JY, Park JY, Kim OY, Ham BM, Kim HJ, Kwon DY, et al. Metabolic profiling of plasma in overweight/obese and lean men using ultra performance liquid chromatography and Q-TOF mass spectrometry (UPLC-Q-TOF MS). J Proteome Res 2010;9: 4368—75.

[97] Wang C, Feng R, Sun D, Li Y, Bi X, Sun C. Metabolic profiling of urine in young obese men using ultra performance liquid chromatography and Q-TOF mass spectrometry (UPLC/Q-TOF MS). J Chromatogr B Anal Technol Biomed Life Sci 2011;879:2871—6.

[98] Mihalik SJ, Michaliszyn SF, de las Heras J, Bacha F, Lee S, Chace DH, et al. Metabolomic profiling of fatty acid and amino acid metabolism in youth with obesity and type 2 diabetes: evidence for enhanced mitochondrial oxidation. Diabetes Care 2012;35: 605—11.

[99] Calvani R, Miccheli A, Capuani G, Tomassini Miccheli A, Puccetti C, Delfini M, et al. Gut microbiome-derived metabolites characterize a peculiar obese urinary metabotype. Int J Obes (Lond) 2010;34:1095—8.

[100] Fahy E, Sud M, Cotter D, Subramaniam S. Lipid maps online tools for lipid research. Nucleic Acids Res 2007;35:W606—12.

[101] Wallace C, Newhouse SJ, Braund P, Zhang F, Tobin M, Falchi M, et al. Genome-wide association study identifies genes for biomarkers of cardiovascular disease: serum urate and dyslipidemia. Am J Hum Genet 2008;82:139—49.

[102] Kettunen J, Tukiainen T, Sarin AP, Ortega-Alonso A, Tikkanen E, Lyytikainen LP, et al. Genome-wide association study identifies multiple loci influencing human serum metabolite levels. Nat Genet 2012;44:269—76.

[103] Shin SY, Fauman EB, Petersen AK, Krumsiek J, Santos R, Huang J, et al. An atlas of genetic influences on human blood metabolites. Nat Genet 2014;46:543—50.

[104] Illig T, Gieger C, Zhai GJ, Romisch-Margl W, Wang-Sattler R, Prehn C, et al. A genome-wide perspective of genetic variation in human metabolism. Nat Genet 2010;42: 137—41.

[105] Vaxillaire M, Cavalcanti-Proenca C, Dechaume A, Tichet J, Marre M, Balkau B, et al. The common P446L polymorphism in GCKR inversely modulates fasting glucose and triglyceride levels and reduces type 2 diabetes risk in the DESIR prospective general french population. Diabetes 2008;57:2253—7.

[106] Prokopenko I, Langenberg C, Florez JC, Saxena R, Soranzo N, Thorleifsson G, et al. Variants in mtnr1b influence fasting glucose levels. Nat Genet 2009;41:77—81.

[107] Kathiresan S, Melander O, Guiducci C, Surti A, Burtt NP, Rieder MJ, et al. Six new loci associated with blood low-density lipoprotein cholesterol, high-density lipoprotein cholesterol or triglycerides in humans. Nat Genet 2008;40:189—97.

[108] Petersen AK, Zeilinger S, Kastenmuller G, Romisch-Margl W, Brugger M, Peters A, et al. Epigenetics meets metabolomics: an epigenome-wide association study with blood serum metabolic traits. Hum Mol Genet 2014;23:534—45.

[109] Suhre K, Shin SY, Petersen AK, Mohney RP, Meredith D, Wagele B, et al. Human metabolic individuality in biomedical and pharmaceutical research. Nature 2011;477: 54—60.

[110] Krumsiek J, Suhre K, Evans AM, Mitchell MW, Mohney RP, Milburn MV, et al. Mining the unknown: a systems approach to metabolite identification combining genetic and metabolic information. PLoS Genet 2012;8:e1003005.

[111] Schellenberger J, Park JO, Conrad TM, Palsson BO. BIGG: a biochemical genetic and genomic knowledgebase of large scale metabolic reconstructions. BMC Bioinforma 2010;11:213.

[112] Caspi R, Altman T, Billington R, Dreher K, Foerster H, Fulcher CA, et al. The MetaCyc database of metabolic pathways and enzymes and the BioCyc collection of pathway/ genome databases. Nucleic Acids Res 2014;42:D459—71.

[113] Karp PD, Paley S, Altman T. Data mining in the MetaCyc family of pathway databases. Methods Mol Biol 2013;939:183—200.

[114] Caspi R, Altman T, Dreher K, Fulcher CA, Subhraveti P, Keseler IM, et al. The MetaCyc database of metabolic pathways and enzymes and the BioCyc collection of pathway/ genome databases. Nucleic Acids Res 2012;40:D742—53.

[115] Caspi R, Karp PD. Using the MetaCyc pathway database and the BioCyc database collection. editoral board. In: Baxevanis. Andreas D, et al., editors. Current protocols in bioinformatics; 2007. Chapter 1: Unit 1.17.

[116] Karp PD, Riley M, Paley SM, Pellegrini-Toole A. The MetaCyc database. Nucleic Acids Res 2002;30:59—61.

[117] de Matos P, Alcantara R, Dekker A, Ennis M, Hastings J, Haug K, et al. Chemical entities of biological interest: an update. Nucleic Acids Res 2010;38:D249—54.

[118] Degtyarenko K, de Matos P, Ennis M, Hastings J, Zbinden M, McNaught A, et al. ChEBI: a database and ontology for chemical entities of biological interest. Nucleic Acids Res 2008;36:D344—50.

[119] Scholz M, Fiehn O. SetupX—a public study design database for metabolomic projects. Pac Symp Biocomput 2007:169—80.

[120] Little JL, Williams AJ, Pshenichnov A, Tkachenko V. Identification of "known unknowns" utilizing accurate mass data and ChemSpider. J Am Soc Mass Spectrom 2012;23:179—85.

[121] Cui Q, Lewis IA, Hegeman AD, Anderson ME, Li J, Schulte CF, et al. Metabolite identification via the Madison Metabolomics Consortium Database. Nat Biotechnol 2008; 26:162—4.

[122] Skogerson K, Wohlgemuth G, Barupal DK, Fiehn O. The volatile compound binbase mass spectral database. BMC Bioinformatics 2011;12:321.

[123] Latendresse M, Paley S, Karp PD. Browsing metabolic and regulatory networks with BioCyc. Methods Mol Biol 2012;804:197—216.

[124] Horai H, Arita M, Kanaya S, Nihei Y, Ikeda T, Suwa K, et al. MassBank: a public repository for sharing mass spectral data for life sciences. J Mass Spectrom JMS 2010; 45:703—14.

[125] Choi C, Munch R, Leupold S, Klein J, Siegel I, Thielen B, et al. SYSTOMONAS—an integrated database for systems biology analysis of Pseudomonas. Nucleic Acids Res 2007;35:D533—7.

[126] Andersen MR, Nielsen ML, Nielsen J. Metabolic model integration of the bibliome, genome, metabolome and reactome of *Aspergillus niger*. Mol Syst Biol 2008;4:178.

[127] Vastrik I, D'Eustachio P, Schmidt E, Gopinath G, Croft D, de Bono B, et al. Reactome: a knowledge base of biologic pathways and processes. Genome Biol 2007;8:R39.

[128] Joshi-Tope G, Gillespie M, Vastrik I, D'Eustachio P, Schmidt E, de Bono B, et al. Reactome: a knowledgebase of biological pathways. Nucleic Acids Res 2005;33:D428—32.

[129] Kopka J, Schauer N, Krueger S, Birkemeyer C, Usadel B, Bergmuller E, et al. GMD@ CSB.DB: the Golm Metabolome Database. Bioinformatics 2005;21:1635—8.

[130] Salek RM, Haug K, Steinbeck C. Dissemination of metabolomics results: role of MetaboLights and COSMOS. Gigascience 2013;2:8.

[131] Salek RM, Haug K, Conesa P, Hastings J, Williams M, Mahendraker T, et al. The metabolights repository: curation challenges in metabolomics. Database J Biol Databases Curation 2013;2013:bat029. http://dx.doi.org/10.1093/database/bat029.

[132] Haug K, Salek RM, Conesa P, Hastings J, de Matos P, Rijnbeek M, et al. MetaboLights—an open-access general-purpose repository for metabolomics studies and associated meta-data. Nucleic Acids Res 2013;41:D781—6.

[133] Steinbeck C, Conesa P, Haug K, Mahendraker T, Williams M, Maguire E, et al. Metabolights: towards a new cosmos of metabolomics data management. Metabolomics Off J Metabolomic Soc 2012;8:757—60.

[134] Zhu ZJ, Schultz AW, Wang J, Johnson CH, Yannone SM, Patti GJ, et al. Liquid chromatography quadrupole time-of-flight mass spectrometry characterization of metabolites guided by the METLIN database. Nat Protoc 2013;8:451—60.

[135] Tautenhahn R, Cho K, Uritboonthai W, Zhu Z, Patti GJ, Siuzdak G. An accelerated workflow for untargeted metabolomics using the METLIN database. Nat Biotechnol 2012;30:826—8.

[136] Sana TR, Roark JC, Li X, Waddell K, Fischer SM. Molecular formula and METLIN personal metabolite database matching applied to the identification of compounds generated by LC/TOF-MS. J Biomol Tech JBT 2008;19:258—66.

[137] Smith CA, O'Maille G, Want EJ, Qin C, Trauger SA, Brandon TR, et al. METLIN: a metabolite mass spectral database. Ther Drug Monit 2005;27:747—51.

2

Modern Transcriptomics and Small RNA Diversity

Kasey C. Vickers

Department of Medicine, Vanderbilt University, School of Medicine, Nashville, TN, USA

1. INTRODUCTION

Due to significant advances in high-throughput RNA sequencing (RNAseq) methods and informatics support after 2010, the field of modern genomics exploded with life and many new RNA species were identified in an ever deepening mammalian transcriptome. Nevertheless, to fully grasp the complexity of modern genomics, one must recognize the early history of genome-scale gene expression analysis. Since the mid-1990s, conventional hybridization arrays have been used to generate mRNA expression profiles and essentially created an entirely new field of transcriptomics that was accompanied by new statistical approaches and bioinformatics support. The term "transcriptomics" was the first of now many "-omics" that are found in the literature and was coined in 1996 as a term to classify the complete set of mRNA expression values in specific cells [1]. The first reported profile of gene expression was published in 1991 by J. Craig Venter's group, which released a database of expressed sequence tags generated by automated Sanger sequencing [2]. Dr Venter later became world recognized for his work on the human genome project. Over the next two decades, single and dual color hybridization gene expression arrays dominated whole-genome mRNA expression profiling. Microarray technologies set forth many of the systems biology and holistic scientific approaches that became favored over more conventional reductionist gene-by-gene biology. These techniques allowed investigators to survey complete pathways at genome scale for both candidate and blinded gene expression changes in biological contexts and diseases. Nonetheless, by 2014 gene expression arrays were being phased out in academic core labs and science in general for sequencing-by-synthesis techniques (next-generation sequencing, NGS), which arrived in 2008. Short-read massive parallel sequencing, the science behind NGS, has facilitated the rapid development of a diverse set of high-throughput DNA (DNAseq) and RNAseq strategies that are being applied to sequence whole genomes, classify DNA variance and single nucleotide polymorphisms, and profile small and long RNA transcripts. RNAseq is a class of methods that are designed to profile both coding (mRNA) and non-coding RNAs. The two most popular approaches are total RNAseq for gene (mRNA) expression profiling and small RNAseq (smRNAseq) for non-coding smRNA profiling, namely microRNAs (miRNA) analysis.

RNAseq methods have many advantages over conventional microarray platforms, including the ability to quantify gene expression at higher genomic resolution. Moreover, RNAseq methods provide absolute expression values compared with less reliable relative signals generated by microarrays. This gives more confidence in data quality, particularly for highly and lowly abundant transcripts, as RNAseq actually counts extreme transcripts opposed to fluorescent detection with microarrays. Likewise, background noise (signal) that is often a problem with microarrays is not an issue with RNAseq due to increased signal-to-noise ratios. Gene expression arrays rely on specific hybridization to immobilized probes, which creates high levels of background noise arising from cross-hybridization or sub-optimal hybridization kinetics. For smRNAs, namely miRNAs, the RNAseq has many scientific advantages over microarray technologies [3]. For example, miRNA microarrays are often limited by probe design and specificity. miRNAs of the same family often only differ by one or two nucleotides, which can generate non-specific binding with microarrays. In addition, microarray probe melting temperatures can vary wildly and often require high-temperature hybridizations to overcome non-specific binding and cross-hybridization issues. Another key feature, in which RNAseq has a major advantage, is its ability to identify and characterize novel transcripts and unannotated RNA species. After 2013, many gene expression cores in academic institutions began to transition away from microarray platform support studies for both scientific and business reasons, and promote projects based on RNAseq technologies for mRNA and miRNA. Due to diminishing customer demand and ever-decreasing sample volumes, many cores are not able to financially support expression arrays, and to do so would drive microarray costs higher than RNAseq per sample costs. As with many technological advances, the costs associated with RNAseq were initially prohibitive; however, by 2014 the price had dropped significantly, making it a direct, albeit superior, competitor to microarray technology. The costs of downstream informatics and data storage now represent greater platform challenges. Nonetheless, with the combination of competitive pricing and superior informative data, RNAseq technologies are the tools of choice for expression profiling.

2. SMALL NON-CODING RNAs

Most RNAs can be grouped into three functional classes: translational, regulatory, or other. Many long non-coding RNAs (lncRNAs) with established function, for example, transfer RNAs (tRNA) in protein translation, are cleaved into small non-coding RNAs (sncRNAs) which have a diverse set of alternative functions, including post-transcriptional gene regulation. Long RNAs are often classified as RNAs longer than 200 nucleotides (nts) in length and refer to as mRNAs, anti-sense RNAs, pseudogenes, and lncRNA. Many functional RNAs are much shorter (100–300 nts) than long (>1 kb) mRNAs or lncRNAs and are referred to here as intermediate RNAs. These include tRNAs, ribosomal RNAs (rRNA), Y RNAs, small nuclear RNAs (snRNA), small nucleolar RNAs (snoRNA), and many others. Most interestingly, it is recognized that most, if not all, classes of long and intermediate RNAs are further processed to produce non-coding smRNAs that are generally less than 40 nts in length. Here we detail the diversity of smRNA classes in the mammalian transcriptome and highlight their biological functions; however, many smRNAs have not been extensively studied and their physiological relevance remains to be determined. Nevertheless, there is tremendous depth and complexity to smRNAs in both cells and extracellular fluids, and they have enormous potential as novel biomarkers and drug targets in cardiometabolic diseases.

2.1 MicroRNAs

The most widely studied non-coding smRNAs are miRNAs that are 19–22 nts in length and post-transcriptionally regulate mRNA targets through complementary binding sites. miRNAs were first discovered by Victor Ambros, Gary Ruvkun, and colleagues in 1993 when they discovered that lin-4S (short) in *Caenorhabditis elegans* (*C. elegans*) was produced by lin-4L (long) and was complementary to sequences in the LIN-14 mRNA, a molecule lin-4s was found to repress [4,5]. It was almost 7 years later when Gary Ruvkun and colleagues found a highly conserved miRNA (let-7) in *C. elegans* and confirmed earlier studies that a non-coding smRNA regulatory pathway exists that suppresses gene expression through post-transcriptional mechanisms [6,7]. After 2000, research on miRNAs exploded, and by 2014 they have been studied in every biological context in plants and animals, as evidenced by over 29,000 miRNA papers in Pubmed Central (pubmed.com, 2014), with over 7000 papers released in 2013 alone. Over time miRNAs have proven to be critical regulators of many biological processes and contribute significantly to metabolic homeostasis and cardiovascular function [8—10]. For example, miR-27b has been reported as a regulatory hub in lipid metabolism and regulates key metabolic genes, including peroxisome proliferator-activated receptor gamma and glycerol-3-phosphate acyltransferase [11]. Many miRNAs, including miR-27b, miR-33a/b, and miR-144, have been found to regulate cholesterol transport, particularly through the regulation of ATP-binding cassette transporter A1 (ABCA1) and cholesterol efflux [12—20]. Moreover, numerous miRNAs have been found to regulate key molecular mechanisms associated with cardiovascular disease, including inflammation (miR-21 and miR-223) and atherosclerosis (miR-126) [21—26]. In diabetes research, miR-144, miR-375, miR-7, miR-103/107, and miR-802 have been reported to regulate glycemic control through various mechanisms [27—34]. Overall, the field of miRNAs in cardiometabolic disease is quite extensive and these small, but powerful regulators of translation and mRNA stability are well established in cardiovascular research. As such, specific miRNAs are now being explored as drug targets to prevent and treat cardiometabolic disease, including miR-208 and miR-103/107 [35]. Although miRNAs have been extensively studied in humans since 2001, modern RNAseq methods have identified many novel miRNAs that are deposited in the annotated miRNA database (miRBase.org). In 2013, miRBase contained over 1500 human miRNAs; however, many submissions may ultimately prove not to be real miRNAs or the miRNAs will be found to be processed from other non-coding RNAs, and thus will be annotated as another type of non-coding smRNA. As such, defining a real miRNA is increasingly difficult. Generally, if an sncRNA (1) has uniform 5′ processed ends (5′ phosphate), (2) possesses Dicer-like 3′ 2 nt overhang processed ends, (3) originates from a folded hairpin structure, and (4) sequenced reads map to both arms (5′ and 3′) of the pre-miRNA, then it is considered a bona fide miRNA. Subsequent evidence that the putative miRNA is loaded into the RNA-induced silencing complex (RISC) and represses target gene (mRNA) translation adds confidence to the miRNA annotation. Still, there are exceptions to even these simple rules. miRtrons are processed from short-length introns and bypass the nuclear DGCR8-Drosha microprocessor [36,37]. Another class of miRNAs originates from nascent transcripts of promoters of protein coding genes enriched at transcriptional start sites (TSSs, TSS-miRNA), and like miRtrons, also bypass Drosha cleavage [38]. In the cytoplasm, some mature miRNAs (e.g., miR-451) have been reported to be processed independent of Dicer, a key RNase III miRNA processing enzyme [39]. The majority of mature miRNAs are loaded onto Argonaute (AGO) family member [1—4] ribonucleoprotein complexes, namely RISC—60% AGO2-RISC and 30% AGO1-RISC [40—42]. It is in this complex where miRNAs recognize target sites within 3′ untranslated regions (3′ UTR) and coding region sites of mRNA targets using partial complementarity through critical seed regions (bases two to seven on the 5′ end of the miRNA) [43—46]. Based on the small requirement of seed-based complementary binding, miRNAs are promiscuous and regulate potentially hundreds of genes. Likewise, depending upon the length of the mRNA's 3′ UTR, each gene is likely regulated by dozens of miRNAs. Therefore, miRNAs work in a complex network of direct and indirect effects that have proven to be difficult to unravel. Nevertheless, miRNAs are critical players in metabolic homeostasis and will continue to be studied as a means to improve cardiometabolic function. In addition to miRNAs, AGO(1—4)-RISC also loads other smRNAs, including tRNA-derived fragments (tDRs); therefore, AGO-RISC loading is not a defining characteristic for verification of a real miRNA, but most miRNAs are loaded onto AGO-RISC. Future investigation into AGO-RISC bound smRNAs and their molecular targets will likely provide a greater fundamental understanding of post-transcriptional gene regulation by miRNAs and non-miRNAs smRNAs in cardiometabolic disease.

2.2 tRNA-Derived smRNAs

Most non-coding smRNA species are named for their parent RNA from which they were derived. For example, one of the most abundant and well-studied classes of non-miRNA smRNAs are tRNA-derived smRNAs (tDRs). Parent tRNAs have the ability to produce a variety of functional smRNAs that generally fall into two distinct classifications—tRNA-derived fragments (tRFs, about 22 nts) and tRNA-derived halves (tRHs, about 33 nts) [47—52]. For most tDRs, their molecular mechanisms of action and physiological relevance remain to be determined; however, both tRFs and tRHs likely contribute to post-transcriptional gene regulation or general protein synthesis repression in some capacity [47,53—55]. During cell stress, parent tRNAs are cleaved by the RNase III enzyme angiogenin to produce tRHs that repress protein translation through multiple mechanisms [47,51,54,56—59]. tRHs have been reported to bind to and sequester translational elongation factors (eIFs, YB-1), interfere with translational machinery, and destabilize mRNAs through sequence-independent recognition [47,51,54,56,59]. Angiogenin activity is regulated by nuclear sequestration and its cognate inhibitor, ribonuclease/angiogenin inhibitor 1 (RNH1) [52,56]. During stress, angiogenin leaves the nucleus, dissociates from RNH1, and localizes to cytoplasmic stress granules where it cleaves mature tRNAs. As such, tRHs and angiogenin support cell survival by attenuating global protein synthesis in states of stress. Most interestingly,

RNA methylation has been found to protect tRNAs from cleavage, and thus regulate the tDR signature produced [55]. Different cell stresses, for example, hypoxia or energy imbalance, result in the production of different tDRs subsets [48,51,52,55,57,60]. Due to their robust response to cell stress, particularly oxidative stress and ischemia, and their roles as regulators of gene expression, tDRs likely contribute substantially to cardiometabolic homeostasis and systemic responses to metabolic disease. Similar to miR-NAs, tDRs are stably present in blood and other fluids where they likely contribute to cell-to-cell communication mediated by extracellular vesicles or lipoproteins [58]. The two subclasses (tRFs and tRHs) have distinct biogenesis mechanisms and likely serve different biological functions. tRFs can be produced by various cleavage events, including RNaseZ processing of 3' uridine tRFs from the 3' trailer sequence of premature tRNAs [47]. Single-stranded smRNAs can also be produced from the 5' leader sequence of tRNAs by RNaseP [47]. Mature tRNAs can also undergo cleavage by Dicer which produces a double-stranded smRNA that yields a 5' tRF and a 3' CCA tRF. The 3' terminal end of a mature tRNA undergoes CCA terminal additions in the nucleus. tDRs, namely tRFs, have been reported to act like miRNAs and repress putative mRNA targets through complementary binding; however, they may also repress protein translation through other unknown mechanisms [47,53]. Collectively, we know very little about the roles of tDRs in pathophysiology. In some samples, tDRs (tRFs and tRHs) are the most abundant class of smRNAs found in smRNAseq datasets, more so than even miRNAs. Nevertheless, unlike miRNAs, the biological functions of tDRs largely remain undefined.

2.3 Small Nuclear RNA

In the nucleus, there is a diverse array of intermediate length and small RNA (snRNA), most of which are sequestered to specific nuclear sub-organelles. snRNA is a class of RNA species, but also refers to a set of specific structural RNAs found in the nucleus. Similar to tRNAs, smRNAs (23−30 nts) can also be processed from long snRNAs. Recently, smRNAs cleaved from U1, U2, and U12 snRNAs were found in complex with AGO proteins in the cytoplasm, suggesting that cleaved smRNAs likely act as miRNAs and post-transcriptioanlly regulate gene expression in the cytoplasm [53,61]. snoRNAs and Cajal body-specific RNAs (scaRNA) are two snRNA class subspecies with different biological functions than snRNAs. snoRNAs are enriched in the nucleolus and scaRNA are restricted to nuclear Cajal bodies. snoRNAs are of intermediate length (60−300 nts) and are often hosted in introns of both ncRNAs and mRNAs, and thus are transcribed by RNA Pol II; however, some evidence of Pol III transcription has been reported [62]. snoRNAs are divided into two classes based on RNA motifs that aid in their respective functions. C/D-box snoRNAs (SNORD) contain structural motif boxes C (RUGAUGA) and D (CUGA) that facilitate 2'O-ribose RNA methylation. Primary targets include ribosomal RNAs (rRNA), transfer RNAs (tRNA), and other smRNAs [63−66]. Highly methylated RNA has been reported to be protected from RNAse hydrolysis and may be a form of RNA-mediated stability or gene regulation [55,67]. H/ACA-box snoRNAs (SNORA) harbor H (ANANNA) and ACA (ACA) boxes that aid in pseudouridylation of targeted RNAs. scaRNAs contain structural motifs similar to both SNORDs and SNORAs, and thus, facilitate the pseudouridylaton and methylation of target spliceosomal RNAs. Studies are ongoing to identify direct nuclear RNA targets of scaRNAs, SNORDs, and SNORAs; however, snoRNAs without identified targets are known as orphan snoRNAs. Most interestingly, snoRNAs can also be exported out of the nucleus, processed by the RNAseIII enzyme Dicer, associate with AGO, and function as "miRNAs" in post-transcriptional gene regulatory networks [53,63,68]. For example, the SNORA ACA45 is cleaved into a 21 nt smRNA that targets and represses CDC2L6 (cyclin-dependent kinase 19, CDK19), and thus likely regulates transcription due to CDK19 association with the mediator activation complex [68]. This snoRNA-derived smRNA was originally identified as mmu-miR-1839. In addition to ACA45, many other snoRNA-derived smRNAs (sdRNA) have been identified from both parent SNORDs and SNORAs [63,68]. sdRNAs from SNORAs are about 20−24 nts in length and are processed from the 3' end of the parent snoRNA. Conversely, SNORD-derived sdRNAs have a bimodal size distribution (17−19, >27 nts in length) and are predominantly processed from the 5' end [63]. In this study, the expression of a majority of the sdRNAs were dependent upon Drosha; however, many, if not all, were dependent upon Dicer processing [63]. Most interestingly, sdRNA expression is likely a highly regulated process as opposed to degradation of parent snoRNAs, as the most abundant sdRNAs originate from lowly abundant snoRNAs [63]. While the central function of parent snoRNAs in the nucleolus is to guide RNA modifications of rRNA, tRNA, and smRNAs, cleaved products of these intermediate length RNAs likely serve as post-transcriptional regulators of gene expression in the cytoplasm. Similar to miRNAs and tDRs, sdRNAs are also present in serum/plasma and have been explored as potential biomarkers for cancer [69,70].

Most interestingly, over 10 canonical bona fide miRNAs have recently been found to be stably present in nuclei, specifically nucleoli, which are the sub-organelle that are highly-enriched in snoRNAs and produces sdRNAs; however, the export of sdRNAs and the import of miRNAs into the nucleolus have not been linked [71]. RNAs generally require a 5' cap to be shuttled between the cytoplasm and nucleus; however, miRNAs that lack 5' caps may not leave the nucleus and might be processed in the nucleus. Nucleolar miRNA abundance was found to be not dependent upon Dicer; however, CMR1 was found to be required, as CMR1 facilitates the localization of snoRNAs to the nucleolus [71]. Moreover, importin 8 has been found to contribute to the translocation of cytoplasmic miRNAs to the nucleus [72]. A key database for snoRNAs is snoRNA-LBME-db (www-snora.biotoul.fr), and human annotations and coordinates can be found using the UCSC Genome Browser wgRNA track [63,73].

2.4 Y RNA-Derived miRNAs

The Ro ribonucleoprotein (RNP) complex consists of non-coding Y RNAs and two autoantigenic proteins Ro and La [74,75]. In humans, there are four types of Y RNAs (hY1, hY3, hY4, and hY5) which are 80−115 nts in length and are largely cytoplasmic where they bind to and inhibit Ro, and thus, attenuate RNA degradation [76]. Parent Y RNAs are transcribed by Pol III from a single cluster on chromosome 7. Y RNAs have a 5' triphosphate, non-modified terminal 3' ribose, and Y RNA nucleotides for the most part are not modified [77]. Parent Y RNAs sequester Ro RNPs to the cytoplasm, and similar to tRNAs, are cleaved during cell stress, an event that facilitates nuclear enrichment of Ro proteins [78]. Y RNAs have also been demonstrated to contribute to DNA replication and cellular RNA control, particularly RNA degradation upon cell stress [79,80]. Human Y RNAs and Y RNA pseudogenes [81] are stem-looped structures, similar in size to pre-miRNAs, and yield a diverse set of sncRNAs that likely post-transcriptionally regulate mRNA targets like miRNAs [77]. Moreover, Y RNA-derived smRNAs are referred to as miY RNA due to their similarity in size, structure, and possible function to bona fide miRNAs. Two miRNAs, hsa-miR-1975 (miR-hY5-3p) and hsa-miR-1979 (miR-hY3-3p), were originally thought to be miRNAs, but now are recognized as miY RNAs as they are processed from the 3' end of hY5 and hY3, respectively [58,82]. In addition to these miY RNAs, MID-19433 (miR-hY1-5p) and MID-19434 (miR-hY3-5p) have been identified in various cells and processed from 3' regions of hY1 and hY3, respectively [83]. Furthermore, miY RNAs have also been reported to be processed from the 5' regions of parent Y RNA, including hsa-miR-1975*, which is now referred to as miR-hY5-5p [58]. Although very similar to miRNAs, Y RNA nuclear and cytoplasmic processing steps are independent of the miRNA pathway [82]. Although there are differing reports, it does appear that miY RNAs can associate with AGO2 of the RISC [82]. Most interestingly, miY RNAs have been found in extracellular vesicles and may represent novel biomarkers or contribute to cell-to-cell communication [84]. The RNP complex has been implicated in autoimmune diseases, namely systemic lupus erythematosus, in which autoantibodies form against Ro and La proteins [85,86]. miY RNAs may also contribute to autoimmunity through post-transcriptionally regulating specific mRNA targets; however, this remains to be determined.

2.5 Vault RNA-Derived smRNAs

Vaults are tiny barrel-shaped structures in the cytoplasm consisting of one vault RNA and three proteins—major vault protein, vault poly (ADP-ribose) polymerase (VPRAP), and telemorase-associated protein 1 (TEP1) [87,88]. Although the biological functions of vaults are unknown, these small (35 × 65 nm) hollow structures may serve as cellular cargo carriers for proteins and other small molecules [88−90]. smRNAs derived from vault RNA (svRNA) have been found to be dependent upon Dicer processing, and svRNAs have been reported to be associated with AGO2-RISC and silence mRNA targets, including CYP3A4 that likely contributes to multidrug resistance [91]. svRNAs, particularly from the internal stem loop of parent vault RNAs, have been reported to be packaged and exported in extracellular vesicles (e.g., exosomes) from inflammatory cells [84]. Determining the relationship between parent vault RNA function and targeting of svRNAs to mRNAs, as well as the role of svRNA in intercellular communication remains to be determined.

2.6 Endogenous siRNAs

The human genome is roughly 3 Gbp with only about 1% encoding the 22,330 protein coding genes. About 44% of the genome is composed of repetitive sequences. Recent evidence suggests that a wide variety of non-coding smRNAs are being produced from these repetitive regions [92]. In plants, a variety of endogenous siRNAs have been reported, including *trans*-acting siRNAs (tasiRNA), *cis*-acting siRNAs (casiRNA), and natural antisense transcript-derived siRNAs (natsiRNA). Mammals also have endo-siRNAs; however, they appear to be restricted to oocytes and stem cells, likely due to the interferon response to dsRNA precursors in somatic cells [93,94]. Mammalian endo-siRNAs (21 nts) have been reported to originate from retrotransposons of long tandem repeats, long interspersed elements, or short interspersed elements [95−97]. Moreover, endo-siRNAs may also arise from dsRNA hairpins resulting from convergent transcription at a specific loci or nascent mRNAs forming dsRNA complexes with pseudogenes in oocytes [95,96]. Nonetheless, endo-siRNAs processing is Drosha independent and Dicer dependent, and endo-siRNAs are found in complex with AGO2 in mammals [95,96,98]. Similar to endo-siRNAs, PIWI-interacting RNAs (piRNA) also arise from transposons and repetitive regions of the genome. piRNAs are also restricted to germ line cells in mammals and are slightly longer than endo-siRNAs (25–31 nts) [99]. Most interestingly, endo-siRNAs and piRNAs likely suppress transposons in germ line cells to prevent mutations from generational propagation. A detailed review of piRNAs can be found here [100,101].

2.7 Chromosome-Associated smRNAs

As detailed above, the diversity of smRNAs has rapidly expanded with the widespread use of high-throughput smRNA sequencing and evidence now suggests that many different elements of the genome produce noncoding smRNAs. Due to the diversity of their origin, a collection of smRNAs derived from specific regions of the genome are referred to as simply chromatin-associated smRNAs (CAsRNA). Centromeres are chromosomal regions that link sister chromatids. Centromere repeat-associated siRNAs (crasiRNA) originate from long dsRNAs produced by bidirectional promoters, and possibly satellite transcripts, which are increased during cell stress [102,103]. CrasiRNAs (34–42 nts in length) associate with centromatic proteins and contribute to heterchromatin formation and chromatin structure [102]. These smRNAs are likely Dicer dependent, as Dicer loss-of-function studies resulted in loss of kinetocore proteins and heterochromatin structure abnormalities; however, this remains to be definitively tested [104]. Non-coding smRNAs can also be produced by telomeres at chromosome ends. Telomere-specific smRNAs (tel-sRNA, 24 nts) are found in mammals, but unlike other smRNAs, tel-sRNAs processing is Dicer independent [105]. Although their functions have not been extensively studied, tel-sRNAs likely contribute to telomere length control and heterochromatin formation [105,106]. Most interestingly, tel-sRNAs are 2′-O-methylated at their 3′ ends that may interfere with column-based RNA isolation methods and smRNA library preparations.

During transcription, Pol II pauses, sputters, and restarts which generates small transcripts immediately downstream of TSSs. These short nascent transcripts are then processed into transcription initiation RNAs (tiRNAs) [107,108]. tiRNAs (18 nts) are one species in a class of promoter-associated RNAs (paRNA) that are highly enriched in the nucleus [109]. Small paRNAs (PASR, 20–200 nts) were identified in 2007 and are processed from 5′ capped nascent transcripts independent of Dicer [110,111]. PASRs can also be generated from bidirectional (promoter) transcription and arise from transcripts upstream of activated gene TSSs [92,112]. For example, PROMoter uPstream Transcripts (PROMPTs) have been identified as transcriptional regulators [113]. Although their biological functions have not been widely studied, PASRs likely contribute to transcriptional regulation through targeting antisense transcripts and epigenetic silencing complexes [114,115]. sncRNAs are also produced at splice junctions. Splice-site associated smRNAs (spliRNA), which are also 18 nts in length, are produced from regions immediately upstream of splice donor sites of internal exons [116]. Collectively, little is known about the biological diversity and functional relevance of CAsRNAs. Nevertheless, they hold great potential like many other non-coding smRNAs to better understand the human genome and gene regulation.

2.8 rRNA-Derived smRNAs

Not to be overlooked, smRNAs can also be produced from ribosomal RNAs (rRNA). Due to their critical role in peptide formation, rRNAs are found in all cells and organisms, and are the most abundant class of RNA by a wide-margin, representing about 80% of total RNA in a cell. These ancient RNAs form subunits of ribosomes and are critical to protein translation. rRNA genes are clustered into tandem repeats divided by internally transcribed spacers. Strikingly, specific miRNAs have been found to be processed from these transcribed, but rapidly degraded spacers—miR-712 (mouse) and miR-663 (humans) [117]. Human rRNA is constructed of large and small subunits that add up to about 7.2 kb in length. Nonetheless, we now appreciate that long rRNA can be cleaved into distinct rRNA-derived smRNA species (rsRNA) which have been found to be loaded into AGO2-RISC, and thus are predicted to post-transcriptionally regulate gene expression [118]. Together tRNAs and rRNAs make up >95% of total cellular RNA and both produce sncRNAs that can act as miRNAs and post-transcriptionally regulate gene expression or general protein translation. Like many of the other smRNAs described, the functional roles of srRNAs are unknown, but like tRNAs could potentially promote cellular response to stress.

2.9 Conclusion

Since 2008, there has been a tremendous growth in the diversity of non-coding RNAs due to advances in high-through smRNA sequencing. In addition to the non-coding smRNAs described above, there are even more examples of non-coding smRNAs that have not been extensively classified or characterized. These include miRNA-offset RNAs (moR) and short hairpin RNAs (shRNA) [98,116,119]. Likewise, there are many other types of smRNAs found in prokaryotes and other classes, and it remains to be determined if there is homology in mammals or similar biological processes. With few exceptions, all non-coding smRNAs are processed from long or intermediate transcripts. miRNAs, the most-widely studied smRNA, are themselves processed from long primary transcripts (pri-miRNA, most 1—10 kb) in the nucleus and hairpin precursors (pre-miRNA, 70—90 nt) in the cytoplasm [120]. Although the deep transcriptome holds enormous potential to identify novel bio-markers, drug targets, or even new biological pathways and regulator networks, the study of transcriptome smRNA diversity is in its infancy. As such, the biological roles of these smRNAs in cardiometabolic disease are almost entirely unknown. One exception is that hepatic levels of srRNAs have been found to be altered and correlate with diabetes [118].

As described above, the current method of choice to profile the deep transcriptome is high-throughput smRNAseq. Most interestingly, mining smRNAseq datasets reveals that miRNAs are only a small fraction of smRNAs found in these datasets. Many of these other smRNAs have only recently been identified and their biological functions are unknown. Furthermore, a significant fraction of smRNAs found in cells and tissues are completely novel and have not been classified or annotated. Current NGS platforms for smRNAseq are producing gigabases of sequencing data per run that results in hundreds of megabases per sample that allows for outstanding coverage depth into the smRNA signature and facilitates novel discovery of lowly abundant species. Therefore, the future is extremely bright in the quest to identify mediators or drug targets of cardiometabolic disease.

References

[1] Pietu G, Mariage-Samson R, Fayein NA, Matingou C, Eveno E, Houlgatte R, et al. The Genexpress IMAGE knowledge base of the human brain transcriptome: a prototype integrated resource for functional and computational genomics. Genome Res February 1999;9(2):195—209. PubMed PMID: 10022985. Pubmed Central PMCID: 310711.

[2] Adams MD, Kelley JM, Gocayne JD, Dubnick M, Polymeropoulos MH, Xiao H, et al. Complementary DNA sequencing: expressed sequence tags and human genome project. Science June 21, 1991;252(5013):1651—6. PubMed PMID: 2047873.

[3] Git A, Dvinge H, Salmon-Divon M, Osborne M, Kutter C, Hadfield J, et al. Systematic comparison of microarray profiling, real-time PCR, and next-generation sequencing technologies for measuring differential microRNA expression. RNA May 2010;16(5): 991—1006. PubMed PMID: 20360395. Pubmed Central PMCID: 2856892.

[4] Lee RC, Feinbaum RL, Ambros V. The C. elegans heterochronic gene lin-4 encodes small RNAs with antisense complementarity to lin-14. Cell December 3, 1993;75(5): 843—54. PubMed PMID: 8252621. Epub 1993/12/03. eng.

[5] Wightman B, Ha I, Ruvkun G. Posttranscriptional regulation of the heterochronic gene lin-14 by lin-4 mediates temporal pattern formation in C. elegans. Cell December 3, 1993;75(5):855—62. PubMed PMID: 8252622.

[6] Reinhart BJ, Slack FJ, Basson M, Pasquinelli AE, Bettinger JC, Rougvie AE, et al. The 21-nucleotide let-7 RNA regulates developmental timing in Caenorhabditis elegans. Nature February 24, 2000;403(6772):901—6. PubMed PMID: 10706289.

[7] Slack FJ, Basson M, Liu Z, Ambros V, Horvitz HR, Ruvkun G. The lin-41 RBCC gene acts in the C. elegans heterochronic pathway between the let-7 regulatory RNA and the LIN-29 transcription factor. Mol Cell April 2000;5(4):659—69. PubMed PMID: 10882102.

[8] Fernandez-Hernando C, Suarez Y, Rayner KJ, Moore KJ. MicroRNAs in lipid metabolism. Curr Opin Lipidol April 2011;22(2):86–92. PubMed PMID: 21178770. Pubmed Central PMCID: 3096067.

[9] Zampetaki A, Mayr M. MicroRNAs in vascular and metabolic disease. Circ Res February 3, 2012;110(3):508–22. PubMed PMID: 22302757.

[10] Vickers KC, Rye KA, Tabet F. MicroRNAs in the onset and development of cardiovascular disease. Clin Sci February 2014;126(3):183–94. PubMed PMID: 24102098. Pubmed Central PMCID: 3873876.

[11] Vickers KC, Shoucri BM, Levin MG, Wu H, Pearson DS, Osei-Hwedieh D, et al. MicroRNA-27b is a regulatory hub in lipid metabolism and is altered in dyslipidemia. Hepatology February 2013;57(2):533–42. PubMed PMID: 22777896. Pubmed Central PMCID: 3470747.

[12] Zhang M, Wu JF, Chen WJ, Tang SL, Mo ZC, Tang YY, et al. MicroRNA-27a/b regulates cellular cholesterol efflux, influx and esterification/hydrolysis in THP-1 macrophages. Atherosclerosis May 2014;234(1):54–64. PubMed PMID: 24608080.

[13] de Aguiar Vallim TQ, Tarling EJ, Kim T, Civelek M, Baldan A, Esau C, et al. MicroRNA-144 regulates hepatic ATP binding cassette transporter A1 and plasma high-density lipoprotein after activation of the nuclear receptor farnesoid X receptor. Circ Res June 7, 2013;112(12):1602–12. PubMed PMID: 23519696.

[14] Ramirez CM, Rotllan N, Vlassov AV, Davalos A, Li M, Goedeke L, et al. Control of cholesterol metabolism and plasma HDL levels by miRNA-144. Circ Res 2013; 112(12):1592–601. PubMed PMID: 23519695.

[15] Gerin I, Clerbaux LA, Haumont O, Lanthier N, Das AK, Burant CF, et al. Expression of miR-33 from an SREBP2 intron inhibits cholesterol export and fatty acid oxidation. J Biol Chem October 29, 2010;285(44):33652–61. PubMed PMID: 20732877. Pubmed Central PMCID: 2962463.

[16] Horie T, Ono K, Horiguchi M, Nishi H, Nakamura T, Nagao K, et al. MicroRNA-33 encoded by an intron of sterol regulatory element-binding protein 2 (Srebp2) regulates HDL in vivo. Proc Natl Acad Sci USA October 5, 2010;107(40):17321–6. PubMed PMID: 20855588. Pubmed Central PMCID: 2951399.

[17] Marquart TJ, Allen RM, Ory DS, Baldan A. miR-33 links SREBP-2 induction to repression of sterol transporters. Proc Natl Acad Sci USA July 6, 2010;107(27): 12228–32. PubMed PMID: 20566875. Pubmed Central PMCID: 2901433.

[18] Najafi-Shoushtari SH, Kristo F, Li Y, Shioda T, Cohen DE, Gerszten RE, et al. MicroRNA-33 and the SREBP host genes cooperate to control cholesterol homeostasis. Science June 18, 2010;328(5985):1566–9. PubMed PMID: 20466882. Pubmed Central PMCID: 3840500.

[19] Rayner KJ, Suarez Y, Davalos A, Parathath S, Fitzgerald ML, Tamehiro N, et al. MiR-33 contributes to the regulation of cholesterol homeostasis. Science June 18, 2010; 328(5985):1570–3. PubMed PMID: 20466885. Pubmed Central PMCID: 3114628.

[20] Rayner KJ, Fernandez-Hernando C, Moore KJ. MicroRNAs regulating lipid metabolism in atherogenesis. Thromb Haemost April 2012;107(4):642–7. PubMed PMID: 22274626.

[21] Kumar S, Kim CW, Simmons RD, Jo H. Role of flow-sensitive microRNAs in endothelial dysfunction and atherosclerosis: mechanosensitive athero-miRs. Arterioscler Thromb Vasc Biol 2014;34(10):2206–16. PubMed PMID: 25012134.

[22] Schober A, Nazari-Jahantigh M, Wei Y, Bidzhekov K, Gremse F, Grommes J, et al. MicroRNA-126-5p promotes endothelial proliferation and limits atherosclerosis by suppressing Dlk1. Nat Med April 2014;20(4):368–76. PubMed PMID: 24584117.

[23] Fan X, Wang E, Wang X, Cong X, Chen X. MicroRNA-21 is a unique signature associated with coronary plaque instability in humans by regulating matrix metalloproteinase-9 via reversion-inducing cysteine-rich protein with Kazal motifs. Exp Mol Pathol April 2014;96(2):242–9. PubMed PMID: 24594117.

[24] Bauernfeind F, Rieger A, Schildberg FA, Knolle PA, Schmid-Burgk JL, Hornung V. NLRP3 inflammasome activity is negatively controlled by miR-223. J Immunol October 15, 2012;189(8):4175—81. PubMed PMID: 22984082. Epub 2012/09/18. eng.

[25] Zhuang G, Meng C, Guo X, Cheruku PS, Shi L, Xu H, et al. A novel regulator of macrophage activation: miR-223 in obesity-associated adipose tissue inflammation. Circulation June 12, 2012;125(23):2892—903. PubMed PMID: 22580331.

[26] Tabet F, Vickers KC, Cuesta Torres LF, Wiese CB, Shoucri BM, Lambert G, et al. HDL-transferred microRNA-223 regulates ICAM-1 expression in endothelial cells. Nat Commun 2014;5:3292. PubMed PMID: 24576947.

[27] Trajkovski M, Hausser J, Soutschek J, Bhat B, Akin A, Zavolan M, et al. MicroRNAs 103 and 107 regulate insulin sensitivity. Nature June 30, 2011;474(7353):649—53. PubMed PMID: 21654750.

[28] Poy MN, Eliasson L, Krutzfeldt J, Kuwajima S, Ma X, Macdonald PE, et al. A pancreatic islet-specific microRNA regulates insulin secretion. Nature November 11, 2004;432(7014):226—30. PubMed PMID: 15538371.

[29] Lynn FC. Meta-regulation: microRNA regulation of glucose and lipid metabolism. Trends Endocrinol Metab November 2009;20(9):452—9. PubMed PMID: 19800254.

[30] Poy MN, Hausser J, Trajkovski M, Braun M, Collins S, Rorsman P, et al. miR-375 maintains normal pancreatic alpha- and beta-cell mass. Proc Natl Acad Sci USA April 7, 2009;106(14):5813—8. PubMed PMID: 19289822. Pubmed Central PMCID: 2656556.

[31] Karolina DS, Armugam A, Tavintharan S, Wong MT, Lim SC, Sum CF, et al. MicroRNA 144 impairs insulin signaling by inhibiting the expression of insulin receptor substrate 1 in type 2 diabetes mellitus. PLoS One 2011;6(8):e22839. PubMed PMID: 21829658. Pubmed Central PMCID: 3148231.

[32] Rottiers V, Naar AM. MicroRNAs in metabolism and metabolic disorders. Nat Rev Mol Cell Biol April 2012;13(4):239—50. PubMed PMID: 22436747. Pubmed Central PMCID: 4021399.

[33] Latreille M, Hausser J, Stutzer I, Zhang Q, Hastoy B, Gargani S, et al. MicroRNA-7a regulates pancreatic beta cell function. J Clin Invest June 2, 2014;124(6):2722—35. PubMed PMID: 24789908. Pubmed Central PMCID: 4038573.

[34] Kornfeld JW, Baitzel C, Konner AC, Nicholls HT, Vogt MC, Herrmanns K, et al. Obesity-induced overexpression of miR-802 impairs glucose metabolism through silencing of Hnf1b. Nature February 7, 2013;494(7435):111—5. PubMed PMID: 23389544.

[35] Li Z, Rana TM. Therapeutic targeting of microRNAs: current status and future challenges. Nat Rev Drug Discov 2014;13(8):622—38. PubMed PMID: 25011539.

[36] Okamura K, Hagen JW, Duan H, Tyler DM, Lai EC. The mirtron pathway generates microRNA-class regulatory RNAs in Drosophila. Cell July 13, 2007;130(1):89—100. PubMed PMID: 17599402. Pubmed Central PMCID: 2729315.

[37] Ruby JG, Jan CH, Bartel DP. Intronic microRNA precursors that bypass Drosha processing. Nature July 5, 2007;448(7149):83—6. PubMed PMID: 17589500. Pubmed Central PMCID: 2475599.

[38] Zamudio JR, Kelly TJ, Sharp PA. Argonaute-bound small RNAs from promoter-proximal RNA polymerase II. Cell February 27, 2014;156(5):920—34. PubMed PMID: 24581493.

[39] Yang JS, Maurin T, Robine N, Rasmussen KD, Jeffrey KL, Chandwani R, et al. Conserved vertebrate mir-451 provides a platform for Dicer-independent, Ago2-mediated microRNA biogenesis. Proc Natl Acad Sci USA August 24, 2010;107(34): 15163—8. PubMed PMID: 20699384. Pubmed Central PMCID: 2930549.

[40] Kim VN, Han J, Siomi MC. Biogenesis of small RNAs in animals. Nat Rev Mol Cell Biol February 2009;10(2):126—39. PubMed PMID: 19165215.

[41] Azuma-Mukai A, Oguri H, Mituyama T, Qian ZR, Asai K, Siomi H, et al. Characterization of endogenous human Argonautes and their miRNA partners in RNA silencing. Proc Natl Acad Sci USA June 10, 2008;105(23):7964—9. PubMed PMID: 18524951. Pubmed Central PMCID: 2430345.

[42] Wang D, Zhang Z, O'Loughlin E, Lee T, Houel S, O'Carroll D, et al. Quantitative functions of Argonaute proteins in mammalian development. Genes Dev April 1, 2012;26(7):693—704. PubMed PMID: 22474261. Pubmed Central PMCID: 3323880.

[43] Baek D, Villen J, Shin C, Camargo FD, Gygi SP, Bartel DP. The impact of microRNAs on protein output. Nature September 4, 2008;455(7209):64—71. PubMed PMID: 18668037. Pubmed Central PMCID: 2745094.

[44] Bartel DP. MicroRNAs: target recognition and regulatory functions. Cell January 23, 2009;136(2):215—33. PubMed PMID: 19167326.

[45] Friedman RC, Farh KK, Burge CB, Bartel DP. Most mammalian mRNAs are conserved targets of microRNAs. Genome Res January 2009;19(1):92—105. PubMed PMID: 18955434. Pubmed Central PMCID: 2612969.

[46] Guo H, Ingolia NT, Weissman JS, Bartel DP. Mammalian microRNAs predominantly act to decrease target mRNA levels. Nature August 12, 2010;466(7308):835—40. PubMed PMID: 20703300. Pubmed Central PMCID: 2990499.

[47] Haussecker D, Huang Y, Lau A, Parameswaran P, Fire AZ, Kay MA. Human tRNA-derived small RNAs in the global regulation of RNA silencing. RNA April 2010; 16(4):673—95. PubMed PMID: 20181738. Pubmed Central PMCID: 2844617.

[48] Thompson DM, Lu C, Green PJ, Parker R. tRNA cleavage is a conserved response to oxidative stress in eukaryotes. RNA October 2008;14(10):2095—103. PubMed PMID: 18719243. Pubmed Central PMCID: 2553748.

[49] Cole C, Sobala A, Lu C, Thatcher SR, Bowman A, Brown JW, et al. Filtering of deep sequencing data reveals the existence of abundant Dicer-dependent small RNAs derived from tRNAs. RNA December 2009;15(12):2147—60. PubMed PMID: 19850906. Pubmed Central PMCID: 2779667.

[50] Elbarbary RA, Takaku H, Uchiumi N, Tamiya H, Abe M, Takahashi M, et al. Modulation of gene expression by human cytosolic tRNase Z(L) through 5'-half-tRNA. PLoS One 2009;4(6):e5908. PubMed PMID: 19526060. Pubmed Central PMCID: 2691602.

[51] Fu H, Feng J, Liu Q, Sun F, Tie Y, Zhu J, et al. Stress induces tRNA cleavage by angiogenin in mammalian cells. FEBS Lett January 22, 2009;583(2):437—42. PubMed PMID: 19114040.

[52] Saikia M, Krokowski D, Guan BJ, Ivanov P, Parisien M, Hu GF, et al. Genome-wide identification and quantitative analysis of cleaved tRNA fragments induced by cellular stress. J Biol Chem December 14, 2012;287(51):42708—25. PubMed PMID: 23086926. Pubmed Central PMCID: 3522271.

[53] Burroughs AM, Ando Y, de Hoon MJ, Tomaru Y, Suzuki H, Hayashizaki Y, et al. Deep-sequencing of human Argonaute-associated small RNAs provides insight into miRNA sorting and reveals Argonaute association with RNA fragments of diverse origin. RNA Biol January—February 2011;8(1):158—77. PubMed PMID: 21282978. Pubmed Central PMCID: 3127082.

[54] Ivanov P, Emara MM, Villen J, Gygi SP, Anderson P. Angiogenin-induced tRNA fragments inhibit translation initiation. Mol Cell August 19, 2011;43(4):613—23. PubMed PMID: 21855800. Pubmed Central PMCID: 3160621.

[55] Schaefer M, Pollex T, Hanna K, Tuorto F, Meusburger M, Helm M, et al. RNA methylation by Dnmt2 protects transfer RNAs against stress-induced cleavage. Genes Dev August 1, 2010;24(15):1590—5. PubMed PMID: 20679393. Pubmed Central PMCID: 2912555.

[56] Gong B, Lee YS, Lee I, Shelite TR, Kunkeaw N, Xu G, et al. Compartmentalized, functional role of angiogenin during spotted fever group rickettsia-induced endothelial

barrier dysfunction: evidence of possible mediation by host tRNA-derived small noncoding RNAs. BMC Infect Dis 2013;13:285. PubMed PMID: 23800282. Pubmed Central PMCID: 3699377.

[57] Thompson DM, Parker R. Stressing out over tRNA cleavage. Cell July 23, 2009;138(2): 215−9. PubMed PMID: 19632169.

[58] Dhahbi JM, Spindler SR, Atamna H, Yamakawa A, Boffelli D, Mote P, et al. 5′ tRNA halves are present as abundant complexes in serum, concentrated in blood cells, and modulated by aging and calorie restriction. BMC Genomics 2013;14:298. PubMed PMID: 23638709. Pubmed Central PMCID: 3654920.

[59] Maute RL, Schneider C, Sumazin P, Holmes A, Califano A, Basso K, et al. tRNA-derived microRNA modulates proliferation and the DNA damage response and is down-regulated in B cell lymphoma. Proc Natl Acad Sci USA January 22, 2013; 110(4):1404−9. PubMed PMID: 23297232. Pubmed Central PMCID: 3557069.

[60] Yamasaki S, Ivanov P, Hu GF, Anderson P. Angiogenin cleaves tRNA and promotes stress-induced translational repression. J Cell Biol April 6, 2009;185(1):35−42. PubMed PMID: 19332886. Pubmed Central PMCID: 2700517.

[61] Chen CJ, Heard E. Small RNAs derived from structural non-coding RNAs. Methods September 1, 2013;63(1):76−84. PubMed PMID: 23684746.

[62] Ikegami K, Lieb JD. Integral nuclear pore proteins bind to Pol III-transcribed genes and are required for Pol III transcript processing in C. elegans. Mol Cell September 26, 2013;51(6):840−9. PubMed PMID: 24011592. Pubmed Central PMCID: 3788088.

[63] Taft RJ, Glazov EA, Lassmann T, Hayashizaki Y, Carninci P, Mattick JS. Small RNAs derived from snoRNAs. RNA July 2009;15(7):1233−40. PubMed PMID: 19474147. Pubmed Central PMCID: 2704076.

[64] Esteller M. Non-coding RNAs in human disease. Nat Rev Genet December 2011; 12(12):861−74. PubMed PMID: 22094949.

[65] Li Z, Ender C, Meister G, Moore PS, Chang Y, John B. Extensive terminal and asymmetric processing of small RNAs from rRNAs, snoRNAs, snRNAs, and tRNAs. Nucleic Acids Res August 2012;40(14):6787−99. PubMed PMID: 22492706. Pubmed Central PMCID: 3413118.

[66] Valleron W, Laprevotte E, Gautier EF, Quelen C, Demur C, Delabesse E, et al. Specific small nucleolar RNA expression profiles in acute leukemia. Leukemia September 2012;26(9):2052−60. PubMed PMID: 22522792.

[67] Stepanov GA, Semenov DV, Kuligina EV, Koval OA, Rabinov IV, Kit YY, et al. Analogues of artificial human box C/D small nucleolar RNA as regulators of alternative splicing of a pre-mRNA target. Acta Naturae January 2012;4(1):32−41. PubMed PMID: 22708061. Pubmed Central PMCID: 3372990.

[68] Ender C, Krek A, Friedlander MR, Beitzinger M, Weinmann L, Chen W, et al. A human snoRNA with microRNA-like functions. Mol Cell November 21, 2008;32(4):519−28. PubMed PMID: 19026782. Epub 2008/11/26. eng.

[69] Liao J, Yu L, Mei Y, Guarnera M, Shen J, Li R, et al. Small nucleolar RNA signatures as biomarkers for non-small-cell lung cancer. Mol Cancer 2010;9:198. PubMed PMID: 20663213. Pubmed Central PMCID: 2919450.

[70] Gee HE, Buffa FM, Camps C, Ramachandran A, Leek R, Taylor M, et al. The small-nucleolar RNAs commonly used for microRNA normalisation correlate with tumour pathology and prognosis. Br J Cancer March 29, 2011;104(7):1168−77. PubMed PMID: 21407217. Pubmed Central PMCID: 3068486.

[71] Bai B, Liu H, Laiho M. Small RNA expression and deep sequencing analyses of the nucleolus reveal the presence of nucleolus-associated microRNAs. FEBS Open Bio 2014;4:441−9. PubMed PMID: 24918059. Pubmed Central PMCID: 4050192.

[72] Wei Y, Li L, Wang D, Zhang CY, Zen K. Importin 8 regulates the transport of mature microRNAs into the cell nucleus. J Biol Chem April 11, 2014;289(15):10270−5. PubMed PMID: 24596094. Pubmed Central PMCID: 4036152.

[73] Karolchik D, Kuhn RM, Baertsch R, Barber GP, Clawson H, Diekhans M, et al. The UCSC genome browser database: 2008 update. Nucleic Acids Res January 2008; 36(Database issue):D773–9. PubMed PMID: 18086701. Pubmed Central PMCID: 2238835.

[74] Lerner MR, Boyle JA, Hardin JA, Steitz JA. Two novel classes of small ribonucleoproteins detected by antibodies associated with lupus erythematosus. Science January 23, 1981;211(4480):400–2. PubMed PMID: 6164096.

[75] Wolin SL, Steitz JA. The Ro small cytoplasmic ribonucleoproteins: identification of the antigenic protein and its binding site on the Ro RNAs. Proc Natl Acad Sci USA April 1984;81(7):1996–2000. PubMed PMID: 6201849. Pubmed Central PMCID: 345423.

[76] Perreault J, Perreault JP, Boire G. Ro-associated Y RNAs in metazoans: evolution and diversification. Mol Biol Evol August 2007;24(8):1678–89. PubMed PMID: 17470436.

[77] Verhagen AP, Pruijn GJ. Are the Ro RNP-associated Y RNAs concealing microRNAs? Y RNA-derived miRNAs may be involved in autoimmunity. BioEssays September 2011;33(9):674–82. PubMed PMID: 21735459.

[78] Sim S, Weinberg DE, Fuchs G, Choi K, Chung J, Wolin SL. The subcellular distribution of an RNA quality control protein, the Ro autoantigen, is regulated by noncoding Y RNA binding. Mol Biol Cell March 2009;20(5):1555–64. PubMed PMID: 19116308. Pubmed Central PMCID: 2649258.

[79] Gardiner TJ, Christov CP, Langley AR, Krude T. A conserved motif of vertebrate Y RNAs essential for chromosomal DNA replication. RNA July 2009;15(7):1375–85. PubMed PMID: 19474146. Pubmed Central PMCID: 2704080.

[80] Chen X, Taylor DW, Fowler CC, Galan JE, Wang HW, Wolin SL. An RNA degradation machine sculpted by Ro autoantigen and noncoding RNA. Cell March 28, 2013;153(1): 166–77. PubMed PMID: 23540697. Pubmed Central PMCID: 3646564.

[81] Perreault J, Noel JF, Briere F, Cousineau B, Lucier JF, Perreault JP, et al. Retropseudogenes derived from the human Ro/SS-A autoantigen-associated hY RNAs. Nucleic Acids Res 2005;33(6):2032–41. PubMed PMID: 15817567. Pubmed Central PMCID: 1074747.

[82] Nicolas FE, Hall AE, Csorba T, Turnbull C, Dalmay T. Biogenesis of Y RNA-derived small RNAs is independent of the microRNA pathway. FEBS Lett April 24, 2012; 586(8):1226–30. PubMed PMID: 22575660.

[83] Meiri E, Levy A, Benjamin H, Ben-David M, Cohen L, Dov A, et al. Discovery of microRNAs and other small RNAs in solid tumors. Nucleic Acids Res October 2010;38(18): 6234–46. PubMed PMID: 20483914. Pubmed Central PMCID: 2952848.

[84] Nolte-'t Hoen EN, Buermans HP, Waasdorp M, Stoorvogel W, Wauben MH, t Hoen PA. Deep sequencing of RNA from immune cell-derived vesicles uncovers the selective incorporation of small non-coding RNA biotypes with potential regulatory functions. Nucleic Acids Res October 2012;40(18):9272–85. PubMed PMID: 22821563. Pubmed Central PMCID: 3467056.

[85] Lerner MR, Steitz JA. Antibodies to small nuclear RNAs complexed with proteins are produced by patients with systemic lupus erythematosus. Proc Natl Acad Sci USA November 1979;76(11):5495–9. PubMed PMID: 316537. Pubmed Central PMCID: 411675.

[86] Peek R, Pruijn GJ, van der Kemp AJ, van Venrooij WJ. Subcellular distribution of Ro ribonucleoprotein complexes and their constituents. J Cell Sci November 1993; 106(Pt 3):929–35. PubMed PMID: 7508449.

[87] Kedersha NL, Rome LH. Isolation and characterization of a novel ribonucleoprotein particle: large structures contain a single species of small RNA. J Cell Biol September 1986;103(3):699–709. PubMed PMID: 2943744. Pubmed Central PMCID: 2114306.

[88] Berger W, Steiner E, Grusch M, Elbling L, Micksche M. Vaults and the major vault protein: novel roles in signal pathway regulation and immunity. Cell Mol Life Sci January 2009;66(1):43–61. PubMed PMID: 18759128.

[89] Hamill DR, Suprenant KA. Characterization of the sea urchin major vault protein: a possible role for vault ribonucleoprotein particles in nucleocytoplasmic transport. Dev Biol October 1, 1997;190(1):117–28. PubMed PMID: 9331335.

[90] Poderycki MJ, Kickhoefer VA, Kaddis CS, Raval-Fernandes S, Johansson E, Zink JI, et al. The vault exterior shell is a dynamic structure that allows incorporation of vault-associated proteins into its interior. Biochemistry October 3, 2006;45(39): 12184–93. PubMed PMID: 17002318. Pubmed Central PMCID: 2538551.

[91] Persson H, Kvist A, Vallon-Christersson J, Medstrand P, Borg A, Rovira C. The non-coding RNA of the multidrug resistance-linked vault particle encodes multiple regulatory small RNAs. Nat Cell Biol October 2009;11(10):1268–71. PubMed PMID: 19749744.

[92] Kawaji H, Nakamura M, Takahashi Y, Sandelin A, Katayama S, Fukuda S, et al. Hidden layers of human small RNAs. BMC Genomics 2008;9:157. PubMed PMID: 18402656. Pubmed Central PMCID: 2359750.

[93] Yang S, Tutton S, Pierce E, Yoon K. Specific double-stranded RNA interference in undifferentiated mouse embryonic stem cells. Mol Cell Biol November 2001; 21(22):7807–16. PubMed PMID: 11604515. Pubmed Central PMCID: 99950.

[94] Wianny F, Zernicka-Goetz M. Specific interference with gene function by double-stranded RNA in early mouse development. Nat Cell Biol February 2000;2(2):70–5. PubMed PMID: 10655585.

[95] Watanabe T, Totoki Y, Toyoda A, Kaneda M, Kuramochi-Miyagawa S, Obata Y, et al. Endogenous siRNAs from naturally formed dsRNAs regulate transcripts in mouse oocytes. Nature May 22, 2008;453(7194):539–43. PubMed PMID: 18404146.

[96] Tam OH, Aravin AA, Stein P, Girard A, Murchison EP, Cheloufi S, et al. Pseudogene-derived small interfering RNAs regulate gene expression in mouse oocytes. Nature May 22, 2008;453(7194):534–8. PubMed PMID: 18404147. Pubmed Central PMCID: 2981145.

[97] Watanabe T, Takeda A, Tsukiyama T, Mise K, Okuno T, Sasaki H, et al. Identification and characterization of two novel classes of small RNAs in the mouse germline: retrotransposon-derived siRNAs in oocytes and germline small RNAs in testes. Genes Dev July 1, 2006;20(13):1732–43. PubMed PMID: 16766679. Pubmed Central PMCID: 1522070.

[98] Babiarz JE, Ruby JG, Wang Y, Bartel DP, Blelloch R. Mouse ES cells express endogenous shRNAs, siRNAs, and other Microprocessor-independent, Dicer-dependent small RNAs. Genes Dev October 15, 2008;22(20):2773–85. PubMed PMID: 18923076. Pubmed Central PMCID: 2569885.

[99] Aravin AA, Naumova NM, Tulin AV, Vagin VV, Rozovsky YM, Gvozdev VA. Double-stranded RNA-mediated silencing of genomic tandem repeats and transposable elements in the D. melanogaster germline. Curr Biol July 10, 2001;11(13):1017–27. PubMed PMID: 11470406.

[100] Ghildiyal M, Zamore PD. Small silencing RNAs: an expanding universe. Nat Rev Genet February 2009;10(2):94–108. PubMed PMID: 19148191. Pubmed Central PMCID: 2724769.

[101] Malone CD, Hannon GJ. Small RNAs as guardians of the genome. Cell February 20, 2009;136(4):656–68. PubMed PMID: 19239887. Pubmed Central PMCID: 2792755.

[102] Carone DM, Longo MS, Ferreri GC, Hall L, Harris M, Shook N, et al. A new class of retroviral and satellite encoded small RNAs emanates from mammalian centromeres. Chromosoma February 2009;118(1):113–25. PubMed PMID: 18839199.

[103] Valgardsdottir R, Chiodi I, Giordano M, Cobianchi F, Riva S, Biamonti G. Structural and functional characterization of noncoding repetitive RNAs transcribed in stressed human cells. Mol Biol Cell June 2005;16(6):2597–604. PubMed PMID: 15788562. Pubmed Central PMCID: 1142408.

[104] Fukagawa T, Nogami M, Yoshikawa M, Ikeno M, Okazaki T, Takami Y, et al. Dicer is essential for formation of the heterochromatin structure in vertebrate cells. Nat Cell Biol August 2004;6(8):784—91. PubMed PMID: 15247924.

[105] Cao F, Li X, Hiew S, Brady H, Liu Y, Dou Y. Dicer independent small RNAs associate with telomeric heterochromatin. RNA July 2009;15(7):1274—81. PubMed PMID: 19460867. Pubmed Central PMCID: 2704082.

[106] Horard B, Gilson E. Telomeric RNA enters the game. Nat Cell Biol February 2008; 10(2):113—5. PubMed PMID: 18246034.

[107] Core LJ, Lis JT. Transcription regulation through promoter-proximal pausing of RNA polymerase II. Science March 28, 2008;319(5871):1791—2. PubMed PMID: 18369138. Pubmed Central PMCID: 2833332. Epub 2008/03/29. eng.

[108] Core LJ, Waterfall JJ, Lis JT. Nascent RNA sequencing reveals widespread pausing and divergent initiation at human promoters. Science December 19, 2008;322(5909): 1845—8. PubMed PMID: 19056941. Pubmed Central PMCID: 2833333.

[109] Taft RJ, Kaplan CD, Simons C, Mattick JS. Evolution, biogenesis and function of promoter-associated RNAs. Cell Cycle August 2009;8(15):2332—8. PubMed PMID: 19597344.

[110] Seila AC, Calabrese JM, Levine SS, Yeo GW, Rahl PB, Flynn RA, et al. Divergent transcription from active promoters. Science December 19, 2008;322(5909):1849—51. PubMed PMID: 19056940. Pubmed Central PMCID: 2692996.

[111] Kapranov P, Cheng J, Dike S, Nix DA, Duttagupta R, Willingham AT, et al. RNA maps reveal new RNA classes and a possible function for pervasive transcription. Science June 8, 2007;316(5830):1484—8. PubMed PMID: 17510325.

[112] Uesaka M, Nishimura O, Go Y, Nakashima K, Agata K, Imamura T. Bidirectional promoters are the major source of gene activation-associated non-coding RNAs in mammals. BMC Genomics 2014;15:35. PubMed PMID: 24438357. Pubmed Central PMCID: 3898825.

[113] Jacquier A. The complex eukaryotic transcriptome: unexpected pervasive transcription and novel small RNAs. Nat Rev Genet December 2009;10(12):833—44. PubMed PMID: 19920851.

[114] Morris KV, Santoso S, Turner AM, Pastori C, Hawkins PG. Bidirectional transcription directs both transcriptional gene activation and suppression in human cells. PLoS Genet November 2008;4(11):e1000258. PubMed PMID: 19008947. Pubmed Central PMCID: 2576438.

[115] Schwartz JC, Younger ST, Nguyen NB, Hardy DB, Monia BP, Corey DR, et al. Antisense transcripts are targets for activating small RNAs. Nat Struct Mol Biol August 2008;15(8):842—8. PubMed PMID: 18604220. Pubmed Central PMCID: 2574822.

[116] Taft RJ, Simons C, Nahkuri S, Oey H, Korbie DJ, Mercer TR, et al. Nuclear-localized tiny RNAs are associated with transcription initiation and splice sites in metazoans. Nat Struct Mol Biol August 2010;17(8):1030—4. PubMed PMID: 20622877.

[117] Son DJ, Kumar S, Takabe W, Kim CW, Ni CW, Alberts-Grill N, et al. The atypical mechanosensitive microRNA-712 derived from pre-ribosomal RNA induces endothelial inflammation and atherosclerosis. Nat Commun 2013;4:3000. PubMed PMID: 24346612. Pubmed Central PMCID: 3923891.

[118] Wei H, Zhou B, Zhang F, Tu Y, Hu Y, Zhang B, et al. Profiling and identification of small rDNA-derived RNAs and their potential biological functions. PLoS One 2013; 8(2):e56842. PubMed PMID: 23418607. Pubmed Central PMCID: 3572043.

[119] Yi R, Pasolli HA, Landthaler M, Hafner M, Ojo T, Sheridan R, et al. DGCR8-dependent microRNA biogenesis is essential for skin development. Proc Natl Acad Sci USA January 13, 2009;106(2):498—502. PubMed PMID: 19114655. Pubmed Central PMCID: 2626731.

[120] Saini HK, Enright AJ, Griffiths-Jones S. Annotation of mammalian primary microRNAs. BMC Genomics 2008;9:564. PubMed PMID: 19038026. PubMed Central PMCID: 2632650.

3

A Systems-Level Understanding of Cardiovascular Disease through Networks

Charles R. Farber, Larry D. Mesner

Departments of Public Health Sciences and Biochemistry and Molecular Genetics, Center for Public Health Genomics, University of Virginia, Charlottesville, VA, USA

1. INTRODUCTION

Cardiovascular disease (CVD) is a diverse constellation of diseases that affect the heart and blood vessels and is the leading cause of death worldwide [1]. The etiology of CVD is multifactorial, with genetics, the environment and genetic by environmental interactions playing a role [1]. Due to it significant impact on human health, considerable effort has been made on understanding the molecular, cellular, and physiological events that contribute to CVD. These investigations have employed a wide range of approaches including molecular biology, biochemistry, physiology, and genetics. Such studies have identified many of the major genes, signaling events, and pathways that influence CVD. Although the tide is beginning to turn, these investigations have mainly been reductionist in nature, focusing on the role of individual cellular components. However, complex systems are not the sum of their parts and genes, and biological pathways are not independent; rather they interrelate in complex, non-additive, and often unpredictable ways (Figure 1) [2]. Moreover, the actions of any single part of the system will never be representative of the behavior of the entire system. As a result, it is hard to capture the essence of a biological system by studying its single components in isolation.

Experiments that have studied the individual parts of biological systems have provided tremendous insight into many aspects of CVD. It has

GENOME   TRANSCRIPTOME   PROTEOME   METABOLOME   PHENOME

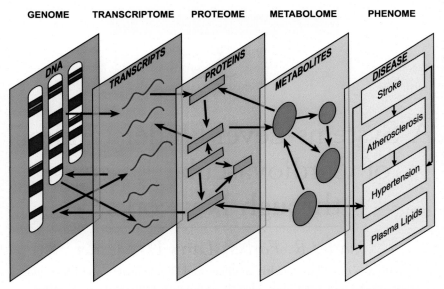

FIGURE 1   Biological systems are complex and composed of discrete scales including the genome, transcriptome, proteome, metabolome, and phenome. The lines between the stages illustrate the effects of genetic perturbations and the various interactions, which exist across all scales. As discussed in the text, biological networks are an efficient and powerful way to study interactions in complex biological systems. *Reprinted with permission from ref. [47].*

begun to be appreciated, however, that in order to comprehensively understand the molecular basis of complex CVD traits, such as atherosclerosis or hypertension, it is critical that the functions of individual genes and proteins be placed in a system's context [3]. In other words, the interactions between cellular components, not just the components themselves, must be identified, studied, and understood. Although a daunting task, the interactions will need to be characterized in all clinically important cell types, in response to genetic and environmental perturbations as well as across time.

One way to capture how cellular components relate and interact is through the generation and study of networks. Networks describe how components of a system relate to one another. Just as schematic diagrams are used to represent electrical circuits (or networks), biological networks describe cellular circuitry. Over the past few years, pioneering work in the area of network theory has revealed that cellular interactions are not random, but instead are governed by a set of organizing principles. These principles are inherent to nearly all biological networks studied to date. Importantly, these properties can be leveraged to generate important insight into cellular and biological function. In this chapter, we provide an overview of the general properties of networks and then describe how they are being used to dissect the molecular basis of disease, with an emphasis on the integration of genetic and network-based approaches for CVD.

2. WHAT IS A NETWORK?

Networks consist of two fundamental entities: nodes and edges. Nodes represent the components of a system under investigation and the edges that connect nodes represent relationships. For example, in a social network nodes are individuals and those who are friends are connected by edges. Nodes in a biological network often represent protein-coding gene products (transcripts and/or proteins), but they can also be metabolites, post-translational modifications, microRNAs/other non-coding RNAs, epigenetic marks, and diseases/phenotypes. The meaning of an edge in a biological network is dependent on the type of network. For instance, in a protein—protein interaction (PPI) network an edge represents a physical interaction between two proteins, whereas an edge in co-expression network signifies that two genes are co-expressed to some degree. Networks can range in size from tens of nodes connected by tens of edges to large genome-wide networks with thousands of nodes and millions of edges.

3. NETWORK ORGANIZING PRINCIPLES

3.1 Scale-Free Topology

The distribution of the number of links (edges) between nodes in a network can either be random or non-random [2]. In a random network, the number of connections per node follows a normal distribution. As a result, most nodes in a random network have a similar number of connections (the mean of the distribution) or "scale," with few nodes possessing a large or small number of links. About 15 years ago, Barabasi and colleagues demonstrated that real world networks are not random [4]. Rather, the degree distribution (the distribution of the number of links per node) of most real-world networks follows a Poisson distribution. His group described such networks as having a "scale-free" topology. In a scale-free network, a small number of nodes are highly connected while most are lowly connected (Figure 2). Barabasi demonstrated that networks such as the World Wide Web had a scale-free topology. When thought of as a network, the Internet consists of websites (nodes) connected by hyperlinks (edges). Barabasi et al. observed that the Internet

(a)   **(b)**

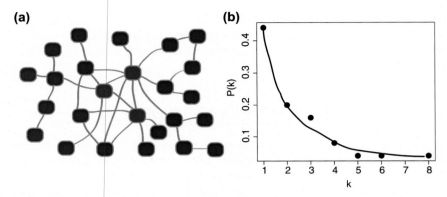

FIGURE 2   Biological networks exhibit scale-free topology. (a) An illustrated example of a scale-free network in which a small number of hubs (green nodes) participate in a large number of edges (black lines), while most (blue nodes) are sparsely connected. (b) The fraction (P(k)) of nodes with k links for the network in A demonstrating its scale-free topology.

consisted of a small number of websites with a high number of links to other sites (e.g., Google) and a much larger number of websites that had few links (e.g., http://cphg.virginia.edu/farber/—our lab's webpage). The consequence of scale-free networks is that any two nodes in a network can be connected to each other through a small number of paths (where one path length is node—edge—node). This idea is conceptualized in the theory of "six degrees of separation" formalized by Frigyes Karinthy in which any two people on the planet are separated by a maximum of six acquaintances (popularized by the parlor game "six degrees of Kevin Bacon").

It has also been demonstrated that most biological networks, independent of their type, are scale free [2,5]. There are a number of theories as to why biological networks have evolved to be scale free. One of the most intriguing hypotheses is that as genomes have duplicated across evolution, any gene that is connected to a duplicated gene will also be connected to the new gene [6,7]. As a result, highly connected genes receive the largest number of new connections as a result of each duplication event; that is, the rich get richer [4]. From a functional standpoint, scale-free networks are efficient and robust to perturbation, at least in part, because most of their nodes are sparsely linked [4].

3.2 Modularity

A second organizing principle of networks is the fact that they are modular and modules are an important concept in the analysis and

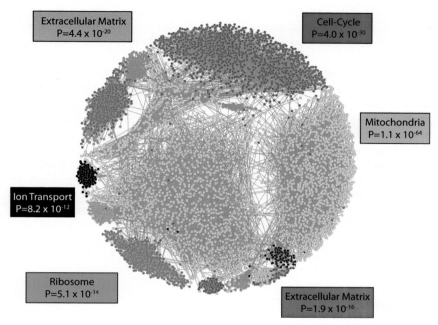

FIGURE 3 Topological and functional network modules. An example of a co-expression network generated from microarray data from a large number of mouse inbred strains [11]. The network consists of 13 topological modules. Nodes belonging to the same module are labeled with the same color. Each topological module is comprised of functional modules. Examples of topological modules are provided on the periphery of the network. For example, the "gray" module in the top left-hand corner of the network is enriched in genes that belong to the gene ontology term "extracellular matrix" and the turquoise module in the lower left-hand corner is enriched in genes associated with ribosomes, and so on. Each box highlighting a functional module is colored to correspond to its given topological module. P values represent the significance of the enrichment of each functional module. It should be noted that most topological modules harbor more than one functional module.

utilization of networks. Three types of modules have been described: topological, functional, and disease modules [8]. A topological module is a group of nodes that are highly interconnected and share more edges with one another than they do with nodes outside the module (Figure 3). In biological networks it has been found that topological modules are often composed of components that are functionally similar. These groups are referred to as functional modules. Functional modules can completely overlap topological modules or they can comprise a subset (i.e., only a subset of genes in a topological module belong to a specific functional module). For example, in co-expression networks generated across diverse cell types and tissues there is always a topological and functional module that is highly enriched for genes involved in the cell cycle [9—11]. The cell-cycle functional module typically includes genes involved in cell-cycle progression and checkpoints and, interestingly, histones. Functional modules are often used to identify genes involved in a particular process. As an example, Ghazalpour et al. generated a co-expression network from liver microarray gene expression data in a mouse genetic population [9]. One topological module, referred to as the Blue module, also contained multiple functional modules comprised of genes involved in extracellular matrix—receptor interactions and the complement and coagulation pathway, as well as others.

A disease module is a collection of genes that are both known to cause disease and are highly interconnected [2]. It has been found in a wide range of networks that genes that cause or influence the same or physiologically similar diseases tend to be more highly connected. This makes sense especially if disease genes are members of complexes. Disease modules can be exploited to identify new disease genes by identifying nodes that are not known to affect a particular disease but are the nearest neighbor of members of a disease module [12].

3.3 Hub Nodes

In a scale-free network, hub nodes are those with the largest number of links to other nodes (Figure 2). Hubs generally refer to those nodes that are in the top 20% based on their number of connections (edges), although this is an arbitrary definition [5]. In biological networks, hubs are thought to be responsible for organizing the behavior of networks. A number of experiments in yeast, flies, and worms indicate that hub proteins in PPI networks are often more essential for life relative to lowly connected proteins, suggesting that disruption of network hub genes results in network failure [13—17]. Because they are essential and not compatible with life, genes that cause Mendelian diseases tend not to encode for hubs in PPI networks [18]. However, it is important to note that the role of hubs is likely to depend on the type of network. For instance, hub proteins in a PPI network may have different properties and characteristics compared to hub genes in a co-expression network. Therefore, the impact of hubs should always be interpreted in the context of the network under investigation.

Above we described network level hubs. Intramodular hubs are a second type of highly connected node. In contrast to network hubs, intramodular hubs are highly connected within, but not typically outside of, a module. Of course, some nodes will be both network and intramodular hubs. It has been suggested that in the context of co-expression networks and complex diseases, the hubs of modules that are associated with disease tend to be more strongly associated with the disease than whole network hubs [19]. This suggests that while perturbation of whole network hubs often leads to catastrophic (and lethal) network failures as discussed earlier, the dysfunction of intramodular hubs may cause localized network failures that influence complex disease states [15,19]. This idea is consistent with recent studies demonstrating that intramodular hub genes play central roles in key biological processes [10,11,20,21]. As a result, it is possible that identification of intramodular hubs is an efficient and powerful approach for elucidating genes contributing to complex disease [19]. This approach has potential advantages over direct genetic analysis. For instance, a gene must be functionally polymorphic to be identified in a genetic mapping study, whereas this would not be required using an intramodular hub discovery strategy. Moreover, networks are modulated by both genetic and environmental inputs allowing for the discovery process to be driven by both sources of variation [22].

3.4 Directed versus Undirected

Although not an organizing principle per se, it is important to discuss network directionality. In any network, the edges that connect nodes can either be directed or undirected. In an undirected network, edges represent relationships that are undefined with respect to causality (i.e., in a node pair, which is upstream and functionally altering the second). For example, most PPIs have no inherent directionality; the presence of the relationship simply means that the two proteins physically interact. However, a PPI between a kinase and its substrate would have inherent directionality, that is, the kinase acts upon (phosphorylates) its protein substrate (although this reaction could also be bidirectional). If directed, then the relationship is said to be oriented, that is one of the nodes is functionally upstream of the other. Some networks, such as metabolic networks, have embedded directionality, while most do not. There are a growing number of approaches that can be used to directionally orient networks [23–26]. Such information can enhance the robustness of biological inferences made using networks [24].

4. TYPES OF NETWORKS

As mentioned above, there are several different types of biological networks. Each type provides a systems-level view of biology from a slightly different perspective and each has its own unique features. Below we describe some of the most popular network types and highlight ways in which they can be used to study CVD.

4.1 Phenotype/Disease Networks

One typically thinks of a biological network as being composed of cellular components; however, this is not always the case. Phenotypic or disease networks do not represent cellular relationships, but instead highlight links between correlated traits or diseases with shared etiologies. When done in an unbiased fashion, such networks can reveal previously unknown relationships between physiological processes or seemingly dissimilar diseases [18].

One of the best ways to generate phenotypic networks is using data collected on recombinant inbred (RI) strains [27]. RI strains have long been used as a tool for genetic mapping in mice, rats, and many other model species. To generate RI strains, genetically divergent parental strains are crossed and then individual inbred lines are created through brother−sister mating for several generations [28]. In the end, the genomes of an RI are inbred (and homozygous) mosaics of the original parents. Since RI strains are inbred they are reproducible and can be phenotyped for many different traits, which makes them ideal for systems studies. One of the best characterized set of RI strains, referred to as the BXD panel, has been typed for hundreds of physiological and millions of molecular traits (http://www.genenetwork.org/webqtl/main.py).

One of the best examples of a phenotypic network for cardiovascular traits is a study by Nadeau et al. using RI mice [29]. This study focused on several traits measuring cardiovascular function such as left ventricular mass (LVM), end diastolic and systolic dimensions, stroke volume, and heart rate, among many others. The correlational relationships among these traits were defined across the RI panel. The idea was that traits that are dependent on one another would co-segregate. The authors were then able to define a model that captured the interdependencies between traits. Many well-known relationships were confirmed in the model. A number of mouse strains with genetic perturbations affecting a specific aspect of cardiovascular function were assayed to test network predictions. Importantly, it was observed that the model could predict pleiotropic relationships, that is, if mice transgenic for gene X had altered LVM, then all the traits connected to LVM were also altered in the transgenics. This study provided a proof-of-principle that constructing phenotypic networks in genetically randomized inbred mice could be used to identify novel and important trait dependencies. This approach has been used subsequently in a number of studies investigating physiological relationships [30,31].

The same or different mutations in individual genes can cause multiple different diseases (e.g., ATM in ataxia−telangiectasia and breast cancer) and a disease can be caused by mutations in multiple genes (e.g., cardiomyopathy has been linked to mutations in many different genes).

Based on this, Goh et al. created networks using data from the Online Mendelian Inheritance in Man (OMIM) [18] database that houses lists of disease gene links. Two networks emerged: the human disease network in which disease nodes were connected if they were caused by mutations in the same gene, and the disease gene network where gene nodes were connected if they caused the same disease. A number of interesting observations emerged from these networks. For instance, proteins associated with the same disorders were much more likely to physically interact and be co-expressed, supporting the utility of PPI and co-expression networks as disease gene discovery tools.

4.2 PPI Networks

PPI networks recapitulate the myriad of physical interactions between proteins on a genome-wide scale [32]. PPIs are biologically meaningful since many functionally similar proteins form complexes and proteins often carry out their function via interactions with other proteins (e.g., heterodimer formation). PPIs can be identified using both low- and high-throughput approaches. For example, the binding partners of individual proteins can be identified by immunoprecipitation. Higher-throughput approaches, such as yeast two-hybrid screening [33] and affinity purification coupled to mass spectrometry [34], have been used to identify PPIs on a genome-wide scale. Large-scale PPI networks have been created for a number of species, including humans [35]. A plethora of online resources exist, such as the Database of Interacting Protein (DIP) [36], the Biomolecular Interaction Network Database (BIND) [37], the Münich Information Center for Protein Sequence (MIPS) protein interaction database [38], the Molecular Interaction database (MINT) [39], the protein Interaction database (IntAct) [40], the Biological General Repository for Interaction Datasets (BioGRID) [41], and the Human Protein Reference Database (HPRD) [42], that have collated and stored PPI information from a variety of sources. Online resources make downloading, visualizing, and analyzing PPI networks relatively straightforward.

PPI networks can be utilized to generate novel disease insight. Disease genes tend to physically interact with other disease genes [18]; therefore, one strategy for novel gene discovery is to identify the PPI network neighbors of known mediators. An example is the work of Zhang et al. who developed methodology based on the network topology of proteins known to be involved in coronary artery disease (CAD) to identify 276 novel CAD candidate genes [43].

A similar approach was used by Li et al. to identify functional modules associated with CAD [44]. These authors started with 266 "seed" genes that had been previously linked to CAD. Using the latest human PPI from the HPRD they expanded the CAD network by identifying additional proteins linked to one or more of the CAD seeds. This network was decomposed into functional modules, each of which was enriched for functionally similar genes. Hub genes were identified revealing putative molecular mechanisms contributing to CAD, such as YY receptor activity, Fc gamma R-mediated phagocytosis, and actin cytoskeleton regulation.

In addition to uncovering new disease genes, mining PPIs have also been proposed as an approach to generate CAD biomarkers. Jin et al. outlined a strategy in which mass spectrometry was used to identify single protein biomarkers for CAD in plasma samples that differed in abundance between CAD patients and controls [45]. By mining the PPI data the group determined that "network" biomarkers consisting of interactions between two to three proteins were better able to distinguish between CAD cases and controls than the levels of a single protein. This suggests that specific PPIs are gained or lost, as a function of CAD, and these network perturbations are robustly predictive of disease even in plasma.

4.3 Co-expression Networks

Of the various types of networks, co-expression networks are the most flexible for interrogating disease. Co-expression generally refers to correlations in transcript levels, but it can also be used to study correlational relationships at all biological scales (e.g., proteins, metabolites, or a combination of transcripts, proteins, and metabolites [46]). Co-expression networks have become popular in part because of technologies (e.g., gene expression microarrays, RNA-seq, and mass-spectrometry) that allow one to interrogate molecular intermediates of different biological scales in a straightforward manner in a relatively large number of samples [47]. Co-expression has also been shown to be biologically meaningful and can be measured in specific cell types in the presence of defined perturbations [48]. As a comparison, most PPI networks represent "generic" interactions that are void of cell type and temporal context, whereas co-expression networks, for example, can be generated using data from specific cell types from normal and disease individuals across the entire spectrum of development.

One of the most widely used algorithms for constructing co-expression networks is Weighted Gene Co-expression Network Analysis (WGCNA) [48,49]. Due to its widespread use and because many co-expression studies that interrogate CVD use WGCNA, it is informative to describe how it works (Figure 4).

The input for a WGCNA is a matrix of data. Gene expression data are mainly used, although as mentioned above it is possible to use any quantitative molecular data collected from the same sample population. In the context of gene expression, these data have historically been

Data from a series of microarray or RNA-seq is used to quantify gene expression changes
due to a set of perturbations (genetic, environmental or both)

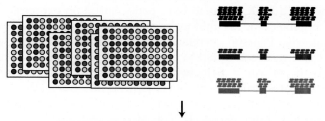

Coexpressed genes share similiar profiles of expression across samples

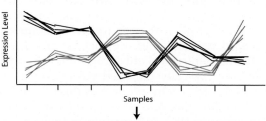

Coexpression relationships are quantified using correlation and then transformed
to a topological overlap measure (TOM)

	G1	G2	G3	G4	G5	G6	G7	G8
G1	1	0.8	0.9	0.1	0.3	0.2	0.5	0.7
G2	0.8	1	0.2	0.3	0.6	0.9	0.1	0.1
G3	0.9	0.2	1	0.9	0.8	0.2	0.1	0.6
G4	0.1	0.3	0.9	1	0.2	0.3	0.7	0.6
G5	0.3	0.6	0.8	0.2	1	0.3	0.2	0.1
G6	0.2	0.9	0.2	0.3	0.3	1	0.6	0.9
G7	0.5	0.1	0.1	0.7	0.2	0.6	1	0.8
G8	0.7	0.1	0.6	0.6	0.1	0.9	0.8	1

Gene coexpression modules are generated by clustering correlated genes and
can be visualized in a number of ways

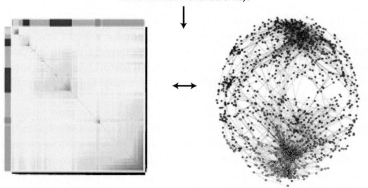

2-D heatmap representation   Multidimensional network representation

FIGURE 4   Overview of the steps involved in the construction of a WGCNA network. *Modified with permission from ref. [47].*

generated using gene expression microarrays, although the use of RNA-seq is becoming more common [50].

The first step in a WGCNA is to filter genes based on expression and variance, since genes that are neither expressed nor vary will be uninformative. Pair-wise Pearson correlations between the remaining genes are calculated and then converted to connection strengths by raising the correlations to the power β. Selection of an appropriate β ensures the network approximates scale-free topology [48]. As described above, scale-free networks consist of many lowly connected genes and a small number of highly connected "hub" genes [4]. This step is critical since most biologically relevant networks are scale free [2]. Next, the connection strengths are transformed to a topological overlap measure (TOM) [51]. Two genes have a high TOM if they share strong connection strengths with the same sets of genes. Genes are then grouped into "modules" based on TOMs using hierarchical clustering. Importantly, WGCNA topological modules are typically enriched for one or many functional modules (e.g., [10,11,52−55]). A by-product of this enrichment is that it is possible to predict the function of a gene of unknown function based on its membership in a specific WGCNA module and the known functions of the genes it is most closely connected to.

It is worth noting that one of the real advantages of co-expression networks is the ability to integrate phenotypic information. Unlike most other networks, co-expression networks of cellular components are collected on individuals across a population. As a result, any other information collected on that population of individuals can be evaluated for associations with network/module behavior. An excellent example is the study by Ghazalpour et al. discussed earlier in which a module (Blue module) was found in liver tissue from over 200 F2 mice. Interestingly, the behavior (i.e., whether the expression of module genes was up or down) of the Blue module correlated with mouse body weight and adiposity, suggesting that Blue module genes influence weight [9]. Importantly, most Blue module genes would not have been found to be associated with body weight by other means such as genetic analysis of differential gene expression alone.

One of the first examples of the use of co-expression networks to inform CVD, studied the transcriptomics of human endothelial cells (ECs) [56]. To model the inflammatory responses of ECs in atherosclerosis, the authors treated ECs with oxidized (an inflammatory stimulus) and non-oxidized phospholipids and measured IL-8 as a measure of response. Across EC donors, IL-8 secretion ranged from no difference to an ~8-fold increase in IL-8 secretion after treatment with oxidized phospholipids. Using WGCNA the authors generated a co-expression network comprised 15 modules. One of the modules was enriched in genes involved in the unfolded protein response (UPR). Functional analysis of the UPR module identified activating transcription factor 4 (*ATF4*) as a transcriptional activator of an unknown gene (later identified as cation transport regulator-like protein 1 (*CHAC1*) [57]) and that IL-8 secretion was partially regulated by the UPR. This study was one of the first to demonstrate that co-expression networks could be used to identify key genes and processes that were partially responsible for a heritable disease associated trait (inflammation in response to oxidized lipids).

5. INTEGRATING GENETICS AND NETWORKS

As we have already illustrated networks can be used to organize a wide array of different data types relevant to CVD collected in diverse experimental settings. One of the areas in which network analysis has been most fruitful is its integration with genetic analysis. Genetic analysis is inherently a reductionist approach. Variants are associated with phenotypes one at a time without considering how variants affect networks at the cellular, tissue, and organismal levels. Several recent studies have begun to combine genetic analysis and networks in a number of clever and productive ways. In this section, we provide a background to the problem and then highlight ways in which network theory has improved our knowledge and understanding of how genetic variation influences CVD. The focus of this section is on CVD, but we also provide examples of applications from other complex diseases that could be used to study CVD.

5.1 Current State of CVD Genetics

Heritability estimates suggest that between 30% and 60% of the risk of developing CVD is due to genetics and robust associations with a number of environmental factors, such as smoking, have been observed [58]. However, we are only beginning to understand the genetic architecture of CVD (the number and frequency of CVD-associated alleles and their mode of inheritance), the identity of the individual variants that contribute to CVD, the genes such variants perturb, the mechanisms connecting such genes to disease, and how genetic and environmental perturbations cause cellular dysfunction. With the advent of genomic technologies considerable attention has been paid on addressing these questions. This work is motivated by the promise that the discovery of novel genes will result in new, more effective, therapeutic modalities capable of treating the diverse range of common abnormalities of the cardiovascular system. There is also the hope that understanding the genomic basis of CVD will allow medicine to move from treatment based on population averages to ones that are "tailored" or "personalized" to individuals [59].

Genetic mapping has always held tremendous promise to provide a more comprehensive understanding of the molecular players and events that lead to CVD. This is due, in large part, to the unbiased nature of genetic discovery. Having the ability to identify novel genes and processes underlying a disease without knowing anything about disease biology has the potential to lead to paradigm shifting discoveries. However, for many years, due to a number of scientific and technical limitations, genetic analysis provided few significant insights. This has recently changed with the arrival of genome-wide association studies (GWASs) in humans and similar approaches in model organisms [60,61].

In a GWAS, tens to hundreds of thousands of individuals are genotyped at millions of single nucleotide polymorphisms (SNPs) [62]. SNPs with allele frequencies that differ between cases and controls or SNPs with allele frequencies that correlate with changes in a CVD-related quantitative trait "tag" regions of the genome that harbor causal variations influencing disease. From a discovery perspective, GWAS in the CVD field have been extremely successful. The most recent and comprehensive GWAS for CAD and plasma lipids are a testament to this success. The CARDIoGRAMplusC4D Consortium GWAS for CAD identified 46 independent genome-wide significant associations in ~63 k cases and ~130 k controls [63]. In one of the largest GWAS for any disease-related traits, the Global Lipids Genetics Consortium identified 157 associations for various plasma lipids (LDL, HDL, triglycerides, and total cholesterol) in ~190 k individuals [64]. These studies implicate a large number of genes with no previously known role in the etiology of CVD. Many of the genes would have never been identified using traditional studies since they belong to pathways that seem quite unrelated to CVD; highlighting the power of unbiased genetic discovery. However, from the perspective of comprehensively understanding the genetic component of CVD, the results of GWAS have been disappointing. Despite the fact that the CAD and plasma lipid GWASs referenced above surveyed the entire genome in hundreds of thousands of people, all of the identified variants together explained less than 15% of variance for their respective disease/traits [63,64]. This is not a phenomena specific to CVD genetics since the same has been observed for nearly all complex diseases [65].

5.2 Systems Genetics

It is clear that GWAS has transformed genetic discovery and in the process has unveiled hundreds of potential therapeutic targets for CVD. However, GWAS alone only provides connections between the genome and phenome (often referred to as the genotype–phenotype map). It ignores all of the information that exists between these two entities. This is important because cellular function is the product of the epigenome, transcriptome, protein, metabolome, microbiome, and so on, and their complex interactions that are poorly understood (Figure 2). In fact, it is likely that many aspects of cellular function cannot be determined just by analyzing patterns of genomic variation and therefore, would never be detected in genetic studies.

To address this limitation, the field of "systems genetics" has emerged [66]. Systems genetics combines transcriptomic, proteomic, and/or metabolomic data with traditional genetic analysis. By doing so, one can identify genetic variation that impacts downstream cellular components to more rapidly identify the subset of variation that contributes to disease [67]. It also allows one to interrogate the interactions between molecular phenotypes and how those interactions impact cellular function and disease. To date, a main focus of system genetics has been to reconstruct cellular networks using large multiscalar datasets to help inform GWAS and other genetic studies.

6. APPLICATIONS OF NETWORK-BASED SYSTEMS GENETICS STUDIES

6.1 Gene and Module Discovery

One of the best examples of the power of systems genetics was a study by Chen et al. In this work, the authors analyzed the genetic regulation of a number of physiological traits including, LDL and HDL cholesterol, free fatty acids, triglycerides, and atherosclerotic lesions, in an F2 mouse intercross [68]. It was found that all of these complex traits were influenced by a strong quantitative trait locus (QTL) on chromosome 1. Using liver and adipose microarray expression data from the same cross, the authors constructed co-expression networks. They observed that the expression of nearly all genes in one module shared between both tissues was linked to the same interval on chromosome 1. Using a modeling approach they predicted that most of the genes in this module were upstream and causal for the change in metabolic phenotypes. This module was referred to as the macrophage-enriched metabolic network (MEMN), given its apparent macrophage-derived origin. Using transgenic and knockout mouse models three of the genes in the MEMN were validated as novel metabolic regulators. Importantly, in a separate study it was found that the MEMN was conserved in human adipose and many of the human genes also showed an apparent causal relationship with metabolic traits [69].

Heinig et al. used a similar, but distinct, approach in rats and humans [70]. Initially, the authors determined what was termed "transcription factor-driven gene networks" by using expression QTL data on seven tissues in the rat and then searched for networks of genes regulated by the same trans locus and enriched for transcription factor binding motifs. One of the strongest predicted networks contained Interferon Regulatory Transcription Factor 7 (*Irf7*) and was enriched for inflammatory response genes leading the authors to name the network the "IRF7-driven inflammatory gene network" (IDIN). An analysis of genetic variation in the inbred rats and siRNA knockdown in macrophages identified G protein-coupled receptor 183 (*Ebi2*) as a negative regulator of the IDIN. The authors extended these findings from rat to humans by performing transcription factor binding site enrichment and co-expression analysis in human monocyte from a couple of different studies with strong overlap observed with rat IDIN members. Multivariate analysis of one of the studies revealed six SNPs in the human orthologous region that were associated with trans-regulation of *IRF7* as well as IDIN genes. Human *EBI2* expression in monocytes was also found to be cis-regulated in both studies. Because of the association of IDIN with macrophages and its enrichment for immune response genes, which includes type I diabetes (T1D), the authors evaluated the human IDIN genes with T1D. SNPs close to IDIN genes were significantly more associated with T1D than non-network genes.

These studies illustrate that multispecies co-expression networks and GWAS together can provide insight into biological processes that the two separately cannot. A significant by-product of these studies was the discovery of novel disease genes.

6.2 Elucidating Gene Function

GWAS can implicate genes in disease; however, it provides no information on gene function and how altered function leads to disease. Elucidating the mechanisms of action for newly minted disease genes is a major bottleneck in translating genetic discoveries into new therapeutics. Addressing this limitation, it has been shown that networks can provide insight on gene function [71,72]. The premise behind this is simple—genes that are connected tend to be functionally similar. Therefore, when assessing disease relevant interactions, one can begin to generate hypotheses regarding the function of an uncharacterized gene just by analyzing the genes of known function with which it is connected to. This "guilt-by-association" approach has the potential to significantly inform genetic studies of complex disease [73].

In an elegant study by McDermott-Roe et al. a network-based approach in rats was used to determine the function of a novel regulator of left ventricular mass (LVM) [71]. In this study, a rat F2 intercross was used to identify a QTL for LVM on Chromosome 3. The QTL was mapped to a 750 kbp region and the only gene in the region whose expression differed as a function of genotype was endonuclease G (*Endog*). *Endog* knockout mice were observed to have increased LVM. To determine the function of *Endog*, a genome-wide co-expression network generated from human heart was investigated. In this network, *Endog* was a member of a module highly enriched for mitochondrial genes and genes involved in oxidative metabolism, suggesting that *Endog* was involved in mitochondrial function. Through functional experiments it was shown that *Endog* was directly regulated by master controllers of mitochondrial function, *Endog* interacted with the mitochondrial genome and loss of *Endog* resulted in a depletion of mitochondria in heart tissue.

Our group recently performed a similar systems genetics study for bone mineral density (BMD) [72]. In this work, an F2 mouse cross was used to identify a BMD QTL on Chromosome 10. The locus contained over 300 genes, yet using gene expression we predicted that Bicaudal C homolog 1 (*Drosophila*) (*Bicc1*) was the causal gene. Mice heterozygous for a null allele of *Bicc1* had lower BMD, validating *Bicc1* as the putative causal gene. *Bicc1* had not been previously implicated in bone cell activities or the regulation of BMD; therefore, we used a previously generated bone co-expression network to infer its function [11]. *Bicc1* was a member of a module enriched for genes involved in the differentiation of bone-forming osteoblasts. As with *Endog* above, we used this information to predict that *Bicc1* was involved in osteoblast differentiation. Consistent with this hypothesis, *Bicc1* siRNA knockdown in primary calvarial osteoblasts impaired their differentiation. *Bicc1* is an RNA-binding protein and to identify potential targets we went back to the network and identified the genes that were the most strongly connected to *Bicc1*. Of these, *Pkd2* stood out and subsequent experimental validation revealed that *Pkd2* knockdown in osteoblasts resulted in their impaired differentiation in a manner identical to that seen when *Bicc1* was knocked down. Using causality modeling *Bicc1* was predicted to regulate *Pkd2* transcript levels and *Pkd2* overexpression was able to rescue the *Bicc1*-dependent osteoblast defects, confirming that *Pkd2* is a *Bicc1* target gene. Furthermore, genetic variants in human *BICC1* and *PKD2* were associated with BMD.

These two studies illustrate how co-expression networks can be used to infer gene function for unannotated genes based on their connections with genes of known function.

6.3 Network Analysis of GWAS Data

In a GWAS, >1 million independent statistical tests are performed. As a consequence, there is a need to correct for this large number of tests, which results in stringent statistical thresholds on the order of $P = 5.5 \times 10^{-8}$. Thus, in a GWAS one expects that many of the nominally significant ($P < 0.05$) SNPs represent true associations that the study was underpowered to identify. Since the advent of GWAS a number of approaches have been developed with the goal of mining GWAS and identifying true associations that do not reach genome-wide significance. Many such approaches include assigning GWAS SNPs to genes and then testing for enrichment of genes with nominally significant P values in particular pathways [74,75]. The underlying assumption is that such pathways play a role in the disease. A complementary approach is to assign GWAS P values to genes and then overlay these onto a network [52,76,77]. The resulting "annotated" network can then be mined for nominally associated genes that are linked to known GWAS genes or for functional modules enriched for genes with nominal associations.

Jensen et al. integrated the results of a GWAS for coronary heart disease (CHD) and a human PPI network [78]. Specifically, they utilized case-−control CHD GWAS data to assign "CHD" P values to human genes. Each gene's P value was a measure of the association between genetic variation in that gene and CHD, with the most significant P values representing the strongest associations. Using a human PPI network the author developed an algorithm to cluster PPIs into >8000 complexes or modules of highly interconnected proteins. They then scored each of the modules based on the strength of member's CHD P values. One module was significantly enriched for genes with strong associations. This module contained 19 genes centered on ADRB1 as the hub, which encodes the β1-adrenergic receptor and is known to play a role in cardiac contractility. The module was also enriched for genes that when knocked out in mice alter the regulation of blood pressure, cardiac function, and/or hypertrophy. It is of interest to note that none of the genes possessed GWAS SNPs that reached genome-wide significance. This study suggests that PPI can inform GWAS and that novel genes in the identified module likely play a role in CHD and warrant further investigation.

7. CONCLUSIONS AND FUTURE PROSPECTS

The field of network theory was reinvigorated just before the turn of the new millennium [4]. Since this time the use of networks to understand complex biological systems has seen an enormous proliferation. This is largely due to the realization that viewing disease from a system's perspective can provide insight not attainable through traditional reductionist approaches. As we have outlined in this chapter, these new insights have significantly advanced our understanding of the molecular mechanisms underlying CVD.

The field of network biology is experiencing new advances almost daily. Such advances include those both in the genomics technologies used to generate the data and in the statistical and analytical approaches to develop and study networks. As a result, the inputs are becoming more refined (e.g., the transition from global transcriptome profiling using microarrays to RNA-seq), as are the tools to analyze these data (e.g., new algorithms to add directionality to network edges). Together these refinements promise to continue to improve the quality of inferences made using networks. An important goal for the field, even while techniques used to generate networks are becoming more complex, is to ensure that non-computational biologists have access to network data to inform their own research programs.

One area that we believe will continue to be influenced to a large degree by network biology is CVD genetics. One of the lessons from GWAS regarding all complex diseases is that they are much more "complex" than expected. The spectrum of perturbations that can affect complex diseases is continually expanding, and it is clear that the complexity can no longer be ignored but should be embraced [79]. Therefore, it is likely that network biology will play an increasingly central role in understanding the complexities of complex disease genetics. As an example, Weiss and colleagues recently proposed a method referred to as a Gene Module Association Study (GMAS) aimed at complementing GWAS. The central tenet of a GMAS is that network modules should be the focus of association studies, not individual genes. The idea is that network modules capture and aggregate many more of the subtle genetic and environmental sources of variation impacting disease that are extremely hard to individually identify.

Network biology is a powerful approach to elucidate interactions between cellular components in a genome-wide and comprehensive manner. It has already led to a number of seminal discoveries in the CVD field, a subset of which were discussed in this chapter. As the tools to generate and investigate networks grow in sophistication, we anticipate that networks will play an increasingly important role in developing a comprehensive understanding of CVD and the development of novel therapies.

LIST OF ABBREVIATIONS

BMD Bone mineral density
CAD Coronary artery disease
CHD Coronary heart disease
CVD Cardiovascular disease
EC Endothelial cell
GMAS Gene module association study

GWAS Genome-wide association study
IDIN IRF7-drive inflammatory gene network
LVM Left ventricular mass
PPI Protein—protein interaction
QTL Quantitative trait locus
RI Recombinant inbred
SNP Single nucleotide polymorphism
TOM Topological overlap measure
UPR Unfolded protein response
WGCNA Weighted Gene Co-Expression Network Analysis

References

[1] Nabel EG. Cardiovascular disease. N Engl J Med 2003;349:60—72.
[2] Barabasi A-L, Oltvai ZN. Network biology: understanding the cell's functional organization. Nat Rev Genet 2004;5:101—13.
[3] Weiss JN, Karma A, MacLellan WR, et al. 'Good enough solutions' and the genetics of complex diseases. Circ Res 2012;111:493—504.
[4] Barabasi A, Albert R. Emergence of scaling in random networks. Science 1999;286: 509—12.
[5] Vidal M, Cusick ME, Barabasi A-L. Interactome networks and human disease. Cell 2011;144:986—98.
[6] Vázquez A, Flammini A, Maritan A, Vespignani A. Modeling of protein interaction networks. Complexus 2003;1(1):38—44.
[7] Pastor-Satorras R, Smith E, Solé RV. Evolving protein interaction networks through gene duplication. J Theor Biol 2003;222:199—210.
[8] Barabasi A-L, Gulbahce N, Loscalzo J. Network medicine: a network-based approach to human disease. Nat Rev Genet 2011;12:56—68.
[9] Ghazalpour A, Doss S, Zhang B, et al. Integrating genetic and network analysis to characterize genes related to mouse weight. PLoS Genet 2006;2:e130.
[10] Farber CR. Identification of a gene module associated with BMD through the integration of network analysis and genome-wide association data. J Bone Miner Res 2010;25: 2359—67.
[11] Calabrese G, Bennett BJ, Orozco L, et al. Systems genetic analysis of osteoblast-lineage cells. PLoS Genet 2012;8:e1003150.
[12] Erlich Y, Edvardson S, Hodges E, et al. Exome sequencing and disease-network analysis of a single family implicate a mutation in KIF1A in hereditary spastic paraparesis. Genome Res 2011;21:658—64.
[13] Jeong H, Mason SP, Barabasi AL, Oltvai ZN. Lethality and centrality in protein networks. Nature 2001;411:41—2.
[14] Carlson MRJ, Zhang B, Fang Z, Mischel PS, Horvath S, Nelson SF. Gene connectivity, function, and sequence conservation: predictions from modular yeast co-expression networks. BMC Genomics 2006;7:40.
[15] Albert R, Jeong H, Barabasi A. Error and attack tolerance of complex networks. Nature 2000;406:378—82.
[16] Albert R, Barabasi AL. Statistical mechanics of complex networks. Rev Mod Phys 2002.
[17] Hahn MW, Kern AD. Comparative genomics of centrality and essentiality in three eukaryotic protein-interaction networks. Mol Biol Evol 2005;22:803—6.
[18] Goh K-I, Cusick ME, Valle D, Childs B, Vidal M, Barabasi A-L. The human disease network. Proc Natl Acad Sci USA 2007;104:8685—90.

[19] Langfelder P, Mischel PS, Horvath S. When is hub gene selection better than standard meta-analysis? PLoS ONE 2013;8:e61505.

[20] Horvath S, Zhang B, Carlson M, et al. Analysis of oncogenic signaling networks in glioblastoma identifies ASPM as a molecular target. Proc Natl Acad Sci USA 2006;103: 17402−7.

[21] Torkamani A, Schork NJ. Identification of rare cancer driver mutations by network reconstruction. Genome Res 2009;19:1570−8.

[22] Stein GS. Molecular networks as sensors and drivers of common human diseases. Nature 2009;461:218−23.

[23] Vinayagam A, Stelzl U, Foulle R, et al. A directed protein interaction network for investigating intracellular signal transduction. Sci Signal 2011;4:rs8.

[24] Schadt EE, Lamb J, Yang X, et al. An integrative genomics approach to infer causal associations between gene expression and disease. Nat Genet 2005;37:710−7.

[25] Chaibub Neto E, Ferrara CT, Attie AD, Yandell BS. Inferring causal phenotype networks from segregating populations. Genetics 2008;179:1089−100.

[26] Aten JE, Fuller TF, Lusis AJ, Horvath S. Using genetic markers to orient the edges in quantitative trait networks: the NEO software. BMC Syst Biol 2008;2:34.

[27] Churchill GA. Recombinant inbred strain panels: a tool for systems genetics. Physiol Genomics 2007;31:174−5.

[28] Silver LM. Mouse genetics: concepts and applications. Oxford University Press; 1995.

[29] Nadeau JH. Pleiotropy, homeostasis, and functional networks based on assays of cardiovascular traits in genetically randomized populations. Genome Res 2003;13: 2082−91.

[30] Jepsen KJ, Hu B, Tommasini SM, et al. Genetic randomization reveals functional relationships among morphologic and tissue-quality traits that contribute to bone strength and fragility. Mamm Genome 2007;18:492−507.

[31] Jepsen KJ, Courtland H-W, Nadeau JH. Genetically determined phenotype covariation networks control bone strength. J Bone Miner Res 2010;25:1581−93.

[32] Cusick ME, Klitgord N, Vidal M, Hill DE. Interactome: gateway into systems biology. Hum Mol Genet 2005;14:R171−81. Spec No. 2.

[33] Fields S, Song O. A novel genetic system to detect protein-protein interactions. Nature 1989;340:245−6.

[34] Dunham WH, Mullin M, Gingras A-C. Affinity-purification coupled to mass spectrometry: basic principles and strategies. Proteomics 2012;12:1576−90.

[35] Stelzl U, Worm U, Lalowski M, et al. A human protein-protein interaction network: a resource for annotating the proteome. Cell 2005;122:957−68.

[36] Xenarios I, Salwínski L, Duan XJ, Higney P, Kim S-M, Eisenberg D. DIP, the Database of Interacting Proteins: a research tool for studying cellular networks of protein interactions. Nucleic Acids Res 2002;30:303−5.

[37] Alfarano C, Andrade CE, Anthony K, et al. The biomolecular interaction network database and related tools 2005 update. Nucleic Acids Res 2005;33:D418−24.

[38] Mewes HW, Ruepp A, Theis F, et al. MIPS: curated databases and comprehensive secondary data resources in 2010. Nucleic Acids Res 2011;39:D220−4.

[39] Licata L, Briganti L, Peluso D, et al. MINT, the molecular interaction database: 2012 update. Nucleic Acids Res 2012;40:D857−61.

[40] Kerrien S, Aranda B, Breuza L, et al. The IntAct molecular interaction database in 2012. Nucleic Acids Res 2012;40:D841−6.

[41] Chatr-Aryamontri A, Breitkreutz B-J, Heinicke S, et al. The BioGRID interaction database: 2013 update. Nucleic Acids Res 2013;41:D816−23.

[42] Prasad TSK, Kandasamy K, Pandey A. Human Protein Reference Database and Human Proteinpedia as discovery tools for systems biology. Methods Mol Biol 2009;577: 67−79.

[43] Zhang L, Li X, Tai J, Li W, Chen L. Predicting candidate genes based on combined network topological features: a case study in coronary artery disease. PLoS ONE 2012;7:e39542.

[44] Li H, Zuo X, Ouyang P, et al. Identifying functional modules for coronary artery disease by a prior knowledge-based approach. Gene 2014;537:260—8.

[45] Jin G, Zhou X, Wang H, et al. The knowledge-integrated network biomarkers discovery for major adverse cardiac events. J Proteome Res 2008;7:4013—21.

[46] Ghazalpour A, Bennett BJ, Shih D, et al. Genetic regulation of mouse liver metabolite levels. Mol Syst Biol 2014;10:730.

[47] Farber CR, Lusis AJ. Integrating global gene expression analysis and genetics. Adv Genet 2008;60:571—601.

[48] Zhang B, Horvath S. A general framework for weighted gene co-expression network analysis. Stat Appl Genet Mol Biol 2005;4. Article17.

[49] Langfelder P, Zhang B, Horvath S. Defining clusters from a hierarchical cluster tree: the Dynamic Tree Cut package for R. Bioinformatics 2008;24:719—20.

[50] Iancu OD, Kawane S, Bottomly D, Searles R, Hitzemann R, McWeeney S. Utilizing RNA-Seq data for de novo coexpression network inference. Bioinformatics 2012;28: 1592—7.

[51] Ravasz E, Barabasi A-L. Hierarchical organization in complex networks. Phys Rev E Stat Nonlin Soft Matter Phys 2003;67:026112.

[52] Farber CR. Systems-level analysis of genome-wide association data. G3 (Bethesda) 2013;3:119—29.

[53] Park CC, Gale GD, de Jong S, et al. Gene networks associated with conditional fear in mice identified using a systems genetics approach. BMC Syst Biol 2011;5:43.

[54] Oldham MC, Konopka G, Iwamoto K, et al. Functional organization of the transcriptome in human brain. Nat Neurosci 2008;11:1271—82.

[55] Miller JA, Oldham MC, Geschwind DH. A systems level analysis of transcriptional changes in Alzheimer's disease and normal aging. J Neurosci 2008;28:1410—20.

[56] Gargalovic PS, Imura M, Zhang B, et al. Identification of inflammatory gene modules based on variations of human endothelial cell responses to oxidized lipids. Proc Natl Acad Sci USA 2006;103:12741—6.

[57] Mungrue IN, Pagnon J, Kohannim O, Gargalovic PS, Lusis AJ. CHAC1/MGC4504 is a novel proapoptotic component of the unfolded protein response, downstream of the ATF4-ATF3-CHOP cascade. J Immunol 2009;182:466—76.

[58] Elder SJ, Lichtenstein AH, Pittas AG, et al. Genetic and environmental influences on factors associated with cardiovascular disease and the metabolic syndrome. J Lipid Res 2009;50:1917—26.

[59] Ginsburg GS, Donahue MP, Newby LK. Prospects for personalized cardiovascular medicine: the impact of genomics. J Am Coll Cardiol 2005;46:1615—27.

[60] Altshuler D, Daly MJ, Lander ES. Genetic mapping in human disease. Science 2008;322: 881—8.

[61] Flint J, Eskin E. Genome-wide association studies in mice. Nat Rev Genet 2012;13: 807—17.

[62] Sale MM, Mychaleckyj JC, Chen W-M. Planning and executing a genome wide association study (GWAS). Methods Mol Biol 2009;590:403—18.

[63] Deloukas P, Kanoni S, Willenborg C, et al. Large-scale association analysis identifies new risk loci for coronary artery disease. Nat Genet 2012;45:25—33.

[64] Consortium GLG. Discovery and refinement of loci associated with lipid levels. Nat Genet 2013;45:1274—83.

[65] Manolio TA, Collins FS, Cox NJ, et al. Finding the missing heritability of complex diseases. Nature 2009;461:747—53.

[66] Civelek M, Lusis AJ. Systems genetics approaches to understand complex traits. Nat Rev Genet 2013;15:34—48.

[67] Nadeau JH, Dudley AM. Genetics. Systems genetics. Science 2011;331:1015—6.

[68] Chen Y, Zhu J, Lum PY, et al. Variations in DNA elucidate molecular networks that cause disease. Nature 2008;452:429—35.

[69] Emilsson V, Thorleifsson G, Zhang B, et al. Genetics of gene expression and its effect on disease. Nature 2008;452:423—8.

[70] Heinig M, Petretto E, Wallace C, et al. A trans-acting locus regulates an anti-viral expression network and type 1 diabetes risk. Nature 2010;467:460—4.

[71] McDermott-Roe C, Ye J, Ahmed R, et al. Endonuclease G is a novel determinant of cardiac hypertrophy and mitochondrial function. Nature 2011;478:114—8.

[72] Mesner LD, Ray B, Hsu Y-H, et al. Bicc1 is a genetic determinant of osteoblastogenesis and bone mineral density. J Clin Invest 2014;124:2736—49.

[73] Wolfe CJ, Kohane IS, Butte AJ. Systematic survey reveals general applicability of 'guilt-by-association' within gene coexpression networks. BMC Bioinf 2005;6:227.

[74] Zhang L, Guo Y-F, Liu Y-Z, et al. Pathway-based genome-wide association analysis identified the importance of regulation-of-autophagy pathway for ultradistal radius BMD. J Bone Miner Res 2010. http://dx.doi.org/10.1002/jbmr.36.

[75] Cantor RM, Lange K, Sinsheimer JS. Prioritizing GWAS results: a review of statistical methods and recommendations for their application. Am J Hum Genet 2010;86:6—22.

[76] Leiserson MDM, Eldridge JV, Ramachandran S, Raphael BJ. Network analysis of GWAS data. Curr Opin Genet Dev 2013;23:602—10.

[77] Jia P, Zhao Z. Network.assisted analysis to prioritize GWAS results: principles, methods and perspectives. Hum Genet 2014;133:125—38.

[78] Jensen MK, Pers TH, Dworzynski P, Girman CJ, Brunak S, Rimm EB. Protein interaction-based genome-wide analysis of incident coronary heart disease. Circ Cardiovasc Genet 2011;4:549—56.

[79] Schadt EE, Sachs A, Friend S. Embracing complexity, inching closer to reality. Sci STKE 2005;2005:pe40.

CHAPTER

4

Using Information about DNA Structure and Dynamics from Experiment and Simulation to Give Insight into Genome-Wide Association Studies

Rodrigo Galindo-Murillo, Thomas E. Cheatham III

Department of Medicinal Chemistry, L.S. Skaggs Pharmacy Institute, University of Utah, Salt Lake City, UT, USA

1. INTRODUCTION

With the development of high-density genotyping analysis, genome-wide association studies (GWASs) are now a common technique to identify and study hundreds of genetic loci associated with commonly observed traits and disease-related pathological traits [1,2]. The genetic variability uncovered identifies the frequency of specific single-nucleotide polymorphisms (SNPs) present in a group of individuals, which facilitates the identification of the haplotype region in the gene where the functional variants reside [3,4]. Once the genotype variant is known, molecular biology analysis (flow cytometry, quantitative PCR, metabolic profiling, RNA sequencing, ChIP sequencing, and so on) helps identify related pathways of action to expose how a particular SNP interacts with, or alters, the molecular machinery of the cell. This exposes the actual effect of the detected variant in the transcription, translation, and/or subsequent production of an active (or inactive) protein. While this analysis is straightforward if the detected SNP is in protein-coding regions of the genome, these SNPs are most often observed outside of the protein-coding regions of cases [5]. Notably, the majority of

83

GWAS-tagged SNPs have been observed in *intronic* regions (about 89%), and this suggests that these sequence aberrations influence the regulating process of gene expression mediated by the non-coding regions [1]. A highly choreographed set of interactions with multiple proteins are involved in gene expression, including the genomic double-stranded DNA in the nucleosome, histone monomers, transcription factors, stabilizing chaperone proteins, counterions, and cofactors [6]. The specific network of interactions and modes of binding of all these players that regulate the expression of genes is an active field of research. Molecular biology tools and techniques allow for a comprehensive analysis of the promotion or inhibitions of the expression of genes, how genes are affected, and, their location inside the genome. Their usefulness resides in studying the complex relations between proteins and genes. To be able to zoom in and study these relations of the genome at the atomic level, one has to switch tools and even the way to address the system one is interested.

Can insight into DNA sequence-specific structure and dynamics elucidated from experiment and simulation provide new understanding of the influence of SNPs on the properties of DNA to provide further mechanistic information? The elucidation of the structure of DNA in the 1950s [7] with the aid of X-ray spectroscopy opened the door for a non-stop stream of studies of the double helix focusing on the understanding of the interplay between the building blocks (phosphate backbone, sugar, and nucleobase). With the coming of nuclear magnetic resonance (NMR) spectroscopy, it was possible to use both tools to characterize DNA sequence-specific structure and dynamics [8,9]. X-ray diffraction spectroscopy has the advantage of showing the precise structural conformation of a biomolecule and highlight sequence-specific differences in structure and dynamics. The drawbacks are the manufacturing of the actual crystal, which is not an easy task and also that the structures are limited to short strands of DNA. Crystals can also present artifacts through packing which can bias the representation of the molecule [10]. NMR spectroscopy is done in solution, which avoids artifacts from crystal packing but is prone to bad or misleading results that make the data interpretation unreliable in some cases [8,9]. The information can now be used to expose sequence-specific differences in structure and dynamics that can help in the study of the sequences of SNPs and their binding interactions with metabolic enzymes or transcription factors. Experimental structures resolved using both X-ray and NMR spectroscopy techniques are available for study and free download in the Protein Data Bank (http://www.rcsb.org/) [11]. The database includes proteins, ligands, and nucleic acids, including in each entry the experimental conditions and details on how the structure was obtained, the quality of the structure, and additional details.

The data provided using these experimental methods illustrate the structural space that double-helical DNA can adopt, and demonstrate that specific structures observed are linked to specific sequence and environment conditions, each of which can modify the DNA structure and dynamics. As effective as these techniques can be, they also have drawbacks that somewhat limit their usefulness. An alternative to experiment is using biomolecular simulation tools on large-scale computational resources; these methods are routinely used to simulate complex systems at the atomic level, allowing a detailed study of motions and dynamics [12]. Molecular dynamics (MD) simulations are a powerful computational tool that allows an atomistic view of the motions, energetics, and structural changes of biopolymers. Since the foundation of the theory driving MD simulations on biological systems in the 1970s by 2013 Nobel laureates Karplus, Levitt, and Warshel, and the recent and significant advances in high-performance computational clusters, parallelization of the codes, and optimization on accelerators, MD simulation has emerged as a vital tool not only to study the structural changes involved in biological processes, but also their energetic profiles [13]. The AMBER engine and family of force fields is one of the tools available to perform MD simulations. It stands for Assisted Modeling Building with Energy Refinement and is used routinely to perform simulations of both proteins and nucleic acids [14–16]. The most widely used force fields for simulation of nucleic acids in the literature are the family of AMBER force fields. The AMBER force field for nucleic acids has been evolving over the years as technological improvements have allowed modelers to extend the simulation time, as extended simulation times continue to reveal new problems with the force fields [15]. The reliable AMBER atomistic force field for nucleic acids that included a proper treatment of the long-range electrostatics was the ff94 force field [17]. The first iteration of ff94 was the ff98 nucleic acids force field which improved the sugar pucker distributions and glycosidic χ torsions [18], which was later tweaked slightly in the same torsions to become the ff99 force field [18]. Further refinements to remove artifacts from anomalous ($\alpha, \gamma = gauche+$, trans) backbone conformations (Figure 1) improved helical DNA twist distributions and the DNA backbone configurations; these are referred to as the parmbsc0 correction to the ff99 force field (ff99-bsc0) [19]. As simulation timescales moved from 10 to 100s of nanoseconds to milliseconds, further deficiencies were noted in the glycosidic χ distributions (within over population of high anti-χ values) and further backbone conformation issues. Work by Šponer and co-workers further improved the force field for the glycosidic χ bond between the sugar and the base (χ_{OL4} modification, [20]) and the distribution of the backbone dihedral between the O3′ oxygen atom (the ε/ζ torsion [21], see Figure 1) which they named

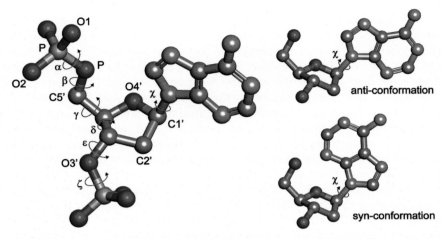

FIGURE 1   Naming convention used for the torsion in the DNA backbone and the nucleobase.

ε/ζ_{OL1}. The accumulation of these small force field adjustments provides a more accurate representation of experimental structures [22]. Still, efforts by our group and others aim to further improve the validity of the nucleic acids force fields. Key lessons learned are the need for long simulations on multiple systems and multiple structures to test the simulation, with or without the use of enhanced sampling (e.g., accelerated MD). Additionally, small changes to a particular force field parameter (i.e., dihedral torsion) can have an accumulative influence in related dihedrals with unintended consequences. For a detailed description and methodology to perform the simulations, we recommend the following references [12,23—25] and tutorials available in http://ambermd.org.

2. THE STRUCTURE AND CHARACTERIZATION OF DNA

Understanding of the structure and dynamics of DNA at a local level actually provides a great deal of information since sequence-specific influences on the structure and dynamics are local, within ±2 to 4 base pairs (except for DNA-protein binding or when sequences of three or more A-T base pairs in poly-A tracts are involved) [26—28]. When working with SNPs that have known sequence, one can infer the overall structure that particular DNA will adopt using structural information provided by experiments and MD simulations, for example, whether the SNP may enhance rigidity or flexibility or whether it may alter the twistability or bendability. Before discussing these properties further, we will briefly visit the naming conventions of DNA.

Double-stranded canonical DNA is built from four basic building blocks called bases or nucleobases: two purines (adenine and guanine) and two pyrimidines (cytosine and thymine). Each of the bases is attached to a pentose monosaccharide sugar, or deoxyribose, and a phosphate group [29]. The three molecules are covalently bonded and form a nucleotide of one variety, depending on the nucleobase, deoxyadenosine (dA), deoxyguanosine (dG), deoxythymidine (dT), and deoxycytocine (dC). In some cases, purines (adenine and guanine) are represented by the letter **R** and pyrimidines (cytosine or thymine) by the letter **Y**. The nucleotides are linked via a phosphate group in the O3′ and O5′ atoms of sequential nucleotides. The direction of the chain starts in a free O5′ oxygen (the 5′ end) and progresses to the free O3′ end. A similar chain of linked nucleotides is brought together in a 3′−5′ direction forming an antiparallel double strand. Pairing of one chain with the other occurs with the formation of hydrogen bonds between the nucleobases in what is called a Watson−Crick pairing (WC). The pairing occurs between dG and dC forms three hydrogen bonds and between dA and dT forms two hydrogen bonds. This WC double-stranded structure under physiological conditions is commonly known as B-DNA. Different base-pairing possibilities (Hoogsteen, Wobble, and so on) are also possible in DNA structures, such as DNA quadruplexes, i-motifs, or in loops, however, with the exception of very transient and low-populated Hoogsteen base pairing [30], these alternative base pairings are not normally seen in the basic genomic material. The backbone of the DNA, which contains the phosphodiester link between two nucleotides, has a formal charge of −1 distributed among all four oxygen atoms. This charge is counterbalanced by positive cations (mainly Na^+, K^+, Mg^{2+}, and Ca^{2+}) that cluster around the DNA in specific locations [31]. The charged nature of the backbone facilitates the process of unwinding and unzipping each of the strands in the duplex, allowing enzymes to manipulate the DNA. An example of this process is the winding of the DNA on the nucleosome where positive residues in the histone interact with the negative phosphate groups in the DNA backbone [32].

In DNA, there are three commonly observed conformations, where the B-DNA conformation is the most abundant [33]. These conformations are based on the structure of the backbone dihedral angles and the sugar ring conformation distribution (or sugar pucker). The backbone on each nucleotide unit starts in the phosphorous atom and ends in the O3′ oxygen atom attached to the sugar molecule (see Figure 1). Starting from the phosphorous atom, there are six dihedral torsions named after the Greek alphabet: from the letter α to ζ. Each of these dihedral torsions adopts values that help reduce the repulsion between the negatively-charged phosphates in the backbone. The dihedral angles involved directly in the sugar ring are called v_1 to v_4, so that the backbone dihedral δ and v_3 are both the same. The ring in the furanose molecule is not planar and can adopt several conformations in what is called sugar puckering. The conformation commonly observed in DNA is with the C2' carbon pointing above the reference plane (known as C2'-endo) and with the C3' carbon above the reference plane (known as C3'-endo). The nucleobase is attached to the sugar ring via a glycosidic bond between the C1' carbon atom in the furanose ring and the N1 nitrogen of the pyrimidines or the N9 nitrogen atom for purines. In DNA, the rotation around this bond, called the χ angle, forms two commonly observed conformations. The syn-conformation has values of $90° \leq \chi \leq 180°$ and sets the nucleobase away from the furanose ring. The anti-conformation, with χ values of $-90° \leq \chi \leq +90°$ positions the nucleobase on top of the ring [29].

These 12 dihedral torsions contribute to the stability of DNA, in addition to stacking interactions of the nucleobases, electrostatics, and the solvent and counterion environment [29]. In order to more easily characterize differences between structures of DNA helices, an internally referenced to the base pair and base-pair step set of helical parameters, measured with respect to a common reference (often to various definitions of the helical axis along the DNA chain) were developed. The nomenclature and definitions were standardized at an EMBO workshop on DNA curvature and bending [34], although the specific mathematical method used were not. Because of this, one has to be careful to use the same definition of the reference helical axis. The base step helical parameters are the measures between adjacent base pairs, and the base-pair helical parameters are measures between two pairing nucleobases (Figure 2). The base step parameters are rise, slide, shift, helical twist, roll, and tilt and the base-pair parameters are stretch, shear, stagger, propeller twist, buckle, inclination, and x-, y-displacements. Each of the commonly observed DNA configurations possesses a unique combination of helicoidal parameters, as well as the aforementioned torsions, that defines its structure and properties. The sequence of an SNP can alter these parameters locally that may, for example, change the overall bend or bendability. This in turn can have several consequences; for example, to alter the ability to easily wrap around the histones [35,36] or the changes in structure or deformability could directly affect the binding of transcription factors [37].

Within the genome, the major observed type of DNA structure is the B-form but nucleic acids present a rich tapestry of motions, deformability, and dynamics that make life possible as we know it. Other commonly recognized, but not as frequent double-stranded DNA configurations are

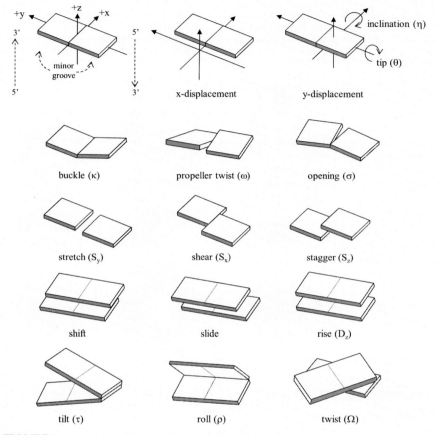

FIGURE 2   Representation of helicoidal parameters.

the A-form and the Z-form, each with specific characteristics in the backbone and helicoidal parameters (Table 1). There are several conformations of DNA that involve multiple strands in the conformation. The H-DNA is formed when a single DNA strand forms the Hoogsteen type of interactions among the major groove of a DNA duplex, forming a triple helical structure. When four strands of DNA are involved, several conformations have been observed, including G-quadruplexes, i-motif, and Holliday junctions. Guanine-rich sequences, also known as G-quadruplexes are found in the telomeric repeats and central regions of the genome [38]. The i-motif structure involves four single strands of C-rich DNA where the nucleobases stack within each other, stabilizing the structure. Holliday junctions occur when two double-stranded DNA chains interchange branches with one another, cross-linking both DNA sets.

TABLE 1   Properties of the Commonly Observed Structures of Duplex DNA in the Genome. Presented Structures Have Been Created Using the Nucleic Acid Builder (NAB [39]) from AmberTools14

	B-DNA	A-DNA	Z-DNA
Helix sense	Right handed	Right handed	Left handed
Base pairs per turn	10	11	12
Vertical rise per base pair	3.4 Å	2.5 Å	3.7 Å
Rotation per base pair	+36°	+33°	−30°
Helical diameter	19 Å	23 Å	18 Å
Helical twist	36°	33°	−30°
Roll	6°	0°	0°
Rise	3.39°	2.56°	3.7°
Inclination	−6.1	21	−6.2
χ angle configuration	Anti	Anti	C: anti, G: syn
Sugar pucker	C2′-endo	C3′-endo	C: C2′-endo, G: C3′-endo
Diameter	20 Å	23 Å	18 Å

3. DNA AND ITS ENVIRONMENT

3.1 Solvation Effects in DNA

Within the cell, the presence of water is of great importance not just to keep solutes dissolved, but to help the stabilization of the secondary and tertiary structure of biomolecules [40]. It also contributes to protein–protein and protein–nucleic acid interactions and specific interactions between biomolecules and ligands [41,42]. Water around the DNA chain reduces the phosphate–phosphate electrostatic repulsion due in part to the dielectric screening of water and to the presence of cations in the medium [29]. Additionally, since the nucleobases show a certain degree of hydrophobicity, they self-assemble into ordered structures in the presence of water [43]. Nucleic acids in general are far more sensitive to hydration effects than proteins due to their highly ionic character [44]. The amount of humidity of DNA plays a key role in its conformation since the secondary structure of DNA is intimately related to the measure of hydration. Multiple experiments and computational studies show two hydration shells around DNA double helixes. The first hydration shell is found interacting with, in affinity order, the phosphate group, the sugar moiety, and the nucleobase. The second hydration shell is indistinguishable from bulk water and exchange constantly between the first shell of hydration. Inside the minor groove exists a well-defined structure of water molecules known as the spine of hydration [45].

At high hydration (low-salt concentration), the B-DNA conformation is favored [45]. About 20 water molecules per nucleotide are present; the sample is conformational stable even if more water molecules are added [29]. When sufficient salt is added, effectively reducing the hydration level, a structural transition occurs toward the A-DNA structure. The B to A transition is also induced by the presence of alcohol and methylation of cytosine residues (the presence of the methyl group disrupts the water structure) [45]. The transformation of B- to Z-DNA involves changing the sense of the duplex causing the major and minor groove to switch and alternating the sugar puckering between C2'-endo and C3'-endo [46]. The conformational change requires a specific sequence of alternating purines and pyrimidines, specifically deoxyguanosine and deoxy-cytidine (e.g., GCGCGC) and high-salt concentration in order to reduce the electrostatic repulsion between the negative charges present in the backbone. Different conditions in the cell could induce switching of these conformations, mainly when DNA is in bound to proteins and ligands. A-DNA conformation is suspected to form as an intermediate step of highly distorted DNA as observed with several DNA–protein complexes [47–49].

3.2 Sequence-Based Deformability and Timescale

Even though there is a perpetual effort to validate, assess, and improve the force fields used for simulation of nucleic acids in general, progress has been made in the study of DNA with its surroundings, DNA, and its sequence-specific structure and the interactions of DNA with proteins. The Ascona B-DNA Consortium (ABC) is an international special interest group focusing on carrying out extensive testing of the AMBER force field for nucleic acids [50–53]. In addition to testing the force fields, the group studies the effect of base sequence on the DNA structure and dynamics using MD simulations. To achieve this, they sampled 136 unique tetranucleotide sequences inside an 18-mer, which translates to all possible nucleobase combinations. An alternative and complementary approach looked at single tetranucleotides embedded in a common sequence [54,55]. Since there are incomplete experimental and structural data on isolated DNA sequences, MD simulations can aid to fill in the information gap of the missing sequences. Using the data, it is possible to predict the structure and flexibility of particular sequences of double-helical B-DNA [56,57]. There are multiple research articles and groups that contribute to the knowledge of how DNA structure shifts and kinks depending on the sequence [54,55], and how deformability and flexibility depend on the sequence [27], although the sampling time and thoroughness of the study are relatively low compared to what has been achieved by the ABC effort. DNA sequence-dependent deformability analysis also has been studied using experimental crystal structures of double-stranded DNA and protein–DNA crystals [58,59], but as mentioned earlier, there are not enough experimental structures with the required sequences to obtain a good representative case [53]. In general, both purine and pyrimidine steps (**R** and **Y**, respectively) are highly dependent on the flanking base steps (tetranucleotide sequence) and not only on the base pair by itself. Purine–purine (**RR**) steps are characterized by higher values of twist and negative shift helicoidal values and pyrimidine–purine (**YR**) steps are characterized by reduced twist and negative slide values. Both dinucleotide sequences (**RR** or **YR**) can show multimodal behavior in a single sequence, which is regulated by the adjacent base pairs. Another ABC finding was a bimodal distribution of the twist helical parameter about a CpG step. Using extensive computational simulations and analyzing experimental structures, Orozco and co-workers found a high-twist–low-twist conformational change, influenced primarily by cation penetration from the minor groove toward the CpG step [60]. Pyrimidine–purine (**YR**) steps were shown to have the most flexibility regarding tilt and twist, increasing the stiffness through RR and RY steps [28]. The most flexible base-pair step is likely TG or CA step. Similar work by Lankaš and our group has revealed deformability at the base-pair level. The GC base pair is, on average, stiffer than AT base pairs in almost all helicoidal parameters (except propeller, twist, and shear, which showed similar values) [27,28]. Another type of distinctive structure found is the one formed with three or more consecutive A-T base pairs in poly-A sequences. These structures are commonly referred to as A-tracts, and this particular sequence, although relatively straight can alter DNA bending at the junctions at the beginning or at the end of the A-tract where the double helix tends to bend toward the minor groove. Particularly interesting are phased A-tracts which are A-tract repeats with an intervening few base-pair sequence break in the A-tract, such as AAAACGTTTT versus TTTTCGAAAA [26], the bending amount corresponding to the length of the A-tract, the sequence at the junctions, and the ionic environment [61,62].

DNA has a multitude of different dynamics, or "DNA breathing" as is sometimes referred, and allows the DNA to be sufficiently flexible to achieve huge conformational changes during biochemical processes such as transcription, replication, and translation. All of these processes have to be performed without losing the genetic encoding, binding specificity, and 3D structure [10,16,54,55]. The dynamics and motions most commonly observed for DNA include:

- bending, twisting, coiling, and supercoiling of the helical chain;
- groove fluctuations;
- fraying of termini base pairs when they lose the Watson—Crick pairing;
- short-lived internal base-pair (not on the termini) openings;
- base flips toward the major groove;
- backbone conformations;
- sugar puckering distributions;
- mismatched base pairing;
- internal base-pair χ flip;
- non-canonical nucleotides.

All of these motions have a specific time range within the DNA molecule that has been characterized with either spectroscopic studies [9] or MD simulations. Regarding the timescale; fast dynamic events such as transient kinks of the grooves, groove fluctuations, and terminal base-pair fraying occur in the nano- to microsecond timescales [30,63,64]. Tertiary DNA structure events such as coiling and protein—DNA recognition have been measured on the micro- to millisecond timescales, internal base-pair opening on the 3—5 μs timescale or longer (except for CpG repeats) and isolated A-T base pairs at ~1 μs [65,66]. Similar measurements of base-pair opening in chemically damaged DNA or where there are nucleotide mismatches or where transient Hoogsteen base-pairing events occur show timescales for opening in the faster micro- to millisecond time range. In standard WC B-DNA, both simulation and experiment suggest a lack of DNA dynamics in the micro- to millisecond time range, the time range where encounter and recognition of DNA occurs, that where DNA damage seems to appear new beacons of motion occur on the sub-millisecond timescale [65,66]. This suggests that motion occurring inside the micro- to millisecond time window may be a means to activate the repair machinery triggered by the abnormal motion.

4. FINAL REMARKS

Through experiment and simulation, the field now has a fairly clear picture of the influence of DNA sequence on its structure and dynamics. This chapter attempts to review this and to provide base insight into DNA structure and dynamics. Although experiments and theory cannot easily provide structural and dynamics at atomic detail for large DNA duplexes, such as those representative of the genomic data, MD approaches can simulate with high level of confidence a range between 1 and ~200 base

TABLE 2   Summary of Commonly Observed DNA Characteristics

A-DNA → B-DNA	• Low salt, low humidity [29,67]
B-DNA → Z-DNA	• High salt, low humidity (observed with 0.7 M MgCl$_2$ or 2.5 NaCl or adding alcohol) [68,69]
A$_4$T$_4$CG repeats	• Bent helix, local bend toward the minor groove at the ApT step [62,70]
T$_4$A$_4$CG repeats	• Straight helixes, local bend toward the major groove at the TpA step [62,70]
CpG steps	• High twist/low twist (HT/LT) bimodal distribution [53] • Cation penetration and tetranucleotide environment [60]
RR and **YR** steps	• Formation of base—phosphate hydrogen bonds (C8—H...O3′) [53,60]
RR steps	• Higher twist and negative shift [53] • Less bendable
YR steps	• Low twist and negative slide [53] • Kinks, more bendable [71]
YR < **RR** < **RY** steps	• Roll flexibility [53] • Stiffness [28]
RY > **RR** > **YR** steps	• Rise flexibility [53]
GC > AT base pairs	• Overall higher stiffness for GC [27,28]

pairs. As the effects of sequence on the structure and dynamics are often local, larger "genome-wide" studies at this level are likely not required. Moreover, this size range is sufficient to model DNA—protein interactions on which SNPs operate. Atomic-level simulations provide insights about the specific interactions involved in binding of biopolymers and what to manipulate chemically to enhance or reduce such interactions. Finally, we include a summary of properties of DNA characteristic and its resulting effect as an aid to do quick characterization of the resulting DNA structure based on the sequence (Table 2).

References

[1] Welter D, MacArthur J, Morales J, Burdett T, Hall P, Junkins H, et al. The NHGRI GWAS Catalog, a curated resource of SNP-trait associations. Nucleic Acids Res 2014;42: D1001—6.

[2] Glazier AM, Nadeau JH, Aitman TJ. Finding genes that underlie complex traits. Science 2002;298:2345—9.

[3] Gabriel SB, Schaffner SF, Nguyen H, Moore JM, Roy J, Blumenstiel B, et al. The structure of haplotype blocks in the human genome. Science 2002;296:2225—9.

[4] Syvänen AC. Accessing genetic variation: genotyping single nucleotide polymorphisms. Nat Rev Genet 2001;2:930—42.

[5] Cooper GM, Shendure J. Needles in stacks of needles: finding disease-causal variants in a wealth of genomic data. Nat Rev Genet 2011;12:628—40.

[6] Alberts B, Johnson A, Lewis J, Morgan D, Raff M, Roberts K, et al. How cells read the genome: from DNA to protein. In: Mol. Biol. Cell; 2008. New York (USA).

[7] Crick FHC, Watson JD. The complementary structure of deoxyribonucleic acid. Proc R Soc A Math Phys Eng Sci 1954;223:80—96.

[8] Zídek L, Stefl R, Sklenár V. NMR methodology for the study of nucleic acids. Curr Opin Struct Biol 2001;11:275—81.

[9] Al-Hashimi HM. NMR studies of nucleic acid dynamics. J Magn Reson 2013;237: 191—204.

[10] Dickerson RE, Goodsell DS, Kopka ML, Pjura PE. The effect of crystal packing on oligonucleotide double helix structure. J Biomol Struct Dyn 1987;5:557—79.

[11] Berman HM. The Protein Data Bank. Nucleic Acids Res 2000;28:235—42.

[12] Galindo-Murillo R, Bergonzo C, Cheatham 3rd TE. Molecular modeling of nucleic acid structure. Curr Protoc Nucleic Acid Chem 2013;54. Unit 7.5.

[13] Karplus M, McCammon JA. Molecular dynamics simulations of biomolecules. Nat Struct Biol 2002;9:646—52.

[14] Case DA, Cheatham 3rd TE, Darden T, Gohlke H, Luo R, Merz KM, et al. The Amber biomolecular simulation programs. J Comput Chem 2005;26:1668—88.

[15] Cheatham 3rd TE, Case DA. Twenty-five years of nucleic acid simulations. Biopolymers 2013;12:969—77.

[16] Cheatham 3rd TE. Simulation and modeling of nucleic acid structure, dynamics and interactions. Curr Opin Struct Biol 2004;14:360—7.

[17] Cornell WD, Cieplak P, Bayly CI, Gould IR, Merz KM, Ferguson DM, et al. A second generation force field for the simulation of proteins, nucleic acids, and organic molecules. J Am Chem Soc 1995;117:5179—97.

[18] Cheatham 3rd TE, Cieplak P, Kollman PA. A modified version of the Cornell et al force field with improved sugar pucker phases and helical repeat. J Biomol Struct Dyn 1999; 16:845—62.

[19] Pérez A, Marchán I, Svozil D, Šponer J, Cheatham 3rd TE, Laughton CA, et al. Refinement of the AMBER force field for nucleic acids: improving the description of alpha/gamma conformers. Biophys J 2007;92:3817−29.

[20] Krepl M, Zgarbová M, Stadlbauer P, Otyepka M, Banáš P, Koča J, et al. Reference simulations of noncanonical nucleic acids with different χ variants of the AMBER force field: quadruplex DNA, quadruplex RNA and Z-DNA. J Chem Theory Comput 2012;8:2506−20.

[21] Zgarbová M, Luque FJ, Šponer J, Cheatham 3rd TE, Otyepka M, Jurečka P. Toward improved description of DNA backbone: revisiting Epsilon and Zeta torsion force field parameters. J Chem Theory Comput 2013;9:2339−54.

[22] Šponer J, Banáš P, Jurečka P, Zgarbová M, Kührová P, Havrila M, et al. Molecular dynamics simulations of nucleic acids from tetranucleotides to the ribosome. J Phys Chem Lett 2014;5:1771−82.

[23] Bergonzo C, Galindo-Murillo R, Cheatham 3rd TE. Molecular modeling of nucleic acid structure: electrostatics and solvation. Curr Protoc Nucleic Acid Chem 2014;55: 7.9.1−7.9.27.

[24] Galindo-Murillo R, Bergonzo C, Cheatham 3rd TE. Molecular modeling of nucleic acid structure: setup and analysis. Curr Protoc Nucleic Acid Chem 2014;56: 7.10.1−7.10.21.

[25] Bergonzo C, Galindo-Murillo R, Cheatham 3rd TE. Molecular modeling of nucleic acid structure: energy and sampling. Curr Protoc Nucleic Acid Chem 2013;54. Unit 7.8.

[26] Sprous D, Young MA, Beveridge DL. Molecular dynamics studies of axis bending in d(G5-(GA4T4C)2-C5) and d(G5-(GT4A4C)2-C5): effects of sequence polarity on DNA curvature. J Mol Biol 1999;285:1623−32.

[27] Lankaš F, Sponer J, Langowski J, Cheatham 3rd TE. DNA deformability at the base pair level. J Am Chem Soc 2004;126:4124−5.

[28] Lankas F, Sponer J, Langowski J, Cheatham 3rd TE. DNA basepair step deformability inferred from molecular dynamics simulations. Biophys J 2003;85:2872−83.

[29] Saenger VW. Principles of nucleic acid structure. New York (USA): Springer-Verlag; 1988.

[30] Nikolova EN, Kim E, Wise AA, O'Brien PJ, Andricioaei I, Al-Hashimi HM. Transient Hoogsteen base pairs in canonical duplex DNA. Nature 2011;470:498−502.

[31] McFail-Isom L, Sines CC, Williams LD. DNA structure: cations in charge? Curr Opin Struct Biol 1999;9:298−304.

[32] Cox M, Nelson DR. Lehninger principles of biochemistry. San Francisco: W.H. Freeman; 2003.

[33] Alberts B, Johnson A, Lewis J, Morgan D, Raff M, Roberts K, et al. DNA, chromosomes, and genomes. In: Mol. Biol. Cell; 2008. New York (USA).

[34] Dickerson RE. Definitions and nomenclature of nucleic acid structure components. Nucleic Acids Res 1989;17:1797−803.

[35] Richmond TJ, Davey CA. The structure of DNA in the nucleosome core. Nature 2003; 423:145−50.

[36] Freeman GS, Lequieu JP, Hinckley DM, Whitmer JK, de Pablo JJ. DNA shape dominates sequence affinity in nucleosome formation. Phys Rev Lett 2014;113:168101.

[37] Jolma A, Yan J, Whitington T, Toivonen J, Nitta KR, Rastas P, et al. DNA-binding specificities of human transcription factors. Cell 2013;152:327−39.

[38] Brooks TA, Kendrick S, Hurley L. Making sense of G-quadruplex and i-motif functions in oncogene promoters. FEBS J 2010;277:3459−69.

[39] Macke TJ, Case DA. Molecular modeling of nucleic acids. Washington (DC): American Chemical Society; 1997.

[40] Westhof E. Water: an integral part of nucleic acid structure. Annu Rev Biophys Biophys Chem 1988;17:125−44.

[41] Woda J, Schneider B, Patel K, Mistry K, Berman HM. An analysis of the relationship between hydration and protein-DNA interactions. Biophys J 1998;75:2170—7.

[42] Schwabe JW. The role of water in protein—DNA interactions. Curr Opin Struct Biol 1997;7:126—34.

[43] Schweitzer BA, Kool ET. Hydrophobic, non-hydrogen-bonding bases and base pairs in DNA. J Am Chem Soc 1995;117:1863—72.

[44] Makarov V, Pettitt BM, Feig M. Solvation and hydration of proteins and nucleic acids: a theoretical view of simulation and experiment. Acc Chem Res 2002;35:376—84.

[45] Drew HR, Wing RM, Takano T, Broka C, Tanaka S, Itakura K, et al. Structure of a B-DNA dodecamer: conformation and dynamics. Proc Natl Acad Sci USA 1981:2179—83.

[46] Dickerson RE. DNA structure from A to Z. Methods Enzymol 1992;211:67—111.

[47] Nekludova L, Pabo CO. Distinctive DNA conformation with enlarged major groove is found in Zn-finger-DNA and other protein-DNA complexes. Proc Natl Acad Sci USA 1994;91:6948—52.

[48] Lu XJ, Shakked Z, Olson WK. A-form conformational motifs in ligand-bound DNA structures. J Mol Biol 2000;300:819—40.

[49] Potaman VN, Sinden RR. DNA: alternative conformations and biology. 2000.

[50] Beveridge DL, Barreiro G, Byun KS, Case DA, Cheatham 3rd TE, Dixit S, et al. Molecular dynamics simulations of the 136 unique tetranucleotide sequences of DNA oligonucleotides. I. Research design and results on d(CpG) steps. Biophys J 2004;87: 3799—813.

[51] Dixit SB, Beveridge DL, Case DA, Cheatham 3rd TE, Giudice E, Lankas F, et al. Molecular dynamics simulations of the 136 unique tetranucleotide sequences of DNA oligonucleotides II: sequence context effects on the dynamical structures of the 10 unique dinucleotide steps. Biophys J 2004;89:3721—40.

[52] Lavery R, Zakrzewska K, Beveridge DL, Bishop TC, Case DA, Cheatham 3rd TE, et al. A systematic molecular dynamics study of nearest-neighbor effects on base pair and base pair step conformations and fluctuations in B-DNA. Nucleic Acids Res 2009;38: 299—313.

[53] Pasi M, Maddocks JH, Beveridge DL, Bishop TC, Case DA, Cheatham 3rd TE, et al. μABC: a systematic microsecond molecular dynamics study of tetranucleotide sequence effects in B-DNA. Nucleic Acids Res 2014;42:12272—83.

[54] Araúzo-Bravo MJ, Fujii S, Kono H, Ahmad S, Sarai A. Sequence-dependent conformational energy of DNA derived from molecular dynamics simulations: toward understanding the indirect readout mechanism in protein-DNA recognition. J Am Chem Soc 2005;127:16074—89.

[55] Fujii S, Kono H, Takenaka S, Go N, Sarai A. Sequence-dependent DNA deformability studied using molecular dynamics simulations. Nucleic Acids Res 2007;35:6063—74.

[56] Dixit SB, Beveridge DL. Structural bioinformatics of DNA: a web-based tool for the analysis of molecular dynamics results and structure prediction. Bioinformatics 2006; 22:1007—9.

[57] Khandelwal G, Lee RA, Jayaram B, Beveridge DL. A statistical thermodynamic model for investigating the stability of DNA sequences from oligonucleotides to genomes. Biophys J 2014;106:2465—73.

[58] Gorin AA, Zhurkin VB, Olson WK. B-DNA twisting correlates with base-pair morphology. J Mol Biol 1995;247:34—48.

[59] Olson WK, Gorin AA, Lu X-J, Hock LM, Zhurkin VB. DNA sequence-dependent deformability deduced from protein-DNA crystal complexes. Proc Natl Acad Sci USA 1998;95:11163—8.

[60] Dans PD, Faustino I, Battistini F, Zakrzewska K, Lavery R, Orozco M. Unraveling the sequence-dependent polymorphic behavior of d(CpG) steps in B-DNA. Nucleic Acids Res 2014;42:11304—20.

[61] Dršata T, Špačková N, Jurečka P, Zgarbová M, Šponer J, Lankaš F. Mechanical properties of symmetric and asymmetric DNA A-tracts: implications for looping and nucleosome positioning. Nucleic Acids Res 2014;42:7383—94.

[62] Stefl R, Wu H, Ravindranathan S, Sklenár V, Feigon J. DNA A-tract bending in three dimensions: solving the dA4T4 vs dT4A4 conundrum. Proc Natl Acad Sci USA 2004;101: 1177—82.

[63] Bothe JR, Nikolova EN, Eichhorn CD, Chugh J, Hansen AL, Al-Hashimi HM. Characterizing RNA dynamics at atomic resolution using solution-state NMR spectroscopy. Nat Methods 2011;8:919—31.

[64] Hansen AL, Nikolova EN, Casiano-Negroni A, Al-Hashimi HM. Extending the range of microsecond-to-millisecond chemical exchange detected in labeled and unlabeled nucleic acids by selective carbon R(1rho) NMR spectroscopy. J Am Chem Soc 2009; 131:3818—9.

[65] Galindo-Murillo R, Roe DR, Cheatham 3rd TE. Convergence and reproducibility in molecular dynamics simulations of the DNA duplex d(GCACGAACGAACGAACGC). Biochim Biophys Acta 2014.

[66] Galindo-Murillo R, Roe DR, Cheatham 3rd TE. On the absence of intrahelical DNA dynamics on the µs to ms timescale. Nat Commun 2014;5:5152.

[67] Ivanov VI, Minchenkova LE, Schyolkina AK, Poletayev AI. Different conformations of double-stranded nucleic acid in solution as revealed by circular dichroism. Biopolymers 1973;12:89—110.

[68] Pohl FM. Polymorphism of a synthetic DNA in solution. Nature 1976;260:365—6.

[69] Pohl FM, Jovin TM. Salt-induced co-operative conformational change of a synthetic DNA: equilibrium and kinetic studies with poly (dG-dC). J Mol Biol 1972;67:375—96.

[70] Hagerman PJ. Sequence-directed curvature of DNA. Nature 1986;321:449—50.

[71] Tolstorukov MY, Jernigan RL, Zhurkin VB. Protein-DNA hydrophobic recognition in the minor groove is facilitated by sugar switching. J Mol Biol 2004;337:65—76.

Genomic Medicine and Lipid Metabolism: LDL Targets and Stem Cell Research Approaches

Theodore P. Rasmussen

Department of Pharmaceutical Sciences, University of Connecticut, Storrs,
CT, USA; University of Connecticut Stem Cell Institute, University of
Connecticut, Storrs, CT, USA; Department of Molecular and Cell Biology,
University of Connecticut, Storrs, CT, USA

1. LDL FUNCTIONS OF THE HEPATOCYTE

Serum cholesterol is derived from two origins—some is absorbed from the diet and the remainder is synthesized endogenously by hepatocytes and released into the bloodstream. All cells produce cholesterol, but in general, non-hepatic cells do not secrete their endogenously synthesized cholesterol into the bloodstream. Within the hepatocyte, de novo cholesterol synthesis begins with the action of the enzyme HMG-CoA reductase. HMG-CoA reductase converts acetyl-CoA to cholesterol via a number of anabolic biochemical steps [1]. HMG-CoA reductase is the first and rate-limiting enzyme in the cholesterol biosynthetic pathway and is the target of statins, which effectively lowers serum cholesterol by decreasing hepatic production of cholesterol [2,3].

The hepatocyte is a unique cell within the human body with regard to cholesterol. It is the only cell type that can produce and secrete cholesterol into the bloodstream yet also removes cholesterol from the blood, which is followed by degradation of cholesterol into bile acids and salts within the hepatocyte. The resulting bile acids and salts are then transported out of the hepatocyte through the action of biliary transporters, where they enter the biliary flow, and are eliminated from the body through the gallbladder and subsequent excretion. Since the hepatocyte is the only cell that can excrete cholesterol from the body, it is central to the maintenance of

Translational Cardiometabolic Genomic Medicine
http://dx.doi.org/10.1016/B978-0-12-799961-6.00005-6

99

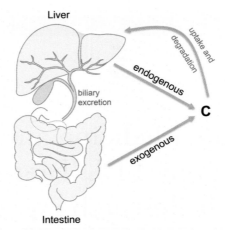

FIGURE 1   **Cholesterol homeostasis overview.** Total cholesterol content in the blood (C) is of two sources: endogenous, which is produced primarily by the liver, and exogenous, which is absorbed through the intestinal tract and deposited in the blood. Hepatocytes of the liver are unique in their ability to take up cholesterol and then excrete it from the body.

cholesterol homeostasis (Figure 1). Disorders of the hepatocyte and the liver can result in elevated cholesterol, with associated cardiovascular risks.

Cholesterol within the blood is carried by a variety of apolipoprotein particles that contain amphipathic apoproteins on their exteriors, and lipids, cholesterol, and fatty acids in their hydrophobic cores. Low-density apolipoprotein (LDL) consists of cholesteryl esters (cholesterol esterified to fatty acids), free cholesterol, fatty acids, and triglycerides complexed with the amphipathic protein apolipoprotein B-100 (ApoB-100), which is produced solely in hepatocytes. In addition, APOB-100 is found in very low-density apolipoproteins (VLDL), intermediate-density apolipoproteins (IDL), and low-density apolipoproteins (LDL). The same *APOB* gene that produces APOB-100 also produces a similar-sized messenger RNA transcript in intestinal (enterocyte) cells, but remarkably, the enterocyte message leads to the production of the lower molecular weight APOB-48. The APOB-48 protein is shorter, because the mRNA of the *APOB* gene is post-transcriptionally edited in enterocytes to yield a stop codon about half way through the reading frame that otherwise would have produced APO-100. This RNA editing event occurs in enterocytes but not hepatocytes, and APOB-48 packages cholesterol and triglycerides of intestinal origin in the form of chylomicrons. This packaging requires the action of microsomal triglyceride transfer protein (MTP). About 75% of the chylomicron's cholesterol is of biliary origin (i.e., of liver origin) and about 25% is dietary. The high level of endogenous cholesterol in chylomicrons occurs because cholesterol is excreted by the bile duct into the intestinal duodenum, and this cholesterol becomes available for re-uptake by intestinal enterocytes downstream from the bile duct. The normal function of APOB-containing chylomicrons and apolipoproteins is to deliver triglycerides to muscle for energy and to fat for storage. The vasculature of muscle and fat is also lined with endothelial cells that bind endogenously produced ApoB-100-containing VLDL particles, which then deliver fatty acids to muscle and fat. This occurs because vascular tissue contains lipoprotein lipase (LPL), which is an enzyme that harvests triglycerides from APOB-100-positive apolipoproteins. As triglycerides are removed, the resulting remnant particles become progressively smaller, and relatively higher in cholesterol content. IDL is formed from VLDL in this way. Some IDL is then further remodeled to yield HDL (which contains APOAI and APOAII). Remaining IDL is then freed to the circulation and then sequestered at the surface of hepatocytes in the Space of Disse. There, hepatic lipase (present on the surface of hepatocytes) further modifies the fatty acid content of remnant IDL particles, and the resulting particles are now remodeled to yield LDL, which can then be released back into the circulation. Thus, LDL is a derivative of remnant apolipoprotein particles which have been produced by a two-step pathway, first at the surface of muscle or fat vasculature to yield IDL, and then at the hepatocyte surface to yield LDL (Figure 2). Overall, LDL is highly significant because 65–75% of total serum cholesterol resides within LDL at any given time. Furthermore, LDL is readily taken up by macrophages (foam cells) in coronary arteries, and these then deposit the cholesterol underneath the thin layer of arterial endothelium to produce atheromatous plaques, which grow over time to cause coronary artery occlusion. Rupture of these plaques exposes thrombus-nucleating proteins such as tissue factor, which then leads rapidly to myocardial

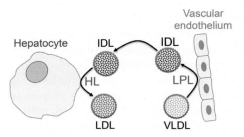

FIGURE 2 **LDL production and uptake.** LDL production is indirect, and occurs in a two-step mechanism. First, very low-density apolipoprotein (VLDL) is recruited to vascular endothelium within muscle and fat, and triglycerides are extracted via the action of lipoprotein lipase (LPL). The resulting particles are now intermediate-density lipoprotein (IDL), and this is processed at the hepatocyte surface by hepatic lipase (HL) to yield LDL.

infarction. Not surprisingly, elevated LDL is the most significant serum cholesterol indicator of the risk for subsequent adverse cardiovascular events.

Although the hepatocyte is the sole producer of LDL in the body (since APOB-100 is a hepatocyte-specific protein), the liver hepatocyte is also highly significant because it is the only cell type that can remove cholesterol from the blood serum for subsequent elimination from the body. One major means of removal of cholesterol is by "reverse choles-terol transport" (RCT) which is primarily mediated by harvesting of cholesterol from HDL. RCT and HDL are discussed extensively in another chapter of this book. A second and very important route of clearance of cholesterol by the liver is the removal of LDL from the bloodstream. Although this route clears less cholesterol than RCT in terms of total amounts of cholesterol, uptake and excretion of LDL-born cholesterol is highly significant from a cardiovascular health standpoint since circu-lating LDL is readily taken up by foam cells in arterial vasculature to yield sub-endothelial plaques from the pool of LDL that is not removed by hepatocytes. Therefore, LDL mechanisms are of keen interest from a cardiovascular health perspective.

LDL is removed from the body by hepatocytes which express a receptor that binds LDL, the LDL receptor (LDLR) [4,5]. The LDLR protein consists of 839 amino acids and is synthesized in the endo-plasmic reticulum and then trafficked to the cell membrane surface via the Golgi apparatus and endosomal trafficking [6]. LDLR is the only receptor that can clear significant amounts of LDL from the serum, although LDLR binds APO-100 with only moderate affinity, possibly leading to the relatively long half-life of LDL in the bloodstream [7]. Once LDL is bound by LDLR, the entire apolipoprotein–receptor complex is endocytosed and encapsulated in lysosomes [8] (Figure 3). The LDLR is then recycled to the hepatocyte surface [9], and the remainder of the LDL particle is degraded in the lysosome by hydrolysis, and free cholesterol is released into the hepatocyte. Levels of LDLR are also regulated by the proteolytic activity of Proprotein Convertase Subtilisin/Kexin type 9 (PCSK9), which selectively degrades LDLR, thus regulating its abundance. Free cholesterol in the hepatocyte can then be degraded, but intrahepatic cholesterol levels also serve as effector molecules to regulate important aspects of cholesterol biosynthesis and LDL uptake. Free cholesterol can then serve as a molecule that inhibits the gene encoding HMG-CoA reductase [10], which is the first and rate-limiting enzyme in de novo cholesterol biogenesis. This is mediated by the binding of cholesterol to nuclear receptors that then become competent to regulate the transcriptional activity of the HMG-CoA reductase gene. Therefore, LDLR-mediated internalization of LDL leads to decreased endogenous cholesterol synthesis. In addition,

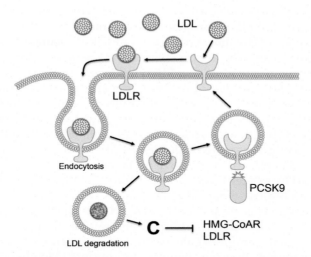

FIGURE 3   **LDLR recycling.** LDLR first binds LDL at the hepatocyte surface, and the entire receptor–apolipoprotein complex is endocytosed in clathrin-coated pits. LDLR then releases LDL for subsequent degradation in endosomes, and free LDLR is then recycled to the cell surface for reuse. However, the LDLR proteolytic enzyme PCSK9 can act on residual LDLR, and thus regulates LDLR protein half-life. Internalized LDL is degraded and releases cholesterol (C), which can then transcriptionally repress production of HMG-CoA reductase and LDLR.

intra-hepatic cholesterol can down-regulate the production of nascent LDLR through similar mechanisms. However, most of the internalized cholesterol within the hepatocyte is either re-excreted into the blood, or excreted into the bile. Cholesterol is degraded via multistep biochemical reactions to yield bile salts and bile acids. The first (and rate-limiting) enzyme in this process is cytochrome P450 7A1 (CYP7A1). Following subsequent biochemical reactions, bile salts and acids are produced and then transported out of the hepatocyte and into the biliary flow via the integral membrane transporter ABCB4.

2. HUMAN SINGLE-GENE GENETIC DISORDERS OF LDL METABOLISM

A classic genetic disorder of LDL metabolism is that of familial hypercholesterolemia (FH). The FH is characterized by very high levels of circulating LDL. A major consequence of this high LDL levels is the early onset of coronary artery disease and frequent cases of heart attacks at a young age. In fact, those in the highest quintile of LDL levels experience a 2.5-fold increase in their risk for myocardial infarction [11]. The situation is worse in FH patients who experience rates of myocardial infarction that are even more severe. From a diagnostic standpoint, FH patients often present with subcutaneous deposits of cholesterol (xanthomas) that are visible and palpable, often around the ankle at the Achilles tendon. In addition, deposits of cholesterol around the periphery of the iris within the eye (corneal arcus) are frequent. FH is typically caused by loss-of-function mutations in one of the two alleles of the *LDLR* gene. Many mutations that are causative of FH have been described, including a large variety of point mutations within the coding and regulatory regions of the *LDLR* gene. Therefore, it is clear that FH has arisen independently by de novo mutagenesis many times. Since FH can be caused by a heterozygous loss-of-function allele of LDLR, FH is a very common human genetic disorder and occurs in as many as one in 500 individuals. This frequency makes FH one of the most common human genetic disorders. In addition, FH is typically observed in the heterozygous state, and most alleles are loss of function. Heterozygous FH is transmitted through families by normal Mendelian genetic mechanisms, and the disease normally appears in each generation since heterozygosity is sufficient to cause the disease. This mode of inheritance has been described as dominant, but is actually best described as an example of haploinsufficiency, since the loss of one of two alleles results in reduced gene dosage that is insufficient to produce adequate amounts of LDLR protein within hepatocytes. In heterozygous FH individuals, LDL accumulates in the blood, leading to the development of atheromatous plaques and cardiovascular disease, xanthomas, and corneal arcus. Occasionally, homozygous FH is observed (at a genotype frequency of less than 1×10^6). Homozygous FH is exceedingly severe, with death due to cardiovascular incidents occurring at a very early age. Heterozygous FH is usually treated with statins, which inhibit HMG-CoA reductase, leading to a reduction in serum LDL levels. Homozygous FH is, unfortunately, refractive to treatment.

Another important protein involved in the removal of LDL from the blood is PCKS9 [12]. PCKS9 is a protease that is involved in regulating the half-life of LDLR since LDLR is a direct substrate of PCKS9-mediated proteolysis. Thus, loss-of-function mutations in the *PCKS9* gene have the effect of making more LDLR available for recycling to the hepatocyte surface. In contrast, other mutations in *PCKS9* lead to increased rates of LDLR proteolysis, and a corresponding decrease in the hepatic ability to remove LDL from the blood. Both types of mutations have been described in human populations. Furthermore, PCKS9 protein is an interesting and attractive drug target, since drugs that inhibit PCKS9 function should lead to increased supplies of LDLR at the hepatocyte surface and a consequent diminution of LDL from the blood.

3. RECENT HUMAN GENOMIC FINDINGS RELATED TO LDL DYSFUNCTION

Recent advances in human genetics research have become a powerful tool when used for the investigation of dyslipidemias, and have led to the identification of dozens of genes that are involved in cholesterol homeostasis and medical conditions involving elevated LDL. Perhaps the most powerful recent method is that of genome-wide association studies (GWAS) that survey individual genomes for single-nucleotide polymorphisms (SNPs) and then associate these with measurable numerical data, such as serum cholesterol profiles. In one such study, 100,000 individuals were typed for both their SNP configurations and associated cholesterol profiles. This study found a surprisingly large number of genes, and suggests that as many as 95 genetic loci may be involved in serum cholesterol management [13]. Many of the hits include expected genes such as *CYP7A1* and *SCARB1*, but many novel genes were also identified, and variants in these genes seem to be correlated with the presence of abnormal cholesterol profiles, including elevated LDL. In another GWAS of elderly patients, cholesterol profiles were utilized as a phenotype, and predominant genomic hits included variants in the *LDLR* gene and polymorphisms in the genes *APOE* and *APOE1*, *HMGCR*, and others [14]. Another recent GWAS involving African-Americans used electronic medical records that included LDL data. This study found a variant in *APOE* which seems to be protective of elevated LDL levels [15]. A number of recent GWASs focused on specific ethnicities indicate that LDL levels vary by ethnic group, suggesting that the regard for population genetics considerations is prudent.

4. MODELING CHOLESTEROL HOMEOSTASIS USING IN VITRO STEM CELL APPROACHES

Although hepatocytes are clearly a central and highly important cell type involved in cholesterol homeostasis, it has been very difficult to obtain informative primary human hepatocytes from human subjects and maintain them in culture. The use of primary human hepatocytes has been difficult for a variety of reasons. The most fundamental technical reason is that primary human hepatocytes rapidly lose their normal enzymatic activities in culture. This includes a rather widespread loss of cytochrome P450 expression soon after culture is commenced to include CYP7A1, the hepatocyte-specific CYP that is solely responsible for the initiation of cholesterol degradation. Additional problems stem from the nature of human liver biopsies. Human liver biopsies (often needle biopsies with relatively small numbers of cells) are usually obtained from patients who are experiencing hepatic disorders such as liver dysfunction or failure or hepatitis infection. Thus, the cells are obtained in small quantities and are usually abnormal. In addition, hepatocyte biopsies are typically not obtained from individuals with elevated cholesterol because this is not necessary in the course of clinical treatment for dyslipidemias. In addition, there is a wide variance in the normal spectrum of hepatocyte functionality from person to person. This is not only true of detoxification activities, but also of cholesterol metabolic activities. Also, in addition to variation in human genetics, non-genetic factors such as age, gender, and dietary history also affect the primary hepatocyte function and serum cholesterol levels. For all these reasons, primary human hepatocytes as research tools present with a host of difficulties for the study of basic cholesterol and apolipoprotein dynamics.

Recently, it has become possible to produce human hepatocytes in vitro from human embryonic stem cells (hESCs) and human-induced pluripotent stem cells (hiPSCs). Embryonic stem cells were first derived from the mouse and shown to be capable of differentiating into all cell types of the adult body [16,17]. The hESC lines are clonal and are derived from a single human embryo at the blastocyst stage, about day 5 in human embryogenesis [18,19]. The hiPSCs are also clonal and are derived by reprogramming somatic cells, such as skin fibroblasts, to a developmental stage that is identical to that of ESCs. This was first achieved with murine somatic cells [20], and soon thereafter with human somatic cells [21–23]. (A complete description of iPS reprogramming methods and the utility of hiPSCs to explore the human genetics of dyslipidemias is presented in a later section.) Both hESCs and hiPSCs are pluripotent, meaning that they can produce all cell types normally present in the adult human body during in vitro cellular differentiation. In addition, both hESCs and hiPSCs are immortal and highly proliferative in cell culture so long as they are maintained under culture conditions that support pluripotency. This feature allows them to be expanded and passaged indefinitely in the pluripotent state, thus allowing the production of a prodigious supply of stem cell-derived hepatocytes (or other desired cell type) upon demand after directed differentiation procedures are employed. Qualities of these stem cell-derived hepatocytes include a uniform and reproducible genotype (due to a ready supply of clonal pluripotent cells). Furthermore, since hESCs and hiPSCs are of non-hepatic origin, they have never experienced insults such as hepatotoxicity or disease during their production. Of major importance is the fact that hiPSCs are derived from a given human subject, and therefore clonal hiPSC lines carry that individual's unique genotype. For this reason, hiPSCs are ideal cellular models of human genetic disease.

The pluripotent state of hESCs and hiPSCs is maintained at both genetic and epigenetic levels. Human blastocyst embryos contain two major cell types, the trophectoderm and the inner cell mass (ICM). The trophectoderm consists of cells that are fated to develop into the embryonic portion of the placenta. In contrast, the ICM contains cells that will subsequently form all cells of the embryo proper. The first cell types established during subsequent embryogenesis from the ICM are those of the embryonic ectoderm, endoderm, and mesoderm (the germ layer cells). All subsequent cells and organs are derivatives of endoderm, ectoderm, and mesoderm. hESCs are derived from the ICM of blastocysts, and the process of hESC derivation yields cells that maintain the developmental potential of the ICM. However, the ICM is a transitory primordium of cells in the intact embryo, but hESCs can be maintained indefinitely and yet maintain the developmental potential of the ICM, and hence reside in a state of permanent pluripotent stasis. In practice, hESC lines have been derived for the most part from excess blastocyst embryos produced in fertility clinics [19]. Hundreds (perhaps thousands) of times as many excess blastocysts have been discarded by fertility clinics in the course of normal operations as assisted reproductive technology clinics have gone into stem cell research.

The hESC genome is managed by a powerful network of cooperating transcription factors that collude to express a transcriptome that is unique to the pluripotent state. The most important of the pluripotency transcription factors are OCT4, SOX2, KLF4, and NANOG [24−26]. In fact, these factors have been shown to bind to several thousand target genes in ESCs by chromatin immunoprecipitation experiments. Together these factors activate a wide variety of downstream genes that maintain the pluripotent state and other genes that inhibit differentiation. MicroRNAs also participate, and LIN28, a microRNA regulatory protein, is notable for its involvement [27,28].

Pluripotent cells are only useful as cell-culture models if they are differentiated into a relevant cell type, such as hepatocytes. Most in vitro differentiation methods have been greatly informed by knowledge gleaned from studies of normal development, usually from model organisms. The hepatocyte is a cell type of endoderm origin. In keeping with this notion, an expansive understanding of hepatocyte specification in vivo during liver organogenesis [29,30] has assisted the design of strategies for the differentiation of pluripotent cells to hepatocyte-like cells (HLCs) in vitro. Hepatogenesis begins during early embryogenesis at the time of gastrulation when definitive endoderm is established. From the anterior portion of the definitive endoderm, foregut is soon formed, and hepatoblast precursors migrate to reside near the heart, which secretes inductive signals including fibroblast growth factors (FGF1 and FGF2) [31]. Other neighboring mesenchymal cells secrete bone morphogenic factor 4 (BMP4), which also assists hepatocyte precursor cells to further assume a hepatocyte fate [32]. WNT signaling also participates in hepatocyte development, and although important, the role and timing are less clear than that of the FGFs and BMP4. Later stages of liver organogenesis rely on the expression cell autonomous factors such HNF1α, FoxA1/A2, and many others within developing hepatoblasts. At the 7-11 somite stage in mice, the liver acquires histological structure and organ design that is identifiable as liver per se, and the expression of several more mature hepatic markers such as albumin, α-fetoprotein, HNF1β, and others. In addition to hepatocyte specification, sinusoidal vasculature, portal and central vein, and bile duct structures are established. Transcriptional and enzymatic activities within hepatocytes themselves are also refined and become specialized, often with segregation of specific biochemical functions to zonal regions of hepatocytes that reside in a columnar arrangement adjacent to sinusoids.

Hepatocyte production from pluripotent stem cells, like development in vivo, first relies upon the induction of endoderm. Several published and effective procedures to produce HLCs from pluripotent cells [30,33] proceed essentially as follows: First, the potent endoderm-inducing factor Activin A is used. Activin A is a potent inducer of definitive endoderm from hESCs and hiPSCs [34]. After Activin A exposure, the resulting endoderm expresses the endoderm marker SOX17. The second step consists of methods to bring about the specification of early hepatic lineages from endoderm. Next, the cells are exposed to FGF2 and BMP4 to establish hepatoblast-like cells from the previously uncommitted endodermal cells [29,33]. Then, the cells are exposed to hepatocyte growth factor, and the resulting cells resemble embryonic hepatocytes as judged by their prodigious secretion of α-fetoprotein. Finally, the cells are induced to a point of greater maturity by exposure to oncostatin M and dexamethasone. The resulting cells now begin to express albumin, although many still continue to express α-fetoprotein, indicating that the final cells are likely to be similar to neonatal hepatocytes rather than fully mature adult hepatocytes. Since the resulting HLCs are not identical to fresh normal hepatoctyes obtained by human biopsy, HLCs, like cultured primary human hepatocytes, also exhibit some deficiencies as cell-culture models. In addition to their immaturity, HLCs also have poor CYP expression, making these cells problematic for most toxicological and some pharmaceutical studies. However, HLCs have been shown to be an excellent cell-culture model for cholesterol homeostasis, because they produce, secrete, and take up cholesterol in vitro in conjunction with apolipoprotein expression.

The cholesterol and lipid processing abilities of HLCs are extensive and recently well characterized [29,33]. These cells produce and secrete cholesterol into the cell-culture medium as solubilized cholesterol (much as a hepatocyte in vivo would excrete cholesterol into the blood). This is likely achieved by the packaging cholesterol into apolipoprotein particles, because HLCs robustly express the apoproteins APOA1, APOA2, APOB, APOC, and APOE. Analysis of the HLCs indicates that there are subsets of cells with regard to the type of APO expression. For instance, some cells are APOA1 positive but albumin negative, while some are both APOA1 and albumin positive. Similar complexity exists in other pairwise analyses of expression of other APOs in HLCs, but the results indicate that HLCs need not be entirely matured in order to produce apolipoproteins, based on albumin and APO expression [33]. For instance, APO expression is evident in HLCs that express either α-fetoprotein alone or α-fetoprotein combined with albumin expression. On a pharmacological basis, HLCs respond to statin treatment in vitro as judged by a marked reduction in secreted solubilized cholesterol after pravastatin treatment. HLCs also express and produce LDLR and bind and endocytose fluorescently labeled LDL.

5. USING hiPSCs FOR GENETIC DISEASE MODELING AND BASIC PERSONALIZED MEDICINE RESEARCH

The researcher has no control over the genetic identity of hESCs. The hESC genome is identical to that which resided in the blastocyst from which it is derived. Since a blastocyst obtained from a fertility clinic obviously has little or no medical information associated with it (and certainly no information with regard to dyslipidemias), it is surprisingly difficult to learn much about human genetic variation and how it might impact a specific disorder such as dyslipidemia through the study of hESCs. In contrast hiPSCs are of tremendous value because the researcher can choose the human somatic cell donor based on clinical phenotype. At the most fundamental level, the human subject in essence donates a genome that is bound for residence in clonal pluripotent cells. In order to more fully understand the remarkable utility of iPS' "disease in a dish" approaches, it is worth knowing about the iPS procedure itself.

Cellular reprogramming was first demonstrated with animal cloning efforts. The first such successful effort at animal cloning was achieved with *Xenopus* frogs, where nuclei of intestinal epithelial cells were reprogrammed by introduction into oocytes, yielding embryos that developed into mature frogs [35]. Only until 30 years later were similar successes reported in mammals, whereby activities present in oocytes successfully reprogrammed mammary epithelial cells to yield a blastocyst which upon embryo transfer and implantation, yielded a cloned live-born offspring (i.e., Dolly the Sheep) [36]. The successes in amphibians and mammals demonstrated that the genetic material of a terminally differentiated cell could be reprogrammed to a very early zygotic state to yield blastocysts, precisely the embryonic stage that is used for the derivation of embryonic stem cells. Indeed, ESCs have been derived from cloned mouse embryos [37]. In this work, reprogramming was conducted by nuclear transfer followed by differentiation of zygotes to blastocysts in vitro, after which ESC derivation from the ICM yielded somatic cell donor-specific pluripotent cell lines. However, this approach is laborious and clearly cannot be accomplished easily in a human system on any reasonable scale. The advent of iPS approaches circumvented this difficulty entirely.

Induced pluripotency begins with the harvest of a tissue biopsy that consists of proliferating somatic cells. The most successful and reliable of these are skin fibroblasts, which can be obtained in sufficient quantity for iPS reprogramming from a 3-mm skin punch biopsy. The skin punch is disaggregated in collagenase and used to establish a primary fibroblast culture. Other somatic cells have also been used, most notably nucleated peripheral blood cells [38–40], which in some cases have been successfully subjected to iPS reprogramming even without establishing a primary culture. Although blood collection is even less invasive than skin punch biopsies, the iPSC success rate with nucleated blood cells is less than for fibroblasts. In either case, the next step is to introduce key reprogramming factors such as OCT4, SOX2, KLF4, c-Myc, n-Myc, NANOG, and LIN28 (in various combinations) into somatic cells. Recently, it has also been found that the transient knock-down by RNA interference of p53 greatly increases the efficiency of iPS although there is some concern that abrogating p53 during iPS may increase levels of DNA damage [41,42]. The classic set of reprogramming factors is OCT4, SOX2, KLF4, and c-Myc, which was first described by Shinya Yamanaka [43]. Now, other combinations are used, most notably substituting n-Myc for c-Myc, which is less likely to yield oncogenic transformations. However, whichever combination is used, iPS methods almost universally employ the action of OCT4 and SOX2. Overall, the collection of reprogramming factors used for iPS is essentially the same set of major transcription factors that maintain the pluripotency transcriptional network in the blastocyst ICM and embryonic stem cells. Reprogramming factors were initially delivered to somatic cells with retroviral and lentiviral vectors, which would integrate into the genome as factor-expressing constructs. After several days, the forced expression of the reprogramming factors in the somatic cells causes the cells to assume the transcriptional and proteomic state of embryonic stem cells, thus conferring upon the cells the qualities of pluripotency and unlimited proliferative potential as long as they are maintained in the undifferentiated state. The resulting iPSCs can then be passaged and utilized as embryonic stem cells. In iPSCs produced by retroviral and lentiviral methods, the integrated transgenes are epigenetically silenced by DNA methylation, but only after the cells have up-regulated the endogenous *OCT4, SOX2, KLF4* genes, etc. Silencing of the ectopic copies of these genes is critical for the ability of these cells to differentiate since *OCT4, SOX2,* and other pluripotency-associated transcription factor genes must be silenced in order for differentiation to proceed. However, significant concerns exist for such virally reprogrammed cells since the integration of foreign DNA into the genome is inherently mutagenic and potentially oncogenic if the transgene integrations disrupt tumor suppressor genes for instance. For this reason, a host of methods have been devised to produce hiPSCs that leave the genome nearly or completely untouched by introduced foreign reprogramming DNA. A number of strategies have utilized transposons (notably piggyBac transposons) to deliver the reprogramming factors [44,45]. After iPS reprogramming is complete, the transposed DNA is removed via transposase-mediated precise excision. More recently, an excellent, simple, and highly efficient method has been developed using episomal vectors [43]. In this strategy, the reprogramming factors are delivered by electroporation to the cell on non-integrating episomes that persist in the cell only long enough to bring about reprogramming. Because the episomes replicate in the cell with only limited efficiency as compared to the rate of cell division of the host cells, the episomes become lost due to cell division and dilution of the residual amount of episomes present in the growing iPSC microcolony. Once hiPSC colonies are formed by this method, and then used to establish clonal cell lines, residual episomes cannot be detected even with as many as 60 cycles of PCR. Such cells are certainly useful for in vitro cell-culture models of cellular functions such as hepatocyte-mediated cholesterol function, but furthermore, these integration-free hiPSCs may meet the safety criteria needed for future stem cell therapies, where an unmutagenized genome is of paramount importance for the safety of these cells from a neoplasia perspective.

In the human subject, cholesterol profiles are the composite outcome of genetics, diet, activity, and other environmental factors. There is also a component of genotype—environment interaction because cholesterol itself (which is of both endogenous and dietary origin) can also act as a regulatory molecule within hepatocytes, where it affects levels of LDLR synthesis, and the activity of HMG-CoA reductase and CYP7A1. In addition, numerous genes have sterol response elements near their promoters, and the transcriptional regulation of these genes can also be affected by intra-hepatocellular levels of cholesterols and derivative sterols. And yet, a substantial contributor to cholesterol metabolism is dictated purely by genetics since the broad sense heritability for LDL levels is 0.36 as determined by twin studies, and the narrow sense heritability for LDL levels has been estimated to be 0.96 [46]. Thus, in the human, overall cholesterol balance, apolipoprotein profiles, and apolipoprotein receptor function are heavily influenced by genetics as well as a significant contribution from "environmental" factors such as diet and exercise.

A basic formula of population genetics states that within a population, the variance of a phenotype is equal to the variance due to genetics plus the variance due to environment ($V_P = V_G + V_E$). This simple equation is useful, but ignores genotype—environment interactions. However, it is a useful starting point for considering the study of HLCs derived from hiPSCs. Cholesterol profiles are readily and routinely measured from fasting blood samples in human subjects, and these show considerable variability from patient to patient. Yet even though these parameters are determined from fasting patients, the impact of diet and exercise is substantial. Such cholesterol profiles constitute the variance of the cholesterol phenotype as a whole in human populations. The cholesterol profile consist of data on levels of total cholesterol, HDL, LDL, and others. With iPS approaches, it is now possible to determine the purely genetic contributions to cholesterol homeostasis for individual persons or a group of individuals. Such determinations are essentially an example of personalized medicine. The way this works is as follows: An individual can be phenotyped for their serum cholesterol, and then enrolled in an iPS medical study in which a sample of somatic cells such as a skin biopsy could be obtained. This could then be used to derive hiPSCs, and these can be differentiated into HLCs. The baseline cholesterol and apolipoprotein profiles of such cells can then be ascertained. These must be entirely due to genetics for two reasons: (1) All patient-specific epigenetic information is erased during the iPS process and (2) the variance due to the environment across patient HLC samples is nearly zero since culture conditions are controlled and identical across cell lines. Thus, in the simple population genetics equation presented above V_E is zero, and $V_P = V_G$. Therefore, such a hypothetical study could accurately determine the magnitude of the variability of cholesterol phenotypes within a human population using an in vitro cell culture approach rather than twin studies. Perhaps an even more valuable use is that an HLC line from an hiPSC line is an exact genetic match to a specific patient. Based on this, if a patient presents with a risky cholesterol phenotype (for instance high LDL), it should be possible to study his or her HLCs and determine the relative contributions of genetics versus environment, and statin therapy or life-style changes could be recommended intelligently and accurately.

The nature of the iPS procedure is that a patient's unique genome is captured into a pluripotent cell context. This is because each cell of the patient's body (including somatic cells such as skin fibroblasts) contains a complete genome, which most significantly contains the sum total of all of the genetic polymorphisms present in that individual. This includes all SNPs, insertion–deletion mutations (indels), variable number tandem repeats (VNTRs), and other larger mutations. During the reprogramming process, essentially all epigenetic information of the donor fibroblast is erased, including patterns of DNA methylation and histone modifications. The terminally differentiated transcriptome of the fibroblast is completely reprogrammed to that of an early pluripotent cell reminiscent of that of the ICM of a blastocyst, and the epigenome of the iPSC is established, which is essentially identical to that of hESCs with bivalent marks (combined trimethylation lysines 4 and 27 of histone H3) which poise developmentally regulated genes for later programmed expression or silencing in specialized cell types as they are established by directed, differentiated strategies in vitro. Thus, the sum total of patient-specific sets of genetic variants (i.e., mutations) can be assessed in cognate cell types in vitro, in a process that can aptly be termed "genome capture."

A specific stem cell-related application of iPS technology is the construction of cell-culture models for specific genetic diseases, or so-called "disease in a dish" models (Figure 4). For instance, hiPSC lines have recently been derived from individuals suffering from FH [47]. In one study, an hiPSC line was derived from an individual suffering from homozygous FH (i.e., an individual with no effective LDLR expression). The genotype of the hiPSC line was confirmed to match that of the patient, and in vitro assays of LDLR expression in HLCs derived from the hiPSCs confirmed that LDLR expression was defective, and the resulting cells failed to properly bind and internalize fluorescently labeled LDL. Additional "disease in a dish" model of FH and those corresponding to other cholesterol-related disorders are undoubtedly in progress.

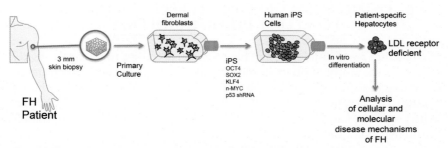

FIGURE 4 **Induced pluripotency to model human genetic disorders.** A "disease in a dish" approach to investigate familial hypercholesterolemia (FH) is shown. First, a skin biopsy is obtained from the FH patient and used to derive a primary culture of fibroblasts. These are then subjected to iPS-mediated reprogramming to yield pluripotent cells. These are then differentiated to HLCs, which contain the same genome as the fibroblast donor patient. These can then be studied by a host of cellular and molecular approaches to learn more about the disease mechanism in the cognate cell type.

Any two randomly chosen humans differ by as many as 3 million changes in their genomes. Most of these are inconsequential, but this high degree of human genetic variability complicates comparisons between normal and affected individuals in both the clinic and "disease in a dish" models. Simply, there is no such thing as a normal or wild-type human. In the recent past, the only point of comparison of appropriately differentiated mutant cells from hiPSCs (e.g., HLCs) is that of an hiPSC line derived from a clinically unaffected individual. Unfortunately, it can be very difficult to ascertain whether an in vitro phenotype is due to a suspected causative mutation or differences in the genetic background between the two cell lines. Now, with recent genome editing technologies such as CRISPR and TALEN endonucleases, it has become possible to produce control or affected cell lines from hiPSCs that differ by only a single genetic change that is causative of a cellular phenotype. Such isogenic pairs of cells promise to greatly increase the utility of "disease in a dish" models. There are two ways in which informative isogenic cell lines can be produced. The first is to derive hiPSCs from an affected individual, in which the suspected causative mutation is known. A genome editing strategy (for instance, CRISPRs) is then used to repair the mutation to a wild-type sequence. The result is two cell lines, one unedited (mutant) and one that is repaired, or rescued (wild type). Subsequently, the cell lines can be differentiated and then subjected to comparative studies, and any cellular or molecular differences can now be attributed to the mutation with confidence, rather than attributed to possible influences of genetic background. The second reciprocal approach is to create a mutation of interest de novo from a normal hiPSC or hESC line. This again yields a pair of isogenic cell lines that differ only by the mutation of interest.

6. USING STEM CELL-DERIVED HEPATOCYTES TO STUDY PHARMACOLOGY

Cells that recapitulate normal biochemical activities in vitro are clearly useful for the study of pharmacological modes of action and as cell-based screening tools for the evaluation of new drug candidates. Because HLCs synthesize, secrete cholesterol (likely in the form of apolipoprotein particles), and bind and endocytose LDL, these cells are useful research tools for pharmacological investigations. Furthermore, HLCs respond to statin treatment as evidenced by the reduced secretion into cell-culture medium of solubilized cholesterol (likely solubilized by apolipoproteins). Convenient fluorimetric assays are available, which can detect total cellular cholesterol secretion or fluorescent LDL uptake. Such assays have already been shown to be of utility for the detection of statin action. These assays should also be of use for the detection or other cholesterol-modulating drug activities. Notably, such fluorimetric assays in HLCs should serve as the basis for the identification of new drugs with cholesterol modulatory activities since HLCs can be differentiated in situ in multi-well plate formats prior to conducting of cell-based screens. In addition to screens, HLCs can be used to study the modes of action of existing drugs at molecular and cellular levels. The approach is simple in essence, whereby dose—response studies can first be conducted to find minimal effective concentrations of specific candidate drugs followed by molecular and genetic assays, such as changes in the transcriptome (using microarrays of RNAseq) and analyses of responses and mechanisms at the level of protein abundance and interactions.

7. FUTURE DIRECTIONS

In the future, perhaps not too distant, it may be possible to use HLCs derived from hiPSCs for liver repair by stem cell therapy. This would be tremendously useful from a medical standpoint because many patients in liver transplant waiting lists fail to receive a liver transplant due to a shortage of available transplantable liver lobes, and liver transplantation has relatively high morbidity and a life-long requirement for immuno-suppressant drugs. The first step of such an approach would be to obtain somatic cell biopsies from a patient and perform iPS on these cells using a non-integrating factor delivery approach such as episomal iPSCs. From the resulting patient-matched hiPSCs, HLCs could be derived in sufficient quantities to yield transplantable cells. At present HLCs derived in vitro have excellent cholesterol-related activities, but poor detoxifica-tion (CYP) activities. However, studies from mice have shown that transplantation of human HLCs (derived by partial reprogramming of human fibroblasts into endoderm followed by differentiation) into the liver microenvironment has resulted in further maturation of HLCs in vivo to yield cells with robust CYP activity [48]. The same may occur in human HLC transplants. These approaches may be useful to replace hepatocytes lost to Hepatitis C, for instance, when combined with antiviral therapy.

In the case of life-long cholesterol disorders such as FH, it may be possible to cure this disease by a two-step approach. Firstly, patient-specific hiPSCs would be derived. Secondly, since most *LDLR* genetic defects in humans are single point mutations, the point mutations could be repaired in vitro by gene editing mediated by TALENS or CRISPRs. The repaired hiPSCs could then be differentiated to HLCs and trans-planted to the FH patient in sufficient quantities to cure the disease. This is essentially gene therapy ex vivo and should be free of immunological events that have previously plagued gene therapy trials, caused by immunological reactions to viral delivery. Furthermore, since hiPSCs contain the exact genome of the patient, the transplant should be autologous and free from rejection.

It is becoming increasingly apparent that human genetic variability both within and between populations has a major impact on drug efficacy and toxicity. People with rare alleles that affect drug metabolism often respond poorly or not at all to standard drug dosages or experience unexpected toxicological responses. Collectively, these genetically encoded predispositions to poor efficacy or adverse drug events are fairly common, and mortality due to drug side effects is a leading cause of death in the United States. In keeping with this, numerous recent studies from the field of pharmacogenomics have for the most part cataloged the major human genetic polymorphisms that lead to such undesired outcomes. For the most part, these polymorphisms lie in a relatively small collection of genes that function in drug metabolism and transport, most notably CYP genes, phase II drug metabolism genes, and transporters. Collectively, these represent fewer than 100 genes, and it would be possible to create a bank of hiPSCs that contain major polymorphisms for each of these genes. During the phase of safety assessment, drugs could be checked for genotype-specific safety problems in hepatocytes, cardiomyocytes, and neurons, all of which can be readily produced from pluripotent hiPSCs.

References

[1] Goldstein JL, Brown MS. Regulation of the mevalonate pathway. Nature 1990;343: 425–30.
[2] Koh KK, Sakuma I, Quon MJ. Differential metabolic effects of distinct statins. Atherosclerosis 2011;215:1–8.
[3] Reiner Z. Statins in the primary prevention of cardiovascular disease. Nat Rev Cardiol 2013;10:453–64.
[4] Brown MS, Goldstein JL. Lipoprotein receptors in the liver. Control signals for plasma cholesterol traffic. J Clin Invest 1983;72:743–7.
[5] Yamamoto T, Davis CG, Brown MS, Schneider WJ, Casey ML, Goldstein JL, et al. The human LDL receptor: a cysteine-rich protein with multiple Alu sequences in its mRNA. Cell 1984;39:27–38.
[6] Gent J, Braakman I. Low-density lipoprotein receptor structure and folding. Cell Mol Life Sci CMLS 2004;61:2461–70.
[7] Toyota Y, Yamamura T, Miyake Y, Yamamoto A. Low density lipoprotein (LDL) binding affinity for the LDL receptor in hyperlipoproteinemia. Atherosclerosis 1999;147: 77–86.
[8] Anderson RG, Brown MS, Goldstein JL. Role of the coated endocytic vesicle in the uptake of receptor-bound low density lipoprotein in human fibroblasts. Cell 1977;10: 351–64.
[9] Anderson RG, Brown MS, Beisiegel U, Goldstein JL. Surface distribution and recycling of the low density lipoprotein receptor as visualized with antireceptor antibodies. J Cell Biol 1982;93:523–31.

[10] Osborne TF, Goldstein JL, Brown MS. 5' end of HMG CoA reductase gene contains sequences responsible for cholesterol-mediated inhibition of transcription. Cell 1985; 42:203−12.

[11] Stampfer MJ, Krauss RM, Ma J, Blanche PJ, Holl LG, Sacks FM, et al. A prospective study of triglyceride level, low-density lipoprotein particle diameter, and risk of myocardial infarction. JAMA 1996;276:882−8.

[12] Seidah NG, Benjannet S, Wickham L, Marcinkiewicz J, Jasmin SB, Stifani S, et al. The secretory proprotein convertase neural apoptosis-regulated convertase 1 (NARC-1): liver regeneration and neuronal differentiation. Proc Natl Acad Sci USA 2003;100: 928−33.

[13] Teslovich TM, Musunuru K, Smith AV, Edmondson AC, Stylianou IM, Koseki M, et al. Biological, clinical and population relevance of 95 loci for blood lipids. Nature 2010;466: 707−13.

[14] Trompet S, de Craen AJ, Postmus I, Ford I, Sattar N, Caslake M, et al. Replication of LDL GWAs hits in PROSPER/PHASE as validation for future (pharmaco)genetic analyses. BMC Med Genet 2011;12:131.

[15] Rasmussen-Torvik LJ, Pacheco JA, Wilke RA, Thompson WK, Ritchie MD, Kho AN, et al. High density GWAS for LDL cholesterol in African Americans using electronic medical records reveals a strong protective variant in APOE. Clin Transl Sci 2012;5: 394−9.

[16] Evans MJ, Kaufman MH. Establishment in culture of pluripotential cells from mouse embryos. Nature 1981;292:154−6.

[17] Martin GR. Isolation of a pluripotent cell line from early mouse embryos cultured in medium conditioned by teratocarcinoma stem cells. Proc Natl Acad Sci USA 1981;78: 7634−8.

[18] Amit M, Carpenter MK, Inokuma MS, Chiu CP, Harris CP, Waknitz MA, et al. Clonally derived human embryonic stem cell lines maintain pluripotency and proliferative potential for prolonged periods of culture. Dev Biol 2000;227:271−8.

[19] Thomson JA, Itskovitz-Eldor J, Shapiro SS, Waknitz MA, Swiergiel JJ, Marshall VS, et al. Embryonic stem cell lines derived from human blastocysts. Science 1998;282: 1145−7.

[20] Takahashi K, Yamanaka S. Induction of pluripotent stem cells from mouse embryonic and adult fibroblast cultures by defined factors. Cell 2006;126:663−76.

[21] Park IH, Zhao R, West JA, Yabuuchi A, Huo H, Ince TA, et al. Reprogramming of human somatic cells to pluripotency with defined factors. Nature 2008;451:141−6.

[22] Takahashi K, Tanabe K, Ohnuki M, Narita M, Ichisaka T, Tomoda K, et al. Induction of pluripotent stem cells from adult human fibroblasts by defined factors. Cell 2007;131: 861−72.

[23] Yu J, Vodyanik MA, Smuga-Otto K, Antosiewicz-Bourget J, Frane JL, Tian S, et al. Induced pluripotent stem cell lines derived from human somatic cells. Science 2007; 318:1917−20.

[24] Chambers I, Tomlinson SR. The transcriptional foundation of pluripotency. Development 2009;136:2311−22.

[25] Kim J, Chu J, Shen X, Wang J, Orkin SH. An extended transcriptional network for pluripotency of embryonic stem cells. Cell 2008;132:1049−61.

[26] Pan G, Thomson JA. Nanog and transcriptional networks in embryonic stem cell pluripotency. Cell Res 2007;17:42−9.

[27] Qiu C, Ma Y, Wang J, Peng S, Huang Y. Lin28-mediated post-transcriptional regulation of Oct4 expression in human embryonic stem cells. Nucleic Acids Res 2010;38:1240−8.

[28] Zhong X, Li N, Liang S, Huang Q, Coukos G, Zhang L. Identification of microRNAs regulating reprogramming factor LIN28 in embryonic stem cells and cancer cells. J Biol Chem 2010;285:41961−71.

[29] Si-Tayeb K, Lemaigre FP, Duncan SA. Organogenesis and development of the liver. Dev Cell 2010;18:175–89.

[30] Si-Tayeb K, Noto FK, Nagaoka M, Li J, Battle MA, Duris C, et al. Highly efficient generation of human hepatocyte-like cells from induced pluripotent stem cells. Hepatology 2010;51:297–305.

[31] Serls AE, Doherty S, Parvatiyar P, Wells JM, Deutsch GH. Different thresholds of fibroblast growth factors pattern the ventral foregut into liver and lung. Development 2005; 132:35–47.

[32] Rossi JM, Dunn NR, Hogan BL, Zaret KS. Distinct mesodermal signals, including BMPs from the septum transversum mesenchyme, are required in combination for hepatogenesis from the endoderm. Genes Dev 2001;15:1998–2009.

[33] Krueger WH, Tanasijevic B, Barber V, Flamier A, Gu X, Manautou J, et al. Cholesterol-secreting and statin-responsive hepatocytes from human ES and iPS cells to model hepatic involvement in cardiovascular health. PLoS One 2013;8:e67296.

[34] D'Amour KA, Agulnick AD, Eliazer S, Kelly OG, Kroon E, Baetge EE. Efficient differentiation of human embryonic stem cells to definitive endoderm. Nat Biotechnol 2005; 23:1534–41.

[35] Gurdon JB. The developmental capacity of nuclei taken from intestinal epithelium cells of feeding tadpoles. J Embryol Exp Morphol 1962;10:622–40.

[36] Wilmut I, Schnieke AE, McWhir J, Kind AJ, Campbell KH. Viable offspring derived from fetal and adult mammalian cells. Nature 1997;385:810–3.

[37] Hwang WS, Ryu YJ, Park JH, Park ES, Lee EG, Koo JM, et al. Evidence of a pluripotent human embryonic stem cell line derived from a cloned blastocyst. Science 2004;303: 1669–74.

[38] Giorgetti A, Montserrat N, Aasen T, Gonzalez F, Rodriguez-Piza I, Vassena R, et al. Generation of induced pluripotent stem cells from human cord blood using OCT4 and SOX2. Cell Stem Cell 2009;5:353–7.

[39] Haase A, Olmer R, Schwanke K, Wunderlich S, Merkert S, Hess C, et al. Generation of induced pluripotent stem cells from human cord blood. Cell Stem Cell 2009;5:434–41.

[40] Loh YH, Agarwal S, Park IH, Urbach A, Huo H, Heffner GC, et al. Generation of induced pluripotent stem cells from human blood. Blood 2009;113:5476–9.

[41] Marion RM, Strati K, Li H, Murga M, Blanco R, Ortega S, et al. A p53-mediated DNA damage response limits reprogramming to ensure iPS cell genomic integrity. Nature 2009;460:1149–53.

[42] Takenaka C, Nishishita N, Takada N, Jakt LM, Kawamata S. Effective generation of iPS cells from CD34+ cord blood cells by inhibition of p53. Exp Hematol 2010;38:154–62.

[43] Okita K, Matsumura Y, Sato Y, Okada A, Morizane A, Okamoto S, et al. A more efficient method to generate integration-free human iPS cells. Nat Methods 2011;8:409–12.

[44] Kaji K, Norrby K, Paca A, Mileikovsky M, Mohseni P, Woltjen K. Virus-free induction of pluripotency and subsequent excision of reprogramming factors. Nature 2009;458: 771–5.

[45] Woltjen K, Michael IP, Mohseni P, Desai R, Mileikovsky M, Hamalainen R, et al. piggy-Bac transposition reprograms fibroblasts to induced pluripotent stem cells. Nature 2009;458:766–70.

[46] Abney M, McPeek MS, Ober C. Broad and narrow heritabilities of quantitative traits in a founder population. Am J Hum Genet 2001;68:1302–7.

[47] Rashid ST, Corbineau S, Hannan N, Marciniak SJ, Miranda E, Alexander G, et al. Modeling inherited metabolic disorders of the liver using human induced pluripotent stem cells. J Clin Invest 2010;120:3127–36.

[48] Zhu S, Rezvani M, Harbell J, Mattis AN, Wolfe AR, Benet LZ, et al. Mouse liver repopulation with hepatocytes generated from human fibroblasts. Nature 2014;508:93–7.

6

Discovery of High-Density Lipoprotein Gene Targets from Classical Genetics to Genome-Wide Association Studies

Lita A. Freeman, Alan T. Remaley

Lipoprotein Metabolism Section, Cardiopulmonary Branch, National Heart, Lung and Blood Institute, Bethesda, MD, USA

1. INTRODUCTION

High-density lipoproteins (HDL) are, as their name implies, complexes of lipids and proteins and have generated great interest, because of their inverse association with cardiovascular disease (CVD) [1]. Currently, there are only a limited number of approved drugs that raise HDL, namely niacin and fibrates. It is not clear, however, whether our current drugs that raise HDL significantly reduce cardiovascular events [1]. Novel drugs, such as cholesteryl ester transfer protein (CETP) inhibitors, are now also being actively investigated in late-stage clinical trials, but the results so far have been disappointing for CETP inhibitors and other novel HDL-modifying drugs. It is now clear that we need a better understanding of HDL metabolism and genetics to better predict the clinical outcome of any pharmaceutical intervention.

In this chapter, we first briefly review HDL composition and function. We then focus on potential gene targets for drug development based on what is known about HDL metabolism from classic Mendelian disorders. Finally, we review recent efforts related to genome-wide association studies (GWASs) that have uncovered a plethora of new genes that are involved in HDL metabolism.

Translational Cardiometabolic Genomic Medicine
http://dx.doi.org/10.1016/B978-0-12-799961-6.00006-8 119

1.1 HDL Composition

HDLs are defined as lipoprotein particles with a density range between 1.21 and 1.063 g/mL [2]. The majority of HDL particles in plasma have a spherical form with a diameter ranging between 8 and 11 nm. A relatively small fraction of HDL, typically less than 5%, can also exist as a disk in either pre-β HDL or α4 HDL subfractions as determined by 2D-gel electrophoresis, which can separate HDL into about 14 different size and charge subfractions [3].

About half of the mass of HDL consists of proteins, which is why HDL is denser than the other major lipoproteins. The main protein component of HDL is apolipoprotein A-I (apoA-I), which comprises about 70% of the protein on HDL [2]. It is 28 kDa in molecular weight and contains a tandem array of amphipathic helices, which stabilizes the structure of HDL. HDL also contains a small number of proteins closely related to apoA-I, such as apoA-II, apoA-IV, apoE, and apoCs. All of these proteins also contain amphipathic helices and like apoA-I can readily exchange between HDL and other lipoproteins and/or can exist in a relatively lipid-free state. These other apolipoproteins on HDL also contribute to the structure of HDL and, in the case of apoE, serve as ligands for the uptake by cellular receptors, and can also modulate the activity of lipid-modifying enzymes, as in case of the apoCs. Recent shotgun proteomic studies have revealed that there are about 80 other proteins that are also associated with HDL [1,4]. Most of these proteins are abundant plasma proteins that have a relatively loose association with HDL, and their functional significance is not well understood; however, they have a wide variety of biological activities related to inflammation, coagulation, and complement activation, and may contribute to the anti-atherogenic properties of HDL [4].

Although the cholesterol content of HDL-C is the historic metric for quantifying HDL, it only represents about 20–25% of its mass. The majority of the cholesterol on HDL, about 65–75%, is esterified by the plasma enzyme lecithin:cholesterol acyltransferase (LCAT) and is found in the core of spherical HDL (Figure 1) as cholesteryl esters. The bulk of the remaining free cholesterol is found on the surface of HDL along with phospholipids and sphingomyelin, which together make up about 25–35% of the mass of HDL. The predominant phospholipid is phosphatidylcholine, but all of the major phospholipid classes can be found on HDL. A relatively small amount of triglyceride, particularly on larger HDL species, can be found in the hydrophobic core of HDL. Due to the structural diversity of fatty acids, which are present in most of the major lipid classes found on HDL, recent lipidomic studies have revealed over 200 different species of lipids on HDL [5–7]. Some of the lipids found on HDL are relatively low in abundance, such as sphingosine-1-phosphate, but are potent signaling molecules and likely account for some of the biological effects of HDL [1,7].

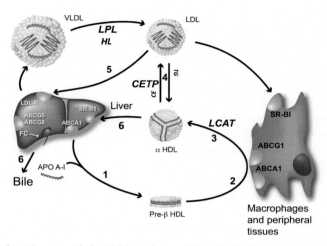

FIGURE 1   **Diagram of the RCT pathway.** (Step 1) The first step begins with the formation of nascent pre-β-HDL, which largely occurs in the liver and to a lesser degree in the intestine, when apoA-I acquires phospholipid and a small amount of free cholesterol (FC) by the ABCA1 transporter. (Step 2) Pre-β-HDL acquires additional lipid, particularly free cholesterol, by ABCA1 transporters in the periphery, including macrophages. (Step 3) Discoidal HDL containing surface PL and FC is transformed into spherical α-migrating HDL after acquiring additional lipid by other transporters and proteins on cell membranes, such as ABCG1 or SR-BI, or by a passive diffusion process. LCAT is involved in this process by converting free cholesterol to cholesteryl esters, which migrate into the core of HDL. (Step 4) Cholesteryl ester (CE) is transferred from HDL to LDL (generated by lipolysis of VLDL) in exchange for triglycerides (TG) by CETP. CETP also transfers CE to VLDL (not shown). (Step 5) The LDL particle is eventually delivered to the liver by the LDL receptor. (Step 6) Alternatively, cholesterol ester in HDL can be delivered directly to the liver after uptake by SR-BI, which then regenerates pre-β-HDL. Cholesterol is then excreted by the liver either as free cholesterol or converted to a bile salt.

1.2 HDL Function

Given its protein and lipid compositional complexity, it is perhaps not surprising that numerous biological functions have been attributed to HDL. Almost every single step in the pathogenesis of atherosclerosis has been described to be affected by HDL mostly in a salutary way (see Toth et al. [1] for a recent review). For example, HDL has been described to have potent anti-inflammatory effects and can suppress the activation of monocytes and the expression of adhesion proteins on endothelial cells. HDL also has antithrombotic effects by either acting on the fibrinolytic pathway or suppressing platelet activation. HDL also transports a wide variety of constituents that can either suppress the oxidation of lipids or promote their catabolism, such as carotene, vitamin E, and paraoxonase. In this way, HDL may prevent the initial formation of macrophage foam cells in the atherosclerotic plaque from the uptake of oxidized lipids.

The best understood anti-atherogenic mechanism for HDL is its participation in the reverse cholesterol transport (RCT) pathway, which is the pathway by which excess cholesterol from the periphery is returned to the liver for excretion (Figure 1). This pathway begins with the biosynthesis of discoidal HDL, which largely occurs in the liver and to a small degree in the intestine. ABCA1 (ATP binding cassette A1 protein), the defective gene in Tangier disease [8–13], is critical in this process by modifying the plasma membrane of cells so that apoA-I can extract phospholipid and cholesterol and form pre-β HDL in a detergent-like extraction mechanism. After the transfer of additional lipid by ABCA1 in peripheral cells, pre-β HDL is converted into larger discoidal particles (α4-HDL). Next, HDL can efflux additional lipid by other cell membrane transporters in the periphery by possibly ABCG1, scavenger receptor class B type I (SR-BI) and or by an aqueous diffusion process [14].

The discoidal species of HDL are then transformed into spherical-shaped HDL once cholesteryl esters formed by LCAT form a central lipid core. This maturation step in HDL is critical because in the absence of LCAT, the smaller discoidal species of HDL are rapidly catabolized, leading to low HDL levels. The esterification of cholesterol has also been proposed to trap cholesterol on HDL until it can be returned to the liver. Once cholesterol is on HDL, it has two possible fates. The first is that it can be directly returned to the liver once HDL interacts with hepatic SR-BI receptors, which promote the selective uptake of cholesteryl esters from the core of HDL, resulting in the regeneration of smaller species of HDL. There is also likely holo-particle uptake of HDL by the liver, but this process has not been well delineated. The other possible fate is that cholesteryl esters on HDL can be exchanged for triglycerides on apoB-containing particles by the CETP and delivered to the liver on very low-density lipoproteins (VLDL) and LDL by hepatic LDL receptors. In humans, about half of cholesterol is believed to be returned by this pathway [15].

Cholesterol delivered to the liver can be stored in the liver, resecreted on lipoproteins, or excreted directly into the bile as either free or unesterified cholesterol or as bile salt after its conversion in the liver. It is likely that most cells rely to some degree on the RCT pathway for cholesterol homeostasis, but macrophages in particular appear to heavily depend on this pathway based on the increased formation of cholesterol-enriched foam cells in Tangier disease patients with a defective ABCA1 transporter. Because macrophages play a key role in the pathogenesis of atherosclerosis, the mobilization of excess cholesterol from macrophages by the RCT pathway is believed to be one of the anti-atherogenic functions of HDL.

2. HDL-MODIFYING GENES IDENTIFIED FROM CLASSIC GENETIC DISORDERS

The first genes identified that modulate HDL were discovered from studies of classic genetic disorders affecting HDL-C levels. Genetic disorders of HDL can be broadly classified as either being hypoalphalipoproteinemia (low HDL-C) [16] or hyperalphalipoproteinemia (high HDL-C). The main differential diagnosis of hypoalphalipoproteinemia, which will first be discussed below, is the following: Tangier disease (ABCA1 mutations), familial LCAT deficiency, and familial ApoA-I deficiency.

2.1 Familial Hypoalphalipoproteinemia

2.1.1 ABCA1 Transporter

The first description of a genetic disorder causing low HDL-C was Tangier disease, an autosomal-recessive disorder first described by Fredrickson and colleagues in 1961 [17]. In addition to very low or un-detectable HDL-C, Tangier disease is characterized by yellow-orange tonsils from carotenoid accumulation, corneal opacities from cholesterol deposits, hepatomegaly, splenomegaly, peripheral neuropathy, and sometimes premature myocardial infarction [18,19]. In an early study in 1964, it was proposed that "a single pair of autosomal alleles exerted major control over plasma HDL concentrations and that absence of the lipoprotein (Tangier disease) results from a double dose of an abnormal mutant at this locus [20]." It was not, however, until 1999 that several groups identified the Tangier disease gene as ABCA1 (originally termed ABC1) [8−13]. The ABCA1 gene had previously been characterized as a protein required for the engulfment of cellular corpses generated by apoptotic cell death [21], but its link to HDL metabolism was not known. The function of ABCA1 was later confirmed in mouse knockouts by Orso [22] and Francone [23] and in mouse overexpression models by several laboratories [24−26]; see also Ref. [27]. The biochemical defect in Tangier disease, the inability to efflux cholesterol from cells to ApoA-I [28,29], was then shown to be corrected after transfection of cells with ABCA1 [30,31].

Later, GWAS and whole-exome studies showed the association of common single-nucleotide polymorphisms (SNPs) in ABCA1 with HDL-C levels [32−34]. Unexpectedly, however, SNPs in ABCA1 that caused low HDL levels were not associated with increased risk for CVD [35]. This may be due to the fact that, at least in Tangier disease, defects in ABCA1 also lead to lower levels of LDL-C, which could have compen-sated for the increased risk from the lower HDL levels. Also, the contribution of ABCA1-dependent cholesterol efflux from macrophages in forming HDL-C is relatively low [36], but macrophage ABCA1 nonetheless protects against atherosclerosis, independent of its effects on plasma lipid levels [37] and thus plasma levels of HDL-C may be a poor indicator of this process. As will be discussed in more detail in the below GWAS section, there is growing realization that the historic metric for quantifying HDL, HDL-C, may be a relatively poor indicator of HDL function. In fact, with regard to cholesterol efflux, a functional assay for HDL has recently been shown to better correlate with CVD events than HDL-C [38].

It is important to note that research on Tangier disease and the ABCA1 transporter has led to a novel therapeutic strategy for raising HDL-C, namely the drug development efforts related to LXR agonists [39]. LXR, which senses intracellular oxysterol levels, is one of the main transcriptional regulators of ABCA1. The up-regulation of ABCA1 in cholesterol-loaded cells helps facilitate the removal of excess cholesterol and raises HDL-C. LXR, however, modulates the expression of many other genes and LXR agonists have been shown to cause fatty liver, which has diminished interest in LXR as a drug target; however, innovative drug discovery efforts designed to circumvent this problem continue [39].

2.1.2 Lecithin: Cholesterol Acyltransferase

Although cholesterol esterification in plasma was observed many years earlier [40], John Glomset, in 1962, was the first to characterize LCAT, the main plasma enzyme that esterifies cholesterol [41]. LCAT has two main enzymatic activities. It has phospholipase-A2-like activity and cleaves fatty acids in the sn-2 position of phospholipids, particularly on phosphatidylcholine. It has also transesterification activity and can transfer the cleaved fatty acid to cholesterol by forming an ester bond with the hydroxyl group in the A-ring of cholesterol. Cholesteryl esters are much more hydrophobic than cholesterol, which accounts for their partitioning of cholesterol from the surface of a lipoprotein particle to the hydrophobic core of lipoproteins once cholesterol becomes esterified. Because ApoA-I is the main protein activator of LCAT, most cholesterol esterification occurs on HDL, although some can also form on LDL [42].

In 1967, Norum and Gjone reported on the first case of a genetic defect in LCAT leading to markedly low levels of HDL-C and cholesteryl esters [43]. Although it is a relatively rare disease, over 100 different mutations in LCAT have been described that cause disease when both alleles are affected [44,45]. Besides low HDL-C, patients with familial LCAT deficiency can present with cloudy corneas, due to accumulation of excess cholesterol, and anemia, due to the abnormal exchange of cholesterol between RBCs and lipoproteins that occurs in this disorder. The most clinically significant problem that patients with LCAT deficiency develop is renal disease, most likely due to the renal deposition of LpX (an abnormal multilamellar lipoprotein particle that forms in the absence of plasma cholesteryl esters). A subset of patients with mutations in LCAT has some preserved function on LDL particles. These patients are classified as having fish eye disease, and except for the cholesterol deposition in the eye they do not have the rest of the clinical manifestations of familial LCAT deficiency, which causes decreased cholesterol esterification on both HDL and LDL.

Drug development efforts related to LCAT have been limited due to the unclear relationship between LCAT deficiency and atherosclerosis (reviewed in Ahsan et al. [45]). Like Tangier disease, patients with LCAT deficiency also have low LDL-C levels, which likely protects them against atherosclerosis despite their low HDL-C levels. A recent 3D-MRI study of carotids from patients with LCAT deficiency has shown, however, that they may have a modest increase in atherosclerosis [46] and arterial stiffness [47]. Genetic mouse models have also been confusing with regard to the role of LCAT in atherosclerosis [45], but in rabbits, which like humans express CETP, a clear atheroprotective benefit of increased LCAT expression was observed [48]. Unexpectedly, in a Mendelian randomization study, LCAT SNPs that decreased HDL-C levels were not associated with an increased incidence of CAD [49]. This finding is also consistent with the main conclusions from recent GWASs that have shown that unlike LDL-C and triglycerides, genetic determinants of HDL-C do not overall appear to be associated with CVD events.

The major effort related to LCAT drug development has been the production of recombinant human LCAT [50]. Similar to the rationale for the use of reconstituted HDL therapy [51], recombinant LCAT is being investigated as an acute treatment for acute coronary syndrome. In a phase I study, it has been shown to be safe and significantly raised HDL-C for several days ([52]; Clinical Trials.gov identifier NCT01554800). Based on its ability to raise HDL-C and otherwise restore lipoprotein metabolism to normal in LCAT-knockout (LCAT-KO) mice [50], recombinant LCAT may also have utility for the treatment of familial LCAT deficiency. There is also one report of a small-molecule activator of LCAT, which in hamsters was found to increase HDL-C and increase other measures of the RCT pathway, such as increasing fecal sterol excretion [53].

2.1.3 Apolipoprotein A-I

Given the importance of ApoA-I in the structure of HDL, it is not unexpected that mutations in ApoA-I can lead to low HDL levels [54]. Patients with heterozygous mutations in ApoA-I have a modest decrease in HDL-C, but homozygous or compound heterozygous mutations can lead to nearly undetectable levels of HDL-C. The clinical presentation of patients with familial ApoA-I deficiency can vary quite significantly, depending on the type of mutation and the degree of HDL-C lowering. Those patients with the most severe mutations can develop xanthelasmas and eruptive-like xanthomas, due to cholesterol accumulation, and are at an increased risk of atherosclerosis [54]. Recently, an apoA-I mutation called ApoA-I Mytilene has been described [55], which decreases the apoA-I production rate by 41% and even in the heterozygote state has been associated with premature CAD [55]. Some types of apoA-I mutations can result in protein aggregation and the formation of amyloidosis [56]. Another unusual mutation called ApoA-I Milano also lowers HDL-C, yet appears to be atheroprotective. This apoA-I variant has a single Arg to Cys substitution, which leads to its more rapid catabolism but may also increase its beneficial antioxidant properties [57].

Despite much effort, useful small-molecule drugs that selectively raise ApoA-I have not been found. The only possible exception is RVX-208, which has been shown in a phase II clinical trial to increase ApoA-I by about 5–10% [58–60]. Whether this increase in HDL-C will be sufficient to reduce cardiovascular events is uncertain based on a lack of significant reduction in atherosclerosis in the Phase II IVUS study (http:// www.escardio.org/about/press/press-releases/esc13-amsterdam/Pages/ hotline-three-assure.aspx). RVX-208 was found to be a BET bromodomain inhibitor, and it affects the transcription of ApoA-I by modulating the methylation of histones [61]. It may, however, have other gene targets, which could possibly explain its ability to lower markers of inflammation, such as C-reactive protein.

Because of the limited success so far in finding small molecules that selectively raise HDL-C, therapeutic strategies based on the intravenous infusion of either purified ApoA-I or recombinant ApoA-I combined with phospholipids have been developed [51]. It is envisioned that an acute treatment with high doses of phospholipid-rich HDL would simulate the formation of pre-β HDL, which would then rapidly promote cholesterol efflux from cells in atherosclerotic plaque, as well as mediate some of the anti-inflammatory properties of HDL. Such a treatment could rapidly stabilize patients with acute coronary syndrome based on pre-clinical animal models. A single infusion of reconstituted HDL in patients with peripheral vascular disease was shown to markedly reduce the lipid content of femoral plaque and decrease inflammatory cell infiltration [62]. Several different preparations of either wild-type ApoA-I or ApoA-I Milano combined with phospholipid have been tested in patients with acute coronary syndrome. A weekly treatment for 4–5 weeks has been shown by IVUS to reduce atherosclerotic plaque in coronary vessels to a greater degree than what has been observed after statin treatment for several years. A phase III trial is now being planned for CSL-112, which is made with purified ApoA-I combined with phosphatidylcholine. Short synthetic peptide mimetics of ApoA-I are also being investigated as alternatives to the full-length ApoA-I protein based on their ability to also promote cholesterol efflux and mediate some of the other beneficial properties of HDL [51].

2.2 Familial Hyperalphalipoproteinemia

The main genetic disorders described to cause hyperalphalipoproteinemia are CETP deficiency and mutations in SR-BI. Other less common genetic causes, which will not be further discussed, are hepatic lipase deficiency [63] and polymorphisms in endothelial lipase [64].

2.2.1 Cholesterol Ester Transfer Protein

Familial homozygous hyperalphalipoproteinemia due to CETP deficiency was first described in a Japanese family in 1984 [65]. The proband had an HDL-C of 181 mg/dL and his sister had an HDL-C of 163 mg/dL [65]. In a second report from Japan, the lack of the normal cholesteryl ester exchange between HDL and VLDL/LDL was found in a proband, who had an HDL-C of 301 mg/dL [66]. In 1989, the affected individuals from families with highly elevated HDL-C were found to lack CETP protein and to have a G-A mutation in a conserved 5′ splice donor of intron 14 of the CETP gene [67,68]. Other CETP mutations found later in other subjects with high HDL-C include the dominant mutation D442G [69], a T insertion at position +3 from the exon 13/intron 14 boundary [70], and a mutation, CETP I405V, found in the Ashkenazi Jewish population [71]. The CETP gene promoter also contains several promoter polymorphisms that are associated with HDL-C levels [72,73]. The relatively common "TaqI polymorphism" (restriction polymorphism Taq11B in intron 1 of CETP gene) [74] has been intensively studied, and its B1 variant has been shown to result in higher plasma CETP concentrations and lower HDL-C levels. Many of the patients with CETP mutations were also described to have relatively low levels of LDL-C, most likely because a significant fraction of cholesterol on LDL is transferred there from HDL as a consequence of CETP activity (Figure 1).

Since increased HDL-C and decreased LDL-C are associated with less CVD, it was hypothesized that inhibition of CETP should provide protection against atherosclerosis. Interestingly, early studies appeared to show an association of decreased levels of CETP and longevity, but this did not always hold up in later studies [75–88]. This raised the concern that although a decrease in CETP activity leads to an increase in HDL-C, this increase may not be beneficial because HDL may become dysfunctional. It may also be that blocking the transfer of cholesteryl esters from HDL to LDL decreases the hepatic uptake of cholesterol as shown in Figure 1, thus decreasing the overall flux of cholesterol by the RCT pathway. The largest study, to date, on the effect of CETP polymorphisms and longevity, however, found a statistically significant association between those polymorphisms linked to high HDL-C and longevity in >10,000 patients from the Copenhagen City Heart Study [89]. It was also associated with decreased cardiovascular events.

Several small-molecule inhibitors of CETP have now been developed and tested in clinical trials [1,90]. The first was torcetrapib, which raised HDL-C as much as 72% and also lowered LDL-C by about 25%. It was found, however, in a phase III clinical trial, to increase all-cause mortality, as well as cardiovascular events, most likely due to off-target effects from hypertension. Dalcetrapib, another CETP-inhibitor, which was more selective in raising HDL-C with a more limited effect in lowering LDL-C, was discontinued when it showed no effect in reducing cardiovascular events [1,91,92]. Several other potent CETP-inhibitors are currently being investigated in late-stage clinical trials. If one of these other agents proves to be effective, it will not be clear, however, whether this is due to their ability to raise HDL-C or lower LDL-C or both.

2.2.2 Scavenger Receptor-BI

The scavenger receptor type B class I (SR-BI) was first identified in 1996 to be a receptor of HDL in mice [93]. The human homolog of SR-BI is also known as CLA1 [94]. SR-BI mediates the selective uptake of HDL cholesteryl esters from HDL [93]. This process is believed to be an important step in the delivery of cholesterol to liver in the RCT pathway (Figure 1) and is also critical for steroid biogenesis in endocrine tissue like the adrenal gland and ovary. Besides HDL, SR-BI can also bind and mediate cholesterol uptake from LDL [93–96], as well as Lp(a) [97]. Its biological relevance in cholesterol homeostasis is not completely clear, but SR-BI also promotes the bidirectional exchange of cholesterol between cells and HDL, and hence may also promote cellular cholesterol efflux [14]. The first genetic knockout of SR-BI in mice [98] was found to have markedly elevated HDL-C but had an increased propensity for atherosclerosis [99], possibly because of decreased hepatic delivery of cholesterol. This was one of the first examples on how not all pathways or interventions that raise HDL-C may be beneficial in reducing atherosclerosis and/or may result in the formation of dysfunctional HDL.

Variations in SCARB1, the gene for CLA1/SR-B1 in humans, have been associated with high HDL-C [100–102], as well as high LDL-C in some cases [100]. In subjects with heterozygous familial hypercholesterolemia, SCARB1 variants were associated with higher levels of TG, total cholesterol, VLDL-C and LDL-C [103]. SCARB1 variants affecting HDL-C were also found by GWAS for the first time in 2010 [33]. Genetic studies of families with rare variants of SCARB1 have confirmed that functional mutations in SR-BI are causative for elevated HDL-C levels [104,105] and certain SCARB1 mutations are associated with subclinical atherosclerosis [106,107] and CVD [107]. Some SCARB1 variants have also been described to be associated with infertility in women most likely because of the role of CLAI/SR-BI in mediating cholesterol uptake in the ovaries for steroid hormone biosynthesis [99,108]. A relatively large family homozygous for a novel SR-BI mutation has been described [104]. The mutation, P297S, resulted in a complete loss of SR-BI function, as measured by selective cholesteryl ester uptake [104]. Heterozygous patients with this mutation were reported to have HDL-C levels of 70.4 mg/dL (1.8 mmol/L) versus 53.4 mg/dL (1.4 mmol/L) in non-carriers ($P < 0.001$); and to have evidence for adrenal insufficiency [104].

Several small-molecule inhibitors of SR-BI have been described [109–113], but there have been limited studies on drugs that modulate SR-BI in humans. Interestingly, probucol, a potent antioxidant that lowers both LDL-C and HDL-C but was discontinued in the USA because of possible cardiac conduction abnormalities, appears to ameliorate some of the fertility problems observed in SR-BI knockout mice [114]. The mechanism for this finding is not clear but suggests that probucol may also have possible utility in improving fertility in women and CVD with certain polymorphisms of SCARB1 associated with high HDL-C [108].

3. HDL-MODIFYING GENES IDENTIFIED FROM GWAS

Besides the genes identified through the study of classic Mendelian disorders of HDL metabolism, many new and known genes associated with HDL have recently been described using GWAS (Table 1). This approach involves the high-throughput determination of SNPs in thousands of individuals, using DNA microarrays, and testing for their association with a particular disease or trait like HDL-C. One of the first GWASs identified 30 genetic loci linked to dyslipidemia [32]. Subsequently, a much larger GWAS of over 100,000 individuals of European descent has been performed. This study identified 95 loci (36 loci previously reported by the GWAS at genome-wide significance and 59 novel GWAS loci) that showed genome-wide significant association ($P < 5 \times 10^{-8}$) with at least one of the four lipid traits tested (HDL-C, LDL-C, TC, and TG) [33]. Thirty-nine novel gene loci demonstrated genome-wide significant association with TC, 22 with LDL-C, 31 with HDL-C, and 16 with TG. Among the 36 known loci, 21 demonstrated genome-wide significant association with another lipid phenotype, in addition to that previously described. The total set of mapped variants found by Teslovich et al. [33] explains 12.4% (TC), 12.2% (LDL-C), 12.1% (HDL-C), and 9.6% (TG) of the total variance in each lipid trait in the Framingham Heart Study, corresponding to about 25–30% of the genetic variance for each trait [33].

TABLE 1   Novel Loci Associated with HDL Cholesterol [33,34], Gene Locus, Associated Trait(s), Effect Sizes, P-Values, and Candidate Gene Names (for Loci in Ref. [33] with More than One Trait, Effect Sizes and P-Values Are Given for HDL-C if HDL-C Was the Main Trait or Labeled NA (Not Available) if HDL-C Was Not the Lead Trait)

Locus	Associated trait(s)	References	Effect size	P-value	Gene name
ABCA8	HDL	33	−0.42	2×10^{-10}	ATP-binding cassette, sub-family A (ABC1), member 8
ACAD11	LDL, HDL	34	−0.034/0.028	$2\times10^{-9}/5\times10^{-9}$	Acyl-CoA dehydrogenase family, member 11
ADH5	HDL	34	0.019	5×10^{-8}	Alcohol dehydrogenase 5 (class III), chi polypeptide
AMPD3	HDL	33	−0.41	5×10^{-8}	Adenosine monophosphate deaminase 3
ANGPTL1	HDL	34	0.021	7×10^{-9}	Angiopoietin-like 1
APOB	TG, HDL	33	NA	NA	Apolipoprotein B
ARL15	HDL	33	−0.49	5×10^{-8}	ADP-ribosylation factor-like 15
ATG7	HDL	34	0.025	5×10^{-8}	Autophagy related 7
C4orf52/ SMIM20	HDL	34	−0.027	5×10^{-8}	Small integral membrane protein 20
C6orf106	HDL	33	−0.49	4×10^{-9}	Chromosome 6 open reading frame 106
CITED2	HDL	33	−0.39	3×10^{-8}	Cbp/p300-interacting transactivator, with Glu/Asp-rich carboxy-terminal domain, 2
CMIP	HDL	33	−0.45	2×10^{-11}	c-Maf inducing protein
COBLL1	HDL	33	0.68	3×10^{-10}	Cordon-Bleu WH2 repeat protein-like 1

Gene	Trait				Description
CPS1	HDL	34	-0.027	9×10^{-10}	Carbamoyl-phosphate synthase 1, mitochondrial
DAGLB	HDL	34	0.024	7×10^{-12}	Diacylglycerol lipase, beta
FAM13A	HDL	34	-0.025	4×10^{-12}	Family with sequence similarity 13, member A
FTO	HDL, TG	34	$-0.020/-0.021$	$7\times10^{-9}/3\times10^{-8}$	Fat mass and obesity associated
GALNT2	HDL, TG	33	-0.61	4×10^{-21}	Polypeptide N-acetylgalactosaminyltransferase 2
GSK3B-NR1/2	HDL	34	0.02	1×10^{-8}	Glycogen synthase kinase 3B; nuclear receptor subfamily 1, group I, member 2
HAS1	HDL	34	-0.029	2×10^{-13}	Hyaluron synthase 1
HDGF-PMVK	HDL	34	0.02	2×10^{-8}	Hepatoma-derived growth factor/phosphomevalonate kinase
IKZF1	HDL	34	0.022	1×10^{-8}	IKAROS family zinc finger 1 (Ikaros)
IRS1	HDL, TG	33	0.46	3×10^{-9}	Insulin receptor substrate 1
KAT5	HDL	34	0.024	3×10^{-8}	K(lysine) acyltransferase 5
KLF14	HDL	33	0.59	1×10^{-15}	Kruppel-like factor 14
LACTB	HDL	33	-0.39	9×10^{-9}	Lactamase, beta
LILRA3	HDL	33	0.83	4×10^{-16}	Leukocyte immunoglobulin-like receptor subfamily A (without TM domain), member 3
LOC55908/C19orf80	HDL	33	-0.64	3×10^{-9}	C19orf80 (chromosome 19 open reading frame 80), aka ANGPTL8, LOC55908

Continued

TABLE 1   Novel Loci Associated with HDL Cholesterol [33,34], Gene Locus, Associated Trait(s), Effect Sizes, P-Values, and Candidate Gene Names (for Loci in Ref. [33] with More than One Trait, Effect Sizes and P-Values Are Given for HDL-C if HDL-C Was the Main Trait or Labeled NA (Not Available) if HDL-C Was Not the Lead Trait)—cont'd

Locus	Associated trait(s)	References	Effect size	P-value	Gene name
LPA	HDL	33	1.95	3×10^{-8}	Lipoprotein, Lp(A)
LRP1	TG, HDL	33	NA	NA	Low density lipoprotein receptor-related protein 1
LRP4	HDL	33	0.78	3×10^{-18}	Low density lipoprotein receptor-related protein 4
MARCH8-ALOX5	HDL, TC	34	$0.026/-0.026$	$2\times10^{-10}/8\times10^{-9}$	Membrane-associated ring finger (C3HC4) 8, E3 ubiquitin protein ligase; arachidonate 5-lipoxygenase
MC4R	HDL	33	-0.42	7×10^{-9}	Melanocortin 4 receptor
MLXIPL	TG, HDL	33	NA	NA	MLX interacting protein-like
MOGAT2-DGAT2	HDL	34	-0.026	1×10^{-8}	Monoacylglycerol O-acyltransferase 2; diacylglycerol O-acyltransferase 2
OR4C46	HDL	34	0.034	2×10^{-10}	Olfactory receptor, family 4, subfamily C, member 46
PABPC4	HDL	33	-0.48	4×10^{-10}	Poly(A) binding protein, cytoplasmic 4 (inducible form)

Gene	Trait	Ref	Effect	p-value	Description
PDE3A	HDL	33	0.4	4×10^{-8}	Phosphodiesterase 3A, cGMP-inhibited
PEPD-CEBPA-CEBPG	TG, HDL	34	$0.022/-0.022$	$3\times10^{-9}/3\times10^{-9}$	PEPD - Peptidase B. CEBP: CCAAT/enhancer binding protein alpha and gamma
PGS1	HDL	33	-0.39	8×10^{-9}	Phosphatidylglycerophosphate synthase 1
PIGV-NR0B2	HDL, LDL, TG	34	$-0.051/0.050/0.037$	$1\times10^{-15}/3\times10^{-12}/1\times10^{-9}$	Phosphatidylinositol glycan anchor biosynthesis, class V; nuclear receptor subfamily 0, group B, member 2
PPP1R3B	HDL, TC, LDL	33	-1.21	6×10^{-25}	protein phosphatase 1, regulatory subunit 3B
RBM5	HDL	34	0.025	9×10^{-12}	RNA binding motif protein 5
RSPO3	HDL, TGa	34	$0.020/-0.020$	$3\times10^{-10}/3\times10^{-8}$	R-spondin 3
SBNO1	HDL	33	0.86	7×10^{-9}	Strawberry notch homolog 1 (Drosophila)
SCARB1	HDL	33	0.61	3×10^{-14}	scavenger receptor class B, member 1
SETD2	HDL	34	-0.030	4×10^{-9}	SET domain containing 2
SLC39A8	HDL	33	-0.84	7×10^{-11}	Solute carrier family 39 (zinc transporter), member 8
SNX13	HDL	34	-0.026	9×10^{-12}	Sorting nexin 13
STAB1	HDL	34	0.029	9×10^{-11}	Stabilin 1
STARD3	HDL	33	-0.48	1×10^{-13}	StAR-related lipid transfer (START) domain containing 3
TMEM176A	HDL	34	$-.036$	2×10^{-8}	Transmembrane protein 176A

Continued

TABLE 1   Novel Loci Associated with HDL Cholesterol [33,34], Gene Locus, Associated Trait(s), Effect Sizes, P-Values, and Candidate Gene Names (for Loci in Ref. [33] with More than One Trait, Effect Sizes and P-Values Are Given for HDL-C if HDL-C Was the Main Trait or Labeled NA (Not Available) if HDL-C Was Not the Lead Trait)—cont'd

Locus	Associated trait(s)	References	Effect size	P-value	Gene name
TRPS1	HDL	33	−0.44	6×10^{-11}	Trichorhinophalangeal Syndrome I
UBASH3B	TC, HDL	33	NA	NA	Ubiquitin associated and SH3 domain containing B
UBE2L3	HDL	33	−0.46	1×10^{-8}	Ubiquitin-conjugating enzyme E2L3
VEGFA	TG, HDL	34	0.029/−0.026	$3\times10^{-15}/2\times10^{-11}$	Vascular endothelial growth factor A
ZBTB42-AKT1	HDL	34	0.02	1×10^{-8}	Zinc finger and BTB domain containing 42; v-akt murine thymoma viral oncogene homolog 1
ZNF648	HDL	33	−0.47	3×10^{-10}	Zinc finger protein 648
ZNF664	HDL, TG	33	0.44	3×10^{-10}	Zinc finger protein 664.

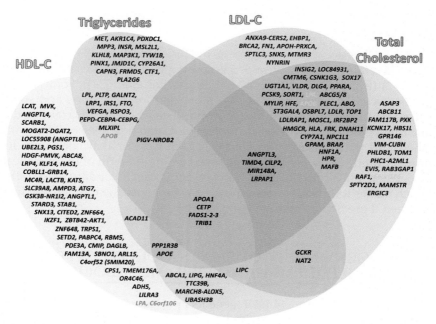

FIGURE 2 **Overlap of loci associated with different lipid traits.** The Venn diagram illustrates the number of loci that show gene association with multiple lipid traits. Genes linked with their primary lipid trait are enclosed in one of the four color-coded ovals labeled with the primary trait at the top of the figure. Loci that show association with two or more traits are displayed in the appropriate segment. Note that the LPA, apoB, and C6orf106 SNPS found to be associated with HDL-C by Teslovich et al. [33] (blue font) differ from the SNPs in these genes associated with LDL-C and/or TC (yellow font) [33,34]. *Modified from Figure 1 in Willer et al. [34].*

A recent meta-analysis of GWASs [34] found even more genes associated with HDL-C. In this study of 188,578 European-ancestry individuals (94,596 genotyped with GWAS arrays (23 studies) and 93,982 genotyped with Metabochip arrays (37 studies)) and 7898 non-European ancestry individuals, a total of 157 gene loci associated with lipid levels at $P < 5 \times 10^{-8}$ (Figure 2), including 62 novel gene loci [34]. Although the effect sizes for these newly identified gene SNPs were relatively small [34], these genes are not necessarily less important than genes with larger effect sizes. The frequency and effect of a given SNP rather than the importance of the gene itself determine the strength of the statistical association and the effect size. For example, LCAT is unquestionably one of the most important enzymes in HDL biogenesis, yet 100,000 individuals were required to link LCAT SNPs to HDL-C at genome-wide significance [33,63].

3.1 Functions of HDL-Modifying Genes Identified from GWAS

The HDL-associated genes found by Teslovich et al. [33] and Willer et al. [34] (Table 1) can be grouped according to their proposed functions (Table 2) (from GeneCards, PubMed and Willer et al. [34]). Some of the genes discussed below may fit into more than one category, and some genes may have additional functions not listed here. A diagram of the identified gene loci that associate with HDL-C and how they overlap with other lipid traits is shown in Figure 2. Each gene is associated with its primary lipid trait, which is largely based on the strength of the statistical association, but is also linked in many cases to other secondary lipid traits, which makes it difficult to assign with any certainty the mechanistic link between these SNPs and the various lipid traits.

4. GENES PREVIOUSLY ASSOCIATED WITH DYSLIPIDEMIA AND LIPID METABOLISM

Teslovich et al. found 14 HDL gene loci previously reported by GWAS at genome-wide significance with previously described associations with dyslipidemia [33]. These genes include ABCA1, APOA1, LCAT, LIPC, LIPG, LPL, CETP, PLTP, HNF4A, MVK, FADS1-2-3, ANGPTL4, GALNT2, and TTC39B. Most of these have well-known functions in lipoprotein metabolism. GALNT2, a glycosyl transferase, and TTC39B, a tetra-tricopeptide repeat motif-containing protein that likely mediates the assembly of multiprotein complexes [115,116], had previously been identified through GWAS as associated with HDL-C [32], but their exact functional role was unclear. To demonstrate functionality, Teslovich et al. knocked down both genes in mice and found an increase in plasma HDL-C for both [33]. Importantly, *SCARB1* was found for the first time to associate with HDL-C in this GWAS [33].

Loci previously considered to be involved in apoB-containing lipoprotein metabolism were also found to be associated with HDL-C levels [33,34]. For example, HDL-C was the lead trait for a SNP at the LPA locus [33], perhaps due to an interaction with SR-BI [97], and for SNPs at the MOGAT2/DGAT2 loci [34]. MOGAT2 facilitates dietary fat absorption from the small intestine and DGAT2 is involved in triglyceride synthesis. HDL-C was also found to be a secondary trait for apoB (rs1042034) [33], LRP1 [33], a receptor for apoE-containing VLDL as well as other ligands, and TRIB1 [33], which was previously associated with CAD.

Both GWAS [33,34] also identified several novel genes associated with HDL-C for the first time, with a possible link to lipoprotein metabolism. LOC55908 [33] (also known as C19orf80, lipasin, ANGPTL8, RIFL,

TABLE 2 Similar to Table 1 but Grouped According to Function: Dyslipidemia and Lipid Metabolism (Table 2(A)), Diabetes and Obesity (Table 2(B)), Lipid Trafficking (Table 2(C)), Chromatin-Modifying Genes, Transcription Factors and RNA Processing (Table 2(D)), Signal Transduction (Table 2(E)), Other (Table 2(F)), and CAD (Table 2(G)). *P*-Values for CAD Are from Willer et al. [34] and Are Italicized if HDL-C Was Not the Lead Trait for that Gene

DYSLIPIDEMIA AND LIPID METABOLISM

Locus	Associated trait(s)	References	Gene function
GALNT2	HDL, TG	33	Participates in first step of O-linked oligosaccharide biosynthesis of proteins such as apoC-III
TTC39B	HDL, TC	33	Knockdown in mice increases HDL-C
SCARB1	HDL	33	Plasma membrane receptor for HDL
LPA	HDL	33	Encodes Apo(a), which modifies apoB in LDL to form Lp(a), an atherogenic lipoprotein
MOGAT2-DGAT2	HDL	34	MOGAT2: facilitates dietary fat absorption; DGAT2: triglyceride synthesis
APOB	TG, HDL	33	Major structural protein of LDL
LRP1	TG, HDL	33	Receptor for apoE-containing VLDL
TRIB1	TG, TC, LDL, HDL	33	Hepatic lipogenesis and glycogenesis; degradation of transcription factors CEBPA and MLXIPL; signaling
LOC55908/C19orf80	HDL	33	Also known as ANGPTL8. Regulates serum TG in mice; associates with LDLC
UBE2L3	HDL	33	Ubiquinates PARK2, a regulator of fat metabolism
PGS1	HDL	33	Biosynthesis of phosphatidylglycerol and cardiolipin

Continued

TABLE 2   Similar to Table 1 but Grouped According to Function: Dyslipidemia and Lipid Metabolism (Table 2(A)), Diabetes and Obesity (Table 2(B)), Lipid Trafficking (Table 2(C)), Chromatin-Modifying Genes, Transcription Factors and RNA Processing (Table 2(D)), Signal Transduction (Table 2(E)), Other (Table 2(F)), and CAD (Table 2(G)). P-Values for CAD Are from Willer et al. [34] and Are Italicized if HDL-C Was Not the Lead Trait for that Gene—cont'd

DYSLIPIDEMIA AND LIPID METABOLISM

Locus	Associated trait(s)	References	Gene function
ACAD11	LDL, HDL	34	β-oxidation of long-chain fatty acids
PMVK	HDL	34	PMVK-cholesterol biosynthesis
ABCA8	HDL	33	Sphingomyelin metabolism in oligodendrocytes
LRP4	HDL	33	Cell-surface endocytic receptor ; binds apoE.

DIABETES, OBESITY, AND ENERGY METABOLISM

Locus	Associated trait(s)	References	Gene function
IRS1	HDL, TG	33	Insulin receptor tyrosine kinase substrate; associated with insulin resistance, diabetes, CAD
KLF14	HDL	33	Transcription factor; diabetes, obesity, CAD
PPP1R3B	HDL, TC, LDL	33	Regulates glycogen synthesis; diabetes
HAS1	HDL	34	Hyaluron accumulates with high-fat feeding and contributes to insulin resistance in mice
COBLL1-GRB14	HDL	33	Diabetes; associated with serum insulin and TG
MC4R	HDL	33	Obesity; hyperphagia in mice
FTO	HDL, TG	34	Obesity; possible hypothalamic regulation of food intake

LACTB	HDL	33	Serine protease; mitochondrial protein; overexpression in mice causes obesity
KAT5	HDL	34	Positive regulator of PPARG transcription involved in adipogenesis
PEPD-CEBPA-CEBPG	TG, HDL	34	PEPD: hydrolyzes peptides with C-terminal proline/OH-proline; CEBPA/G: transcription factors
SLC39A8	HDL	33	Zinc transporter; negatively regulate proinflammatory responses
VEGFA	TG, HDL	34	Obesity, T2D, SBP, DBP, and CAD
RSPO3	HDL, TGa	34	Signalling; regulates VEGF expression; waist–hip ratio
AMPD3	HDL	33	Energy metabolism
ATG7	HDL	34	Intracellular energy metabolism
GSK3B	HDL	34	Energy metabolism
ANGPTL1	HDL	34	Member of ANGPTL family
ZBTB42-AKT1	HDL	34	AKT1: Ser-Thr protein kinase; associated with glucose homeostasis, metabolic syndrome.

LIPID TRAFFICKING

Locus	Associated trait(s)	References	Gene function
STARD3	HDL	33	Cholesterol trafficking
STAB1	HDL	34	Endocytosis of various ligands (including LDLr)

Continued

TABLE 2   Similar to Table 1 but Grouped According to Function: Dyslipidemia and Lipid Metabolism (Table 2(A)), Diabetes and Obesity (Table 2(B)), Lipid Trafficking (Table 2(C)), Chromatin-Modifying Genes, Transcription Factors and RNA Processing (Table 2(D)), Signal Transduction (Table 2(E)), Other (Table 2(F)), and CAD (Table 2(G)). *P*-Values for CAD Are from Willer et al. [34] and Are Italicized if HDL-C Was Not the Lead Trait for that Gene—cont'd

LIPID TRAFFICKING

Locus	Associated trait(s)	References	Gene function
SNX13	HDL	34	Intracellular trafficking
MARCH8	HDL, TC	34	MARCH8: membrane bound E3 ubiquitin ligase.

CHROMATIN-MODIFYING GENES, TRANSCRIPTION FACTORS, RNA PROCESSING

Locus	Associated trait(s)	References	Gene function
HNF4A	HDL, TC	33	Transcription factor - master regulator
NR0B2	HDL, LDL, TG	34	NR0B2 (SHP) transcription factor
CEBPA-CEBPG	TG, HDL	34	CEPBA, CEBPG: transcription factors involved in adipogenesis. CEBPA degraded by TRIB1
CITED2	HDL	33	Transcriptional coactivator; stimulates PPARA and upregulates NR5A1 (SF-1)
KLF14	HDL	33	Transcription factor; diabetes, obesity, CAD
ZNF664	HDL, TG	33	Transcription factor
IKZF1	HDL	34	Transcription factor
KAT5	HDL	34	Positive regulator of PPARG transcription involved in adipogenesis

Locus	Associated trait(s)	References	Gene function
NR1/2	HDL	34	NR1/2: transcription factor; heterodimerizes with RXR; activated by guggulipid; lipid homeostasis
MLXIPL	TG, HDL	33	Transcription factor; binds carbohydrate response elements in gene promoters; degraded by TRIB1
ZBTB42-AKT1	HDL	34	ZBTB42: transcriptional repressor; locus associated with glucose homeostasis, MetS
ZNF648	HDL	33	Transcription factor
TRPS1	HDL	33	Transcriptional repressor; Binds GATA elements
SETD2	HDL	34	Histone H3 methyltransferase
PABPC4	HDL	33	Inducible cytoplasmic polyA-binding protein
RBM5	HDL	34	Nuclear RNA binding protein; component of the spliceosome A complex; FAS gene splicing

SIGNAL PROCESSING

Locus	Associated trait(s)	References	Gene function
TRIB1	HDL	33	Signaling
PDE3A	HDL	33	cAMP and cGMP hydrolysis
CMIP	HDL	33	T-cell signalling
RSPO3	HDL, TGa	34	Signaling; regulates VEGF expression; waist–hip ratio
VEGFA	TG, HDL	34	Obesity, T2D, SBP, DBP and CAD
UBASH3B	TC, HDL	33	Ubiquitin-modifying protein that modulates tyrosine kinase pathways.

Continued

TABLE 2   Similar to Table 1 but Grouped According to Function: Dyslipidemia and Lipid Metabolism (Table 2(A)), Diabetes and Obesity (Table 2(B)), Lipid Trafficking (Table 2(C)), Chromatin-Modifying Genes, Transcription Factors and RNA Processing (Table 2(D)), Signal Transduction (Table 2(E)), Other (Table 2(F)), and CAD (Table 2(G)). *P*-Values for CAD Are from Willer et al. [34] and Are Italicized if HDL-C Was Not the Lead Trait for that Gene—cont'd

SIGNAL PROCESSING

Locus	Associated trait(s)	References	Gene function
DAGLB	HDL	34	Hydrolysis of diacylglycerol to 2-arachidonoyl-glycerol (2-AG), an abundant endocannabinoid
FAM13A	HDL	34	Signal transduction
ALOX5	HDL, TC	34	Leukotriene formation
PIGV	HDL, LDL, TG	34	PIGV: Mannosyl transferase.

OTHER

Locus	Associated trait(s)	References	Gene function
SBNO1	HDL	33	Unknown
ARL15	HDL	33	Unknown
C4orf52/SMIM20	HDL	34	Small integral membrane protein
CPS1	HDL	34	Catalyzes first commited step of urea cycle
TMEM176A	HDL	34	Transmembrane protein
OR4C46	HDL	34	Olefactory receptor
ADH5	HDL	34	Oxidation of long-chain primary alcohols
LILRA3	HDL	33	Unknown

CAD-ASSOCIATED GENES

Locus	Associated trait(s)	References	Effect Size	P-value	Function	CAD effect size
ABCA8	HDL	33	−0.42	2×10^{-10}	Sphingomyelin metabolism in oligodendrocytes	2.0×10^{-2}
APOA1	TG, TC, HDL, LDL	33	NA	NA	Main protein component of HDL. ApoA-I accepts cholesterol effluxed from ABCA1	4.8×10^{-11}
APOE	LDL, TC, HDL	33	NA	NA	Chylomicron and VLDL remnants clearance; defects cause type III hyperlipoproteinemia	2.1×10^{-4}
C6orf106	HDL	33	−0.49	4×10^{-9}	Unknown	1.6×10^{-3}
CETP	HDL,TC,LDL,TG	33	3.39	7×10^{-380}	Exchanges CE in HDL for TG in apoB-Lps	2.2×10^{-3}
CITED2	HDL	33	−0.39	3×10^{-8}	Transcriptional coactivator; stimulates PPARA and upregulates NR5A1 (SF-1)	1.5×10^{-2}
GALNT2	HDL, TG	34	−1.2	14×10^{-41} / 7×10^{-31}	Participates in first step of O-linked oligosaccharide biosynthesis of proteins such as apoC-III	4.1×10^{-2}
IKZF1	HDL	34	0.022	1×10^{-8}	Transcription factor.	3.4×10^{-2}

Continued

TABLE 2   Similar to Table 1 but Grouped According to Function: Dyslipidemia and Lipid Metabolism (Table 2(A)), Diabetes and Obesity (Table 2(B)), Lipid Trafficking (Table 2(C)), Chromatin-Modifying Genes, Transcription Factors and RNA Processing (Table 2(D)), Signal Transduction (Table 2(E)), Other (Table 2(F)), and CAD (Table 2(G)). P-Values for CAD Are from Willer et al. [34] and Are Italicized if HDL-C Was Not the Lead Trait for that Gene—cont'd

CAD-ASSOCIATED GENES

Locus	Associated trait(s)	References	Effect Size	P-value	Function	CAD effect size
IRS1	HDL, TG	33	0.46	3×10^{-9}	Substrate of insulin receptor tyrosine kinase; associated with insulin resistance, diabetes, CAD	3.8×10^{-4}
KAT5	HDL	34	0.024	3×10^{-8}	Positive regulator of PPARG transcription involved in adipogenesis	3.8×10^{-2}
KLF14	HDL	33	0.59	1×10^{-15}	Transcription factor; diabetes, obesity, CAD	3.2×10^{-3}
LPL	TG, HDL	33	NA	NA	Hydrolyzes triglycerides in chylomicrons and VLDL	4.7×10^{-5}
RBM5	HDL	34	0.025	9×10^{-12}	Nuclear RNA-binding protein; component of the spliceosome A complex; role in FAS gene splicing.	7.0×10^{-5}
TRIB1	TG, TC, LDL, HDL	33	NA	NA	Hepatic lipogenesis and glycogenesis; degradation of transcription factors CEBPA and MLXIPL; signalling; CAD	2.8×10^{-5}
VEGFA	TG, HDL	34	0.029/−0.026	3×10^{-15} / 2×10^{-11}	Obesity, T2D, SBP, DBP, and CAD	9×10^{-3}
ZNF664	HDL, TG	33	0.44	3×10^{-10}	Transcription factor.	3.6×10^{-3}

Gm6484 and betatrophin) was found to regulate serum triglyceride levels in mice and to associate with LDL-C as well as HDL-C in later studies [117−119]. Another interesting gene is UBE2L3 (ubiquitin-conjugating enzyme E2L3) [33], also known as UBCH7. UBE2L3 ubiquitinates Parkin (PARK2) [120], which in turn is a lipid-sensitive regulator of fat metabolism [121]. PGS1 (phosphatidylglycerophosphate synthase 1) [33], and ACAD11, involved in β-oxidation of long-chain fatty acids in muscle and heart [34], were also associated with HDL-C. HDL-C was also associated for the first time with PMVK, a gene present at the HDGF/PMVK locus [34], which encodes an enzyme in the cholesterol biosynthesis pathway that converts mevalonate 5-phosphate to mevalonate 5-diphosphate. ABCA8, which stimulates sphingomyelin production in oligodendrocytes [122] but was not known to function in HDL biogenesis, was associated with HDL-C [33], as well as CAD [34].

Finally, HDL-C was the lead trait for LRP4 [33]. LRP4 has been proposed to function as a cell-surface endocytic receptor, binding and internalizing extracellular ligands for degradation by lysosomes. Diseases associated with LRP4 include syndactyly cenani lenz type, which affects bone and kidney and can involve hypothyroidism, and sclerosteosis 2, which affects bone (GeneCards). LRP4 is expressed in brain and liver as well as in other tissues and is known to bind apoE [123], thus identifying LRP4 as a potential novel target in lipoprotein metabolism.

5. GENES PREVIOUSLY ASSOCIATED WITH DIABETES AND OBESITY

Several loci with previous associations to diabetes were found to be linked for the first time with HDL-C as the lead trait [33]. These gene loci include IRS1 [33], a major intracellular substrate of the insulin receptor (IR) tyrosine kinase, which was previously associated with insulin resistance and lipoatrophic diabetes. Another example is the transcription factor KLF14 [33], which is a master regulator of gene expression in adipose tissue and was previously shown to be associated with a variety of metabolic traits, including type 2 diabetes mellitus (T2DM) and HDL-C [124]. Both of these genes are excellent candidates for future HDL-based drug development, as they were both found to be associated with CAD and with HDL-C [33].

Additional genes linked to HDL-C with a role in diabetes but not yet linked to CAD were also identified by GWAS. PPP1R3B [33] (also known as hepatic glycogen-targeting protein phosphatase 1 regulatory subunit), associated with LDL-C and TC as well as HDL-C, is expressed in liver and skeletal muscle tissue, where it regulates glycogen synthesis. PPP1R3B may also be involved in T2DM and maturity-onset diabetes of the young.

Overexpression of the mouse homolog *Ppp1r3B* in mouse liver was shown to decrease plasma HDL-C and TC [33], confirming its role in HDL metabolism. HAS1 (hyaluron synthase 1) was also found to associate with HDL-C [34], perhaps because hyaluron accumulates with high-fat feeding and contributes to insulin resistance in mice [125]. Finally, SNPs within the novel COBLL1-GRB14 locus [33] have recently been linked to serum insulin, HDL-C, and triglyceride levels in adults [126] and serum insulin levels in overweight children [127].

HDL loci with associations to obesity were identified as well by GWAS [33]. MC4R [33] is a membrane-bound receptor that is highly expressed in hypothalamus, hippocampus, and hindbrain and interacts with adrenocorticotropic and MSH hormones. Defects in this gene cause hyperphagia in mice and autosomal-dominant morbid obesity in humans [128] and associate with LDL-C, HDL-C, and triglyceride levels. Similarly, variants in the FTO (fat mass and obesity associated) gene associate with obesity-related phenotypes, and the FTO gene product may act through hypothalamic regulation of food intake [34]. KLF14 [33], as noted above, is a master regulator of gene expression in adipose tissue [124], and the serine protease LACTB, another GWAS identified gene [33], when overexpressed in mice, was found to cause obesity [129]. KAT5, a lysine acyltransferase, is a positive regulator of PPARG transcription involved in adipogenesis [34]. CEBPA and CEBPG are well-known transcription factors that are also involved in adipogenesis, and the PEPD gene, in the same locus, is associated with adiponectin [130]. The role of SLC39A8 [33], a zinc transporter, in HDL metabolism may relate to its ability to negatively regulate proinflammatory responses through zinc-mediated down-modulation of IκB kinase activity in vitro [131]. It is associated with BMI [34], and its association with HDL-C has been confirmed in subsequent studies [132]. Finally, variants in the VEGFA and RSPO3 loci associate with triglycerides and waist/hip ratio, as well as HDL-C [34]. As VEGFA is associated with T2DM, systolic blood pressure (SBP), diastolic blood pressure (DBP), and CAD as well [34], it would seem to be a prime target for therapeutic intervention.

A number of identified energy metabolism genes associated with HDL (AMPD3 [33], ATG7 [34], and GSK3B [34]) might also be future therapeutic targets. ANGPTL1, an ANGPTL family member not known to have a role in lipid metabolism, had a SNP associated with a small but statistically significant effect on HDL-C [34].

6. LIPID TRAFFICKING-RELATED GENES

Lipid trafficking plays a key role in HDL biogenesis and metabolism [29,31,133–136], hence genes involved in trafficking represent interesting possible therapeutic targets. The strongest lipid trafficking candidate to emerge from GWAS is the STARD3 gene, for which HDL-C is the lead and only trait [33]. The STARD3 gene product, also known as MLN64, belongs to a subfamily of lipid trafficking proteins characterized by a C-terminal steroidogenic acute regulatory domain and an N-terminal metastatic lymph node 64 domain. The protein localizes to late endosomal membranes and is involved in cholesterol trafficking [137]. Another intriguing lipid trafficking candidate is STAB1 (Stabilin 1), which associates with HDL [34] and functions to mediate endocytosis of various ligands, including LDLR [34]. Its function in modulating HDL-C levels is not understood but variants in STAB1 associate with waist/hip ratio [34].

Additional HDL-C associated genes that could be involved in lipid trafficking include SNX13 (sorting nexin 13) [34] and MARCH8 [34]. While its role in HDL metabolism is unclear, SNX13 does target endosomal sorting and plays a crucial role in preserving cardiomyocyte survival and cardiac performance [138]. MARCH8 is an actin-binding protein that belongs to the MARCH family of membrane-bound E3 ubiquitin ligases and induces the internalization (endocytosis and sorting to lysosomes) of several cell-surface receptors, including FAS (Fas Cell-Surface Death Receptor) [139–142]. This possible link to FAS function is of interest because markedly decreased HDL-C levels have been found in autoimmune lymphoproliferative syndrome (ALPS) patients with FAS mutations [143].

7. CHROMATIN-MODIFYING GENES, TRANSCRIPTION FACTORS AND RNA PROCESSING RELATED GENES

A number of chromatin-modifying genes and transcription factors were found by GWAS to be associated with HDL-C as the lead trait. HNF4A [32,33], NR0B2 (SHP) [34], CEBPA [34], and CEBPG [34] are well-known transcription factors found to be associated with HDL-C as well as other lipids (Figure 2). These genes are very attractive therapeutic target candidates as they could potentially regulate multiple pathways involved in HDL metabolism.

Several novel transcription factors associated with CAD are of particular interest as potential therapeutic targets. CITED2 [33] is a transcriptional coactivator of the p300/CBP-mediated transcription complex and stimulates PPAR-alpha and also indirectly up-regulates the nuclear hormone receptor NR5A1 (SF-1), which in turn regulates transcription of SCARB1 and SOAT1/ACAT [144,145]. SNPs in CITED2 were found to be associated with CAD events [33], making it an attractive therapeutic target. KLF14 [33] is a master regulator of gene expression in adipose tissue [124], and is also associated with diabetes, TG and CAD, as noted above. Similarly, ZNF664 [33], IKZF1 [34], and KAT5 (K(lysine) acetyltransferase 5) [34] are all transcription factors/chromatin-modifying genes associated with CAD, as well as with HDL-C. IKZF1 is reported to regulate LDLr transcription in some cell types but was associated only with HDL-C by GWAS [34] and KAT5 is a positive regulator of PPARG transcription involved in adipogenesis [34]. Further investigation of these proteins as potential therapeutic targets appears to be warranted.

Other novel transcription factors affecting lipid metabolism and associated with HDL-C for the first time were also identified. NR1I2, associated with HDL-C, heterodimerizes with RXR and is involved in lipid homeostasis [34] and MLXIPL [33] is a transcription factor that binds and activates carbohydrate response element (ChoRE) motifs in the promoters of triglyceride synthesis genes. MLXIPL, also known as CHREBP, was associated with both TG and HDL-C [33,34]. It binds to and is degraded by TRIB1 [146], which is associated with HDL-C, TG, LDL-C, TC, and CAD. Finally, a haplotype at the ZBTB42/AKT1 locus [34] may be associated with glucose homeostasis and metabolic syndrome [147].

Additional chromatin modifiers and transcription factors less well understood but significantly associated with HDL-C include ZNF648 [33], TRPS1 [33], and SETD2 [34]. Little is known about these genes' roles in HDL metabolism. Similarly, the roles of the polyA-binding protein PABPC4 [33] and the nuclear RNA-binding protein RBM5 [34] in lipoprotein metabolism are also unclear; however, as RBM5 is associated with CAD as well as HDL-C it warrants further investigation. Interestingly, RBM5 has been found to be involved in the alternative splicing of the FAS gene [148]. As noted above, the link to FAS gene expression is of interest because of the markedly decreased HDL-C levels found in ALPS patients with FAS mutations [143].

8. SIGNAL TRANSDUCTION-RELATED GENES

Signal transduction genes, like transcription factors and chromatin-modifying genes, have the potential to affect cellular HDL metabolism by multiple pathways. TRIB1 (tribbles pseudokinase 1) polymorphisms are associated with all four different lipid traits (TG, TC, LDL, and HDL) and CAD events [33] and MI (reviewed in Ref. [149]). Its exact function is not understood, but it is believed to signal by the MAPK and CEBP pathways, down-regulating hepatic lipogenesis and glycogenesis via multiple molecular interactions [146,149,150]. Interestingly, TRIB1 promotes degradation of CEBPA, a master transcription factor for adipocyte differentiation, and binds and degrades the hepatic lipogenic master regulator MLXIPL. Both of these transcription factors were found to be associated with HDL and triglycerides [33,34]. TRIB1 has also recently been found to be critical for adipose tissue maintenance and suppression of metabolic disorders by controlling the differentiation of tissue-resident M2-like macrophages [151].

Other signal transduction genes associated with HDL-C include PDE3A [33], a member of the cGMP-inhibited cyclic nucleotide phosphodiesterase family that hydrolyzes both cAMP and cGMP; and CMIP [33], which plays a role in T-cell signaling pathways. Another gene locus in this category is RSPO3, which regulates beta-catenin signaling and, in mice, is required for VEGF expression and endothelial cell proliferation [34]. RSPO3 associates with TG and HDL-C. VEGFA is a signal transduction gene associated with TG and hip/waist ratio, as well as with obesity, T2D, SBP, DBP, and CAD [34]. UBASH3B (ubiquitin associated and SH3 domain containing B) (lead trait TC, secondary trait HDL-C [33]), codes for a ubiquitin-modifying protein that modulates receptor-type tyrosine kinases pathways. Willer et al. identified several other gene SNPs with small effects on HDL-C (DAGLB, FAM13A, ALOX5, and PIGV) [34] that may play a role in signal transduction, but their role in lipid metabolism is not clear except for DAGLB [34], which hydrolyzes DAG to form endogenous cannabinoids and ALOX5 [34], which converts arachidonic acid to leukotrienes.

9. OTHER GWAS-IDENTIFIED HDL-ASSOCIATED GENES

GWAS identified a number of miscellaneous genes associated with HDL-C but with no apparent mechanistic link to lipoprotein metabolism. These include SBNO1 (Strawberry Notch Homolog 1 (*Drosophila*)) [33]; ARL15 (ADP-ribosylation factor-like 15) [33]; C4orf52 [34] [official gene symbol now SMIM20 (small integral membrane protein 20)], CPS1 [34], TMEM176A [34], OR4C46, and ADH5 [34]. The function of LILRA3 (leukocyte immunoglobulin-like receptor, subfamily A (without TM domain), member 3) [33] in HDL metabolism is not known, but its association with HDL-C has been confirmed in subsequent studies [152,153].

10. HDL-MODIFYING GENES ASSOCIATED WITH CAD

Large-scale epidemiological studies have long demonstrated an inverse relationship between HDL-C and CAD, yet demonstrating causality has remained elusive. Table 2(G) lists genes associated with both HDL-C and CAD. Among the many novel genes found by Teslovich et al. [33] to be associated with HDL-C as their primary trait, only IRS1, a major intracellular substrate of the IR tyrosine kinase, and KLF14, a master regulator of gene expression in adipose tissue [124], are statistically linked with CAD. Both genes have also been associated with diabetes and TG levels as a secondary trait, in the case of IRS1 [33], so it is difficult to assign causality to HDL-C even in these cases. Additionally, a novel SNP in apoA-I was also found to be associated with CAD, with TG as the lead trait and HDL as a secondary trait. As discussed above, mutations in apoA-I can cause both low HDL-C and CAD, so this is likely a causal association.

Other novel SNPs associated with CAD and with HDL-C as a trait secondary to other lipid parameters included LPL, TRIB1, and APOE [33]. C6orf106 had two SNPs, one associated with CAD but not HDL-C and one associated with HDL-C but not CAD [33]. Willer et al. confirmed the association between HDL-C levels and CAD events for IRS1 and KLF14 [34], and found additional genes associated with both CAD and HDL-C (CETP, CITED2, ZNF664, IKZF1, KAT5, RBM5, ABCA8, and GALNT2) but the HDL-C effect sizes were not large [34]. As pointed out by Kuivenhaven and Hegele [63], a small effect size for an SNP in the GWAS should not be taken as evidence that the gene does not play a major role in a metabolic pathway or disease.

11. SUMMARY

Genes that modify HDL-C were first identified from the study of classic genetic disorders. Although in many cases these genetic disorders are quite rare, such as Tangier disease, the discovery of the genetic basis of these disorders has led to the identification of key genes in HDL metabolism. This work has also led to the identification of targets for drug development, such as CETP. Although much has been accomplished by the classic genetic approach, in about half of patients with markedly decreased HDL-C, defects in the known genes for HDL-C do not appear to be the cause. Recent advances in high-throughput DNA sequencing, however, will likely lead to new HDL candidate genes, particularly when applied to families with genetic disorders in HDL metabolism.

Many potential gene candidates for modulating HDL-C have been revealed by GWAS. In fact, the pace of discovery of new genes by GWAS has outpaced our ability to confirm their role by complementary approaches, such as routine biochemical studies and testing their effect in genetic animal models. Additional complexity arises from the fact that GWAS identifies SNPS, not genes, so identifying the actual gene(s) in a GWAS locus causative for the associated phenotype poses further challenges. Finally, the number of genes associated with plasma HDL-C levels just below the threshold for genome-wide significance suggests that many additional loci may be discovered in future studies. The strongest novel GWAS gene candidates are those supported by multiple lines of evidence—independent association with more than one lipid and/or metabolic trait, plausible function according to pathway analysis and prior studies, changes in gene expression and chromatin and promoter markers consistent with effect size and direction, concordance with previous eQTL's, and so on [34].

Another ongoing challenge in the GWAS is the seemingly lack of association between many gene loci that affect HDL-C and CAD events [63] despite the abundant animal data showing a clear protective role of HDL in the pathogenesis of atherosclerosis [63]. In contrast to what has been found for HDL-C, a clear relationship between genes that regulated LDL-C and TG and CAD events have been found by the GWAS. This suggests that the strong association of HDL-C with CAD events from large epidemiology studies is possibly due to the known inverse linkage between triglyceride levels and HDL-C. In addition, mutations in genes such as LCAT may lower LDL-C as well as HDL-C, confounding interpretation of CAD association results. An additional important insight from a large GWAS is that diverse subphenotypes with different physiological effects result from different HDL-C-associated gene SNPs, and the specific effects of each SNP may confer differential effects on CAD [34]. More detailed genetic and phenotypic analysis of these subphenotypes could lead to functional groupings of HDL-associated variants, thus reconciling the results of genetic and epidemiological studies [34]. Additionally, for some genes, SNPs common enough to be found in GWAS may not have great functional consequences, but severe, rare mutations in the same genes, or an accumulation of SNPs with small HDL effects in multiple genes, might still be causative for CAD. Lastly, genetic factors are known to account for only half of the variation of HDL-C levels, so it may be that dietary and lifestyle factors or other diseases, such as inflammatory diseases, that affect HDL-C levels accounts for much of the inverse association found between HDL-C and CAD in epidemiological studies. Despite these concerns, the increased risk of atherosclerosis in familial ApoA-I deficiency underscores the importance of HDL in protecting against atherosclerosis, and genetics and genomics will continue to provide insight into the role of HDL in atherosclerosis.

A possible interpretation of the recent GWAS findings is that the cholesterol content of HDL is an inadequate measure of the anti-atherogenic function of HDL and that other compositional or functional tests, such as cholesterol efflux, should be used [1,78,154]. The resolution of some of these questions is not only vital for understanding HDL metabolism and the pathogenesis of CAD but, obviously, is also of great importance to the ongoing efforts for developing drugs that alter HDL metabolism.

Acknowledgment

This research was supported by the Intramural Research Program of the NIH, NHLBI.

References

[1] Toth PP, Barter PJ, Rosenson RS, et al. High-density lipoproteins: a consensus statement from the National Lipid Association. J Clin Lipidol 2013;7:484—525.

[2] Kontush A, Chapman MJ. High density lipoproteins: structure, metabolism, function and therapeutics [Chapter 1]. John Wiley & Sons, Inc; 2012. 3—38.

[3] Asztalos BF, Tani M, Schaefer EJ. Metabolic and functional relevance of HDL subspecies. Curr Opin Lipidol 2011;22:176—85.

[4] Gordon SM. Proteomic diversity in HDL: a driving force for particle function and target for therapeutic intervention. In: Komoda T, editor. The HDL handbook: biological functions and clinical implications. Walton (MA): Elsevier; 2013. p. 293—322.

[5] Wiesner P, Leidl K, Boettcher A, et al. Lipid profiling of FPLC-separated lipoprotein fractions by electrospray ionization tandem mass spectrometry. J Lipid Res 2009;50: 574—85.

[6] Yetukuri L, Soderlund S, Koivuniemi A, et al. Composition and lipid spatial distribution of HDL particles in subjects with low and high HDL-cholesterol. J Lipid Res 2010; 51:2341—51.

[7] Kontush A, Lhomme M, Chapman MJ. Unraveling the complexities of the HDL lipidome. J Lipid Res 2013;54:2950—63.

[8] Brooks-Wilson A, Marcil M, Clee SM, et al. Mutations in ABC1 in Tangier disease and familial high-density lipoprotein deficiency. Nat Genet 1999;22:336—45.

[9] Bodzioch M, Orso E, Klucken J, et al. The gene encoding ATP-binding cassette transporter 1 is mutated in Tangier disease. Nat Genet 1999;22:347—51.

[10] Rust S, Rosier M, Funke H, et al. Tangier disease is caused by mutations in the gene encoding ATP-binding cassette transporter 1. Nat Genet 1999;22:352—5.

[11] Lawn RM, Wade DP, Garvin MR, et al. The Tangier disease gene product ABC1 controls the cellular apolipoprotein-mediated lipid removal pathway. J Clin Invest 1999; 104:R25—31.

[12] Marcil M, Brooks-Wilson A, Clee SM, et al. Mutations in the ABC1 gene in familial HDL deficiency with defective cholesterol efflux. Lancet 1999;354:1341—6.

[13] Remaley AT, Rust S, Rosier M, et al. Human ATP-binding cassette transporter 1 (ABC1): genomic organization and identification of the genetic defect in the original Tangier disease kindred. Proc Natl Acad Sci USA 1999;96:12685—90.

[14] Rosenson RS, Brewer Jr HB, Davidson WS, et al. Cholesterol efflux and atheroprotection: advancing the concept of reverse cholesterol transport. Circulation 2012;125: 1905—19.

[15] Schwartz CC, VandenBroek JM, Cooper PS. Lipoprotein cholesteryl ester production, transfer, and output in vivo in humans. J Lipid Res 2004;45:1594—607.

[16] Santos RD, Asztalos BF, Martinez LR, et al. Clinical presentation, laboratory values, and coronary heart disease risk in marked high-density lipoprotein-deficiency states. J Clin Lipidol 2008;2:237—47.

[17] Fredrickson DS, Altrocchi PH, Avioli LV, Goodman DS, Goodman HC. Tangier disease. Ann Intern Med 1961;55:1016—20.

[18] Schaefer EJ, Santos RD, Asztalos BF. Marked HDL deficiency and premature coronary heart disease. Curr Opin Lipidol 2010;21:289—97.

[19] Puntoni M, Sbrana F, Bigazzi F, et al. Tangier disease: epidemiology, pathophysiology, and management. Am J Cardiovasc Drugs 2012;12:303—11.

[20] Fredrickson DS. The inheritance of high density lipoprotein deficiency (Tangier disease). J Clin Invest 1964;43:228—36.

[21] Luciani MF, Chimini G. The ATP binding cassette transporter ABC1 is required for the engulfment of corpses generated by apoptotic cell death. EMBO J 1996;15:226—35.

[22] Orso E, Broccardo C, Kaminski WE, et al. Transport of lipids from golgi to plasma membrane is defective in Tangier disease patients and Abc1-deficient mice. Nat Genet 2000;24:192—6.

[23] McNeish J, Aiello RJ, Guyot D, et al. High density lipoprotein deficiency and foam cell accumulation in mice with targeted disruption of ATP-binding cassette transporter-1. Proc Natl Acad Sci USA 2000;97:4245—50.

[24] Vaisman BL, Lambert G, Amar M, et al. ABCA1 overexpression leads to hyperalpha-lipoproteinemia and increased biliary cholesterol excretion in transgenic mice. J Clin Invest 2001;108:303—9.

[25] Cavelier LB, Qiu Y, Bielicki JK, et al. Regulation and activity of the human ABCA1 gene in transgenic mice. J Biol Chem 2001;276:18046—51.

[26] Singaraja RR, Bocher V, James ER, et al. Human ABCA1 BAC transgenic mice show increased high density lipoprotein cholesterol and ApoAI-dependent efflux stimulated by an internal promoter containing liver X receptor response elements in intron 1. J Biol Chem 2001;276:33969—79.

[27] Joyce C, Freeman L, Brewer Jr HB, et al. Study of ABCA1 function in transgenic mice. Arterioscler Thromb Vasc Biol 2003;23:965—71.

[28] Francis GA, Knopp RH, Oram JF. Defective removal of cellular cholesterol and phospholipids by apolipoprotein A-I in Tangier disease. J Clin Invest 1995;96:78—87.

[29] Remaley AT, Schumacher UK, Stonik JA, et al. Decreased reverse cholesterol transport from Tangier disease fibroblasts. Acceptor specificity and effect of brefeldin on lipid efflux. Arterioscler Thromb Vasc Biol 1997;17:1813—21.

[30] Remaley AT, Stonik JA, Demosky SJ, et al. Apolipoprotein specificity for lipid efflux by the human ABCAI transporter. Biochem Biophys Res Commun 2001;280:818—23.

[31] Neufeld EB, Stonik JA, Demosky Jr SJ, et al. The ABCA1 transporter modulates late endocytic trafficking: insights from the correction of the genetic defect in Tangier disease. J Biol Chem 2004;279:15571—8.

[32] Kathiresan S, Willer CJ, Peloso GM, et al. Common variants at 30 loci contribute to polygenic dyslipidemia. Nat Genet 2009;41:56—65.

[33] Teslovich TM, Musunuru K, Smith AV, et al. Biological, clinical and population relevance of 95 loci for blood lipids. Nature 2010;466:707—13.

[34] Willer CJ, Schmidt EM, Sengupta S, et al. Discovery and refinement of loci associated with lipid levels. Nat Genet 2013;45:1274—83.

[35] Frikke-Schmidt R, Nordestgaard BG, Stene MC, et al. Association of loss-of-function mutations in the ABCA1 gene with high-density lipoprotein cholesterol levels and risk of ischemic heart disease. JAMA 2008;299:2524—32.

[36] Haghpassand M, Bourassa PA, Francone OL, et al. Monocyte/macrophage expression of ABCA1 has minimal contribution to plasma HDL levels. J Clin Invest 2001;108: 1315—20.

[37] Aiello RJ, Brees D, Bourassa PA, et al. Increased atherosclerosis in hyperlipidemic mice with inactivation of ABCA1 in macrophages. Arterioscler Thromb Vasc Biol 2002;22: 630—7.

[38] Rohatgi A, Khera A, Berry JD, et al. HDL cholesterol efflux capacity and incident cardiovascular events. N Engl J Med 2014;371:2383—93.

[39] Hong C, Tontonoz P. Liver X receptors in lipid metabolism: opportunities for drug discovery. Nat Rev Drug Discov 2014;13:433—44.

[40] Sperry W. Cholesterol esterase in blood. J Biol Chem 1935;111:467—78.

[41] Glomset JA. The mechanism of the plasma cholesterol esterification reaction: plasma fatty acid transferase. Biochim Biophys Acta 1962;65:128—35.

[42] Glomset JA, Janssen ET, Kennedy R, et al. Role of plasma lecithin:cholesterol acyltransferase in the metabolism of high density lipoproteins. J Lipid Res 1966;7:638—48.

[43] Norum KR, Gjone E. Familial serum-cholesterol esterification failure. A new inborn error of metabolism. Biochim Biophys Acta 1967;144:698—700.

[44] Calabresi L, Simonelli S, Gomaraschi M, et al. Genetic lecithin:cholesterol acyltransferase deficiency and cardiovascular disease. Atherosclerosis 2012;222:299—306.

[45] Ahsan L, Ossoli AF, Freeman LA, Vaisman B, Amar MJ, Shamburek RD, et al. Role of lecithin: cholesterol acyltransferase in HDL metabolism and atherosclerosis. In: Komoda T, editor. The HDL handbook second edition: biological functions and clinical implications. Academic Press; 2014. p. 159—94.

[46] Duivenvoorden R, Holleboom AG, van den Bogaard B, et al. Carriers of lecithin cholesterol acyltransferase gene mutations have accelerated atherogenesis as assessed by carotid 3.0-T magnetic resonance imaging [corrected]. J Am Coll Cardiol 2011;58: 2481—7.

[47] van den Bogaard B, Holleboom AG, Duivenvoorden R, et al. Patients with low HDL-cholesterol caused by mutations in LCAT have increased arterial stiffness. Atherosclerosis 2012;225:481—5.

[48] Hoeg JM, Santamarina-Fojo S, Berard AM, et al. Overexpression of lecithin:cholesterol acyltransferase in transgenic rabbits prevents diet-induced atherosclerosis. Proc Natl Acad Sci USA 1996;93:11448—53.

[49] Haase CL, Tybjaerg-Hansen A, Qayyum AA, et al. LCAT, HDL cholesterol and ischemic cardiovascular disease: a Mendelian randomization study of HDL cholesterol in 54,500 individuals. J Clin Endocrinol Metab 2012;97:E248—56.

[50] Rousset X, Vaisman B, Auerbach B, et al. Effect of recombinant human lecithin cholesterol acyltransferase infusion on lipoprotein metabolism in mice. J Pharmacol Exp Ther 2010;335:140—8.

[51] Krause BR, Remaley AT. Reconstituted HDL for the acute treatment of acute coronary syndrome. Curr Opin Lipidol 2013;24:480—6.

[52] Shamburek R. Single-dose ACP-501 (rhLCAT) is well-tolerated, raises HDL-C in those with low HDL with CAD. In: Workshop on HDL structure-function Lake Buena Vista, FL, 2013; 2013.

[53] Chen Z, Wang SP, Krsmanovic ML, et al. Small molecule activation of lecithin cholesterol acyltransferase modulates lipoprotein metabolism in mice and hamsters. Metabolism 2012;61:470—81.

[54] Matsunaga A. Apolipoprotein A-I mutations and clinical investigation. In: Komoda T, editor. The HDL handbook, 2nd edition. Biological functions and clinical implications. Academic Press; 2014. p. 9—35.

[55] Anthanont P, Polisecki E, Asztalos BF, et al. A novel ApoA-I truncation (ApoA-IMytilene) associated with decreased ApoA-I production. Atherosclerosis 2014;235: 470—6.

[56] Traynor CA, Tighe D, O'Brien FJ, et al. Clinical and pathologic characteristics of hereditary apolipoprotein A-I amyloidosis in Ireland. Nephrology (Carlton) 2013;18: 549—54.

[57] Bielicki JK, Oda MN. Apolipoprotein A-I(Milano) and apolipoprotein A-I(Paris) exhibit an antioxidant activity distinct from that of wild-type apolipoprotein A-I. Biochemistry 2002;41:2089—96.

[58] Wong NC, Gordon A, Johansson J, et al. RVX-208 given orally to humans raises plasma ApoA-I and HDL in a phase 1b/2a clinical trial. In: 59th Annual Scientific Session of the American College of Cardiology; 2010.

[59] Bailey D, Jahagirdar R, Gordon A, et al. RVX-208: a small molecule that increases apolipoprotein A-I and high-density lipoprotein cholesterol in vitro and in vivo. J Am Coll Cardiol 2010;55:2580–9.

[60] Rvx 208. Drugs R D 2011;11:207–13.

[61] McLure KG, Gesner EM, Tsujikawa L, et al. RVX-208, an inducer of ApoA-I in humans, is a BET bromodomain antagonist. PLoS One 2013;8:e83190.

[62] Shaw JA, Bobik A, Murphy A, et al. Infusion of reconstituted high-density lipoprotein leads to acute changes in human atherosclerotic plaque. Circ Res 2008;103:1084–91.

[63] Kuivenhoven JA, Hegele RA. Mining the genome for lipid genes. Biochim Biophys Acta 2014;1842:1993–2009.

[64] Voight BF, Peloso GM, Orho-Melander M, et al. Plasma HDL cholesterol and risk of myocardial infarction: a mendelian randomisation study. Lancet 2012;380:572–80.

[65] Saito F. A pedigree of homozygous familial hyperalphalipoproteinemia. Metabolism 1984;33:629–33.

[66] Koizumi J, Mabuchi H, Yoshimura A, et al. Deficiency of serum cholesteryl-ester transfer activity in patients with familial hyperalphalipoproteinaemia. Atherosclerosis 1985;58:175–86.

[67] Brown ML, Inazu A, Hesler CB, et al. Molecular basis of lipid transfer protein deficiency in a family with increased high-density lipoproteins. Nature 1989;342:448–51.

[68] Inazu A, Brown ML, Hesler CB, et al. Increased high-density lipoprotein levels caused by a common cholesteryl-ester transfer protein gene mutation. N Engl J Med 1990;323: 1234–8.

[69] Takahashi K, Jiang XC, Sakai N, et al. A missense mutation in the cholesteryl ester transfer protein gene with possible dominant effects on plasma high density lipoproteins. J Clin Invest 1993;92:2060–4.

[70] Inazu A, Jiang XC, Haraki T, et al. Genetic cholesteryl ester transfer protein deficiency caused by two prevalent mutations as a major determinant of increased levels of high density lipoprotein cholesterol. J Clin Invest 1994;94:1872–82.

[71] Barzilai N, Atzmon G, Schechter C, et al. Unique lipoprotein phenotype and genotype associated with exceptional longevity. JAMA 2003;290:2030–40.

[72] Klerkx AH, Tanck MW, Kastelein JJ, et al. Haplotype analysis of the CETP gene: not TaqIB, but the closely linked -629C → A polymorphism and a novel promoter variant are independently associated with CETP concentration. Hum Mol Genet 2003;12: 111–23.

[73] Frisdal E, Klerkx AH, Le Goff W, et al. Functional interaction between -629C/A, -971G/A and -1337C/T polymorphisms in the CETP gene is a major determinant of promoter activity and plasma CETP concentration in the REGRESS Study. Hum Mol Genet 2005;14:2607–18.

[74] Kuivenhoven JA, de Knijff P, Boer JM, et al. Heterogeneity at the CETP gene locus. Influence on plasma CETP concentrations and HDL cholesterol levels. Arterioscler Thromb Vasc Biol 1997;17:560–8.

[75] Brewer Jr HB. High-density lipoproteins: a new potential therapeutic target for the prevention of cardiovascular disease. Arterioscler Thromb Vasc Biol 2004;24:387–91.

[76] Mohrschladt MF, van der Sman-de Beer F, Hofman MK, et al. TaqIB polymorphism in CETP gene: the influence on incidence of cardiovascular disease in statin-treated patients with familial hypercholesterolemia. Eur J Hum Genet 2005;13:877–82.

[77] Durlach A, Clavel C, Girard-Globa A, et al. Sex-dependent association of a genetic polymorphism of cholesteryl ester transfer protein with high-density lipoprotein cholesterol and macrovascular pathology in type II diabetic patients. J Clin Endocrinol Metab 1999;84:3656–9.

[78] Kingwell BA, Chapman MJ, Kontush A, et al. HDL-targeted therapies: progress, failures and future. Nat Rev Drug Discov 2014;13:445–64.

[79] Marschang P, Sandhofer A, Ritsch A, et al. Plasma cholesteryl ester transfer protein concentrations predict cardiovascular events in patients with coronary artery disease treated with pravastatin. J Intern Med 2006;260:151–9.

[80] Vasan RS, Pencina MJ, Robins SJ, et al. Association of circulating cholesteryl ester transfer protein activity with incidence of cardiovascular disease in the community. Circulation 2009;120:2414–20.

[81] Khera AV, Wolfe ML, Cannon CP, et al. On-statin cholesteryl ester transfer protein mass and risk of recurrent coronary events (from the pravastatin or atorvastatin evaluation and infection therapy-thrombolysis in myocardial infarction 22 [PROVE IT-TIMI 22] study). Am J Cardiol 2010;106:451–6.

[82] Duwensee K, Breitling LP, Tancevski I, et al. Cholesteryl ester transfer protein in patients with coronary heart disease. Eur J Clin Invest 2010;40:616–22.

[83] Ritsch A, Scharnagl H, Eller P, et al. Cholesteryl ester transfer protein and mortality in patients undergoing coronary angiography: the Ludwigshafen Risk and Cardiovascular Health study. Circulation 2010;121:366–74.

[84] Robins SJ, Lyass A, Brocia RW, et al. Plasma lipid transfer proteins and cardiovascular disease. The Framingham Heart Study. Atherosclerosis 2013;228:230–6.

[85] Boekholdt SM, Kuivenhoven JA, Wareham NJ, et al. Plasma levels of cholesteryl ester transfer protein and the risk of future coronary artery disease in apparently healthy men and women: the prospective EPIC (European Prospective Investigation into Cancer and nutrition)-Norfolk population study. Circulation 2004;110:1418–23.

[86] Borggreve SE, Hillege HL, Dallinga-Thie GM, et al. High plasma cholesteryl ester transfer protein levels may favour reduced incidence of cardiovascular events in men with low triglycerides. Eur Heart J 2007;28:1012–8.

[87] Kappelle PJ, Perton F, Hillege HL, et al. High plasma cholesteryl ester transfer but not CETP mass predicts incident cardiovascular disease: a nested case-control study. Atherosclerosis 2011;217:249–52.

[88] Schierer A, Been LF, Ralhan S, et al. Genetic variation in cholesterol ester transfer protein, serum CETP activity, and coronary artery disease risk in Asian Indian diabetic cohort. Pharmacogenet Genomics 2012;22:95–104.

[89] Johannsen TH, Frikke-Schmidt R, Schou J, et al. Genetic inhibition of CETP, ischemic vascular disease and mortality, and possible adverse effects. J Am Coll Cardiol 2012; 60:2041–8.

[90] Rader DJ, deGoma EM. Future of cholesteryl ester transfer protein inhibitors. Annu Rev Med 2014;65:385–403.

[91] Lim GB. Lipids. Dalcetrapib raises HDL-cholesterol level, but does not reduce cardiac risk. Nat Rev Cardiol 2013;10:5.

[92] Barter PJ, Caulfield M, Eriksson M, et al. Effects of torcetrapib in patients at high risk for coronary events. N Engl J Med 2007;357:2109–22.

[93] Acton S, Rigotti A, Landschulz KT, et al. Identification of scavenger receptor SR-BI as a high density lipoprotein receptor. Science 1996;271:518–20.

[94] Murao K, Terpstra V, Green SR, et al. Characterization of CLA-1, a human homologue of rodent scavenger receptor BI, as a receptor for high density lipoprotein and apoptotic thymocytes. J Biol Chem 1997;272:17551–7.

[95] Wang N, Arai T, Ji Y, et al. Liver-specific overexpression of scavenger receptor BI decreases levels of very low density lipoprotein ApoB, low density lipoprotein ApoB, and high density lipoprotein in transgenic mice. J Biol Chem 1998;273: 32920–6.

[96] Rhainds D, Brissette L. The role of scavenger receptor class B type I (SR-BI) in lipid trafficking. defining the rules for lipid traders. Int J Biochem Cell Biol 2004;36:39–77.

[97] Yang XP, Amar MJ, Vaisman B, et al. Scavenger receptor-BI is a receptor for lipoprotein(a). J Lipid Res 2013;54:2450–7.

[98] Rigotti A, Trigatti BL, Penman M, et al. A targeted mutation in the murine gene encoding the high density lipoprotein (HDL) receptor scavenger receptor class B type I reveals its key role in HDL metabolism. Proc Natl Acad Sci USA 1997;94: 12610–5.

[99] Trigatti B, Rayburn H, Vinals M, et al. Influence of the high density lipoprotein receptor SR-BI on reproductive and cardiovascular pathophysiology. Proc Natl Acad Sci USA 1999;96:9322–7.

[100] Acton S, Osgood D, Donoghue M, et al. Association of polymorphisms at the SR-BI gene locus with plasma lipid levels and body mass index in a white population. Arterioscler Thromb Vasc Biol 1999;19:1734–43.

[101] Roberts CG, Shen H, Mitchell BD, et al. Variants in scavenger receptor class B type I gene are associated with HDL cholesterol levels in younger women. Hum Hered 2007;64:107–13.

[102] West M, Greason E, Kolmakova A, et al. Scavenger receptor class B type I protein as an independent predictor of high-density lipoprotein cholesterol levels in subjects with hyperalphalipoproteinemia. J Clin Endocrinol Metab 2009;94:1451–7.

[103] Tai ES, Adiconis X, Ordovas JM, et al. Polymorphisms at the SRBI locus are associated with lipoprotein levels in subjects with heterozygous familial hypercholesterolemia. Clin Genet 2003;63:53–8.

[104] Vergeer M, Korporaal SJ, Franssen R, et al. Genetic variant of the scavenger receptor BI in humans. N Engl J Med 2011;364:136–45.

[105] Brunham LR, Tietjen I, Bochem AE, et al. Novel mutations in scavenger receptor BI associated with high HDL cholesterol in humans. Clin Genet 2011;79:575–81.

[106] Naj AC, West M, Rich SS, et al. Association of scavenger receptor class B type I polymorphisms with subclinical atherosclerosis: the Multi-Ethnic Study of Atherosclerosis. Circ Cardiovasc Genet 2010;3:47–52.

[107] Manichaikul A, Naj AC, Herrington D, et al. Association of SCARB1 variants with subclinical atherosclerosis and incident cardiovascular disease: the multi-ethnic study of atherosclerosis. Arterioscler Thromb Vasc Biol 2012;32:1991–9.

[108] Yates M, Kolmakova A, Zhao Y, et al. Clinical impact of scavenger receptor class B type I gene polymorphisms on human female fertility. Hum Reprod 2011;26:1910–6.

[109] Nieland TJ, Penman M, Dori L, et al. Discovery of chemical inhibitors of the selective transfer of lipids mediated by the HDL receptor SR-BI. Proc Natl Acad Sci USA 2002; 99:15422–7.

[110] Faloon PW, Dockendorff C, Yu M, et al. A small molecule inhibitor of scavenger receptor BI-mediated lipid uptake – probe 3. 2010.

[111] Faloon PW, Dockendorff C, Youngsaye W, et al. A small molecule inhibitor of scavenger receptor BI-mediated lipid uptake – probe 1. 2010.

[112] Faloon PW, Dockendorff C, Germain A, et al. A small molecule inhibitor of scavenger receptor BI-mediated lipid uptake – probe 2. 2010.

[113] Masson D, Koseki M, Ishibashi M, et al. Increased HDL cholesterol and apoA-I in humans and mice treated with a novel SR-BI inhibitor. Arterioscler Thromb Vasc Biol 2009;29:2054–60.

[114] Miettinen HE, Rayburn H, Krieger M. Abnormal lipoprotein metabolism and reversible female infertility in HDL receptor (SR-BI)-deficient mice. J Clin Invest 2001;108: 1717–22.

[115] D'Andrea LD, Regan L. TPR proteins: the versatile helix. Trends Biochem Sci 2003;28: 655–62.

[116] Allan RK, Ratajczak T. Versatile TPR domains accommodate different modes of target protein recognition and function. Cell Stress Chaperones 2011;16:353–67.

[117] Quagliarini F, Wang Y, Kozlitina J, et al. Atypical angiopoietin-like protein that regulates ANGPTL3. Proc Natl Acad Sci USA 2012;109:19751–6.

[118] Gusarova V, Alexa CA, Na E, et al. ANGPTL8/betatrophin does not control pancreatic beta cell expansion. Cell 2014;159:691–6.
[119] Kaestner KH. Betatrophin-promises fading and lessons learned. Cell Metab 2014;20: 932–3.
[120] Wenzel DM, Lissounov A, Brzovic PS, et al. UBCH7 reactivity profile reveals parkin and HHARI to be RING/HECT hybrids. Nature 2011;474:105–8.
[121] Kim KY, Stevens MV, Akter MH, et al. Parkin is a lipid-responsive regulator of fat uptake in mice and mutant human cells. J Clin Invest 2011;121:3701–12.
[122] Kim WS, Hsiao JH, Bhatia S, et al. ABCA8 stimulates sphingomyelin production in oligodendrocytes. Biochem J 2013;452:401–10.
[123] Lu Y, Tian QB, Endo S, et al. A role for LRP4 in neuronal cell viability is related to apoE-binding. Brain Res 2007;1177:19–28.
[124] Small KS, Hedman AK, Grundberg E, et al. Identification of an imprinted master trans regulator at the KLF14 locus related to multiple metabolic phenotypes. Nat Genet 2011;43:561–4.
[125] Kang L, Lantier L, Kennedy A, et al. Hyaluronan accumulates with high-fat feeding and contributes to insulin resistance. Diabetes 2013;62:1888–96.
[126] Manning AK, Hivert MF, Scott RA, et al. A genome-wide approach accounting for body mass index identifies genetic variants influencing fasting glycemic traits and insulin resistance. Nat Genet 2012;44:659–69.
[127] Mancina RM, Burza MA, Maglio C, et al. The COBLL1 C allele is associated with lower serum insulin levels and lower insulin resistance in overweight and obese children. Diabetes Metab Res Rev 2013;29:413–6.
[128] Kebede MA, Attie AD. Insights into obesity and diabetes at the intersection of mouse and human genetics. Trends Endocrinol Metab 2014;25:493–501.
[129] Chen Y, Zhu J, Lum PY, et al. Variations in DNA elucidate molecular networks that cause disease. Nature 2008;452:429–35.
[130] Wu Y, Gao H, Li H, et al. A meta-analysis of genome-wide association studies for adiponectin levels in East Asians identifies a novel locus near WDR11-FGFR2. Hum Mol Genet 2014;23:1108–19.
[131] Liu MJ, Bao S, Galvez-Peralta M, et al. ZIP8 regulates host defense through zinc-mediated inhibition of NF-κB. Cell Rep 2013;3:386–400.
[132] Waterworth DM, Ricketts SL, Song K, et al. Genetic variants influencing circulating lipid levels and risk of coronary artery disease. Arterioscler Thromb Vasc Biol 2010; 30:2264–76.
[133] Neufeld EB, Remaley AT, Demosky SJ, et al. Cellular localization and trafficking of the human ABCA1 transporter. J Biol Chem 2001;276:27584–90.
[134] Neufeld EB, Demosky Jr SJ, Stonik JA, et al. The ABCA1 transporter functions on the basolateral surface of hepatocytes. Biochem Biophys Res Commun 2002;297: 974–9.
[135] Santamarina-Fojo S, Remaley AT, Neufeld EB, et al. Regulation and intracellular trafficking of the ABCA1 transporter. J Lipid Res 2001;42:1339–45.
[136] Wang S, Smith JD. ABCA1 and nascent HDL biogenesis. Biofactors 2014;40:547–54.
[137] van der Kant R, Zondervan I, Janssen L, et al. Cholesterol-binding molecules MLN64 and ORP1L mark distinct late endosomes with transporters ABCA3 and NPC1. J Lipid Res 2013;54:2153–65.
[138] Li J, Li C, Zhang D, et al. SNX13 reduction mediates heart failure through degradative sorting of apoptosis repressor with caspase recruitment domain. Nat Commun 2014;5: 5177.
[139] Bartee E, Mansouri M, Hovey Nerenberg BT, et al. Downregulation of major histocompatibility complex class I by human ubiquitin ligases related to viral immune evasion proteins. J Virol 2004;78:1109–20.

[140] Goto E, Ishido S, Sato Y, et al. c-MIR, a human E3 ubiquitin ligase, is a functional homolog of herpesvirus proteins MIR1 and MIR2 and has similar activity. J Biol Chem 2003;278:14657—68.

[141] Fujita H, Iwabu Y, Tokunaga K, et al. Membrane-associated RING-CH (MARCH) 8 mediates the ubiquitination and lysosomal degradation of the transferrin receptor. J Cell Sci 2013;126:2798—809.

[142] Apaja PM, Lukacs GL. Protein homeostasis at the plasma membrane. Physiology (Bethesda) 2014;29:265—77.

[143] Moraitis AG, Freeman LA, Shamburek RD, Wesley R, Wilson W, Grant CM, et al. Elevated interleukin-10: a new cause of dyslipidemia leading to severe HDL deficiency. J Clin Lipidol, in press.

[144] Cao G, Garcia CK, Wyne KL, et al. Structure and localization of the human gene encoding SR-BI/CLA-1. Evidence for transcriptional control by steroidogenic factor 1. J Biol Chem 1997;272:33068—76.

[145] Ferraz-de-Souza B, Hudson-Davies RE, Lin L, et al. Sterol O-acyltransferase 1 (SOAT1, ACAT) is a novel target of steroidogenic factor-1 (SF-1, NR5A1, Ad4BP) in the human adrenal. J Clin Endocrinol Metab 2011;96:E663—8.

[146] Ishizuka Y, Nakayama K, Ogawa A, et al. TRIB1 downregulates hepatic lipogenesis and glycogenesis via multiple molecular interactions. J Mol Endocrinol 2014;52: 145—58.

[147] Harmon BT, Devaney SA, Gordish-Dressman H, et al. Functional characterization of a haplotype in the AKT1 gene associated with glucose homeostasis and metabolic syndrome. Hum Genet 2010;128:635—45.

[148] Bonnal S, Martinez C, Forch P, et al. RBM5/Luca-15/H37 regulates Fas alternative splice site pairing after exon definition. Mol Cell 2008;32:81—95.

[149] Dugast E, Kiss-Toth E, Soulillou JP, et al. The Tribbles-1 protein in humans: roles and functions in health and disease. Curr Mol Med 2013;13:80—5.

[150] Angyal A, Kiss-Toth E. The tribbles gene family and lipoprotein metabolism. Curr Opin Lipidol 2012;23:122—6.

[151] Satoh T, Kidoya H, Naito H, et al. Critical role of Trib1 in differentiation of tissue-resident M2-like macrophages. Nature 2013;495:524—8.

[152] Singaraja RR, Tietjen I, Hovingh GK, et al. Identification of four novel genes contributing to familial elevated plasma HDL cholesterol in humans. J Lipid Res 2014;55: 1693—701.

[153] Edmondson AC, Braund PS, Stylianou IM, et al. Dense genotyping of candidate gene loci identifies variants associated with high-density lipoprotein cholesterol. Circ Cardiovasc Genet 2011;4:145—55.

[154] Balder JW, Staels B, Kuivenhoven JA. Pharmacological interventions in human HDL metabolism. Curr Opin Lipidol 2013;24:500—9.

The Genetics of Obesity

Rodrigo Alonso, Magdalena Farías, Veronica Alvarez, Ada Cuevas

Obesity and Lipid Units, Department of Nutrition, Clínica Las Condes, Santiago de Chile, Chile

1. INTRODUCTION

Obesity is a chronic, serious, and costly disease. It is associated with an increased risk of several co-morbidities, including type 2 diabetes mellitus (T2DM), high blood pressure, cardiovascular disease, dyslipidemia, fatty liver disease, and some types of cancer [1]. The rate of obesity in the world has nearly doubled since 1980, especially in developing countries [2]. The World Health Organization estimates that more than 1.4 billion of individuals aged 20 years or more are overweight (body mass index, BMI $25-29.9 \, kg/m^2$), and more than 500 million are obese (BMI over $30 \, kg/m^2$). The percentage of children classified as overweight and obese is rising rapidly, especially in low- and middle-income countries, reaching 40 million in 2012 [3].

The pathogenesis of obesity is complex, resulting from an imbalance of energy intake and energy expenditure, and caused by a complex interaction of genes, environmental factors, and human behavior; however, the relative contribution of any of these factors is still not completely understood. Some of the most important environmental factors related with obesity are low physical activity, and the availability and accessibility to high-energy low-cost foods. However, a wide degree of variability in the susceptibility to developing obesity is observed among subjects, or communities exposed to the same environmental factors, suggesting that genetic factors likely play a role in this variability. There are some hypotheses that try to explain the contribution of genetic factors to obesity. Genes are involved in some way in how the energy is captured, stored, and released in the body. According to "thrifty gene" hypothesis, there are individuals with a genetic predisposition to accumulate energy in periods

161

of food availability, and these individuals are more able to survive in periods of food scarcity [4,5]. This "thrifty gene" hypothesis to accumulate fat and use the calories more efficiently was probably an advantage in primitive populations, and currently can be considered a maladaptive process in individuals with high intake of calories and sedentary life style that characterize our current modern society. One of the main criticisms of this hypothesis is that it cannot explain the high heterogeneity of obesity in different populations and within the same population [6]. Determining the genetic basis of a major susceptibility to obesity may help to understand the different underlying mechanisms involved in the pathogenesis of this disorder and, therefore, to improve preventive measurements and to develop new treatments. The genetic contribution to body weight has been established initially through family studies, twins, and parent–offspring relationships studies. According to these studies, it has been estimated that in a population, genetic factors may explain 40–70% of the variation in BMI in a population [7–11].

Progress in the identification of genes related to obesity has not been an easy task. It has been based on two different genetic epidemiological analyses, the candidate gene approach and genome-wide approach (GWAS) including linkage and association studies.

The candidate gene approach is based on the understanding and knowledge of the physiological role of an identified or candidate gene in pathways related to food intake and energy expenditure and in the pathophysiology of the disease. Candidate genes are selected on the basis of previous evidence from animal models, cellular model systems, and linkage studies in extreme cases [12]. However, since the pathophysiological mechanisms underlying obesity are generally still not well known, the use of the hypothesis-driven candidate gene association approach will identify only a relatively small fraction of the genetic risk factors for the disease.

On the other hand, in the genome-wide linkage studies, the analysis of the whole genome should identify new variants associated with a disease. In this approach, a susceptibility gene is discovered because it resides on a chromosomal region that segregates with a specific phenotype [13]. In the Human Obesity Gene Map update from 2005, 15 from 253 loci replicated in more than three studies using this approach [14]; however, a subsequent meta-analysis could not locate any single locus related with obesity or BMI with convincing evidence, indicating that this approach might not always be effective in identifying genetic variants associated with common obesity [15]. On the other hand, GWAS is an approach to look for associations between many specific variations (commonly, intronic single nucleotide polymorphisms, SNPs) and particular traits or diseases. The identification of any locus as part of a GWAS depends upon its effect size and prevalence of the risk allele. Therefore, the loci that have smaller effect size or lower allele frequency in the population will require larger studies in order to detect them. In the GWAS, the SNPs that show the highest level of association with the phenotype of interest are then tested for association in a new population with at least a similar size and design. The whole genome is screened at higher resolution levels compared to linkage studies and has replaced the linkage studies for common diseases like obesity in unrelated individuals [13].

The discovery of some monogenic forms of early-onset and severe obesity has contributed to the knowledge of different mechanisms that regulate body weight. However, the increase in the prevalence of obesity in the past 20 years seems to result from a more complex interaction of lifestyle factors and a polygenic background determining the suscepti-bility of an individual to gain weight.

2. LEPTIN AND HYPOTHALAMIC CONTROL OF APPETITE AND ENERGY REGULATION (FIGURE 1)

Leptin is a 167-amino acid produced predominantly in adipose tissue. Leptin binds to the leptin receptor (LEPR) located throughout the central nervous system, especially in the hypothalamus. There are at least six isoforms of LEPR, some of them involved in the transport of leptin across

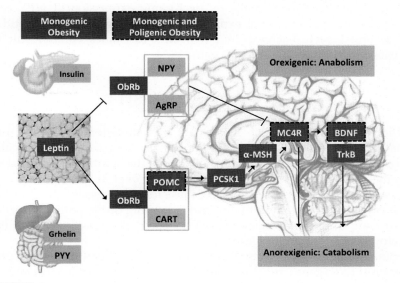

FIGURE 1

ObRb, Long Leptin receptor; AgRP, agouti-related protein; *BDNF*, brain-derived neurotro-phic factor; CART, cocaine-and amphetamine—related transcript; MC4R, melanocortin-4 receptor; α-MSH, α-melanocyte-stimulating hormone; NYY, neuropeptide Y; POMC, pro-opiomelanocortin; PCSK1, proprotein convertase 1; PYY, peptide YY.

the blood—brain barrier and others responsible for leptin signaling in the hypothalamus where it regulates energy homeostasis [16].

The hypothalamus plays an important role in the regulation of body weight, through afferent signals from peripheral issues (i.e., adipose tissue) and efferent signals, balancing the intake of food, energy expenditure, and fat storage [17]. Some hypothalamic neurons produce orexigenic peptides such as agouti-related protein (AgRP) and neuropeptide Y (NPY) and others produce anorexigenic peptides such as pro-opiomelanocortin (POMC) and cocaine- and amphetamine-related transcript (CART). Leptin is produced in adipose tissue, and the binding of this hormone to its receptor, stimulates the expression of CART and POMC, and suppresses the expression of NPY and AgRP in the hypothalamus. Pro-opiomelanocortin is then transformed by means of pro-protein convertase 1 (PCSK1) into α-melanocyte-stimulating hormone (α-MSH), acting as an agonist of the melanocortin-4 receptor (MC4R). The stimulation of this receptor leads to an anorexigenic state reducing appetite, food intake, and increasing energy expenditure due to an enhanced sympathetic tone and release of some pituitary hormones controlling adipose tissue metabolism and metabolic rate [18]. Besides leptin released from adipose tissue, other signals affecting the anorexigenic center in the hypothalamus are insulin from the pancreas, and ghrelin and peptide YY from the gastrointestinal tract. Insulin has been shown to decrease food intake by decreasing expression of NPY and AgRP, and by increasing the expression of anorexigenic neuropeptides POMC and CART. Therefore, brain insulin resistance could result in increased food intake and weight gain [18A].

3. MONOGENIC OBESITY

The monogenic forms of obesity are rare disorders resulting from mutations in a single gene or chromosomal region, and are typically characterized by an early onset (usually before the age of 10 years) and an extreme obesity (BMI standard deviation score over 3) [13,14,19,20].

Despite the very low prevalence of these forms of obesity, the study of these cases has contributed to our understanding of the pathogenesis of obesity and to the knowledge of some of the pathways involved in the development of this disorder (Table 1).

Some examples are obesity due to mutations in leptin (LEP) and leptin receptor (LEPR) genes, resulting in a rapid weight gain since birth and an intense hyperphagia. The LEP gene was first characterized in mouse in 1994 and LEPR gene in 1996 [21,22]. The first report of severe obesity associated with undetectable levels of serum leptin produced by a severe mutation in LEP gene appeared in 1997 [23]. One year later, the first

TABLE 1   Genes Associated with Human Monogenic Obesity

Gene	Encoded protein	Function	Location
LEP	Leptin	Hormone producing satiety	7q31.1
LEPR	Leptin receptor	Receptor for leptin	1p31
MC4R	Melanocortin 4 receptor	Receptor for α-MSH	18q22
POMC	Pro-opiomelanocortin	Anorexigenic hypothalamic peptide	2p23.3
PCSK1	Proprotein convertase 1	Related to conversion of POMC into α-MSH	5q15-21

mutation in *LEPR* gene was described [24]. These cases due to *LEPR* mutations represent 3% of subjects with hyperphagia and severe early-onset obesity [25].

In some cases of severe obesity, treatment with recombinant leptin has shown beneficial effects in normalizing appetite, and reducing body weight and fat mass [26]. Other examples of monogenic obesity are caused by mutations in human *MC4R*, *POMC*, and *PCSK1* genes, among others. All of them are characterized by hyperphagia and are usually associated with other additional phenotypic and endocrine abnormalities [27,28]. Mutations in *MC4R* gene, associated with dominantly inherited obesity were first described in 1998 by two different groups [29,30]. Currently, more than 160 functional mutations (no synonymous, nonsense, deletion, or frame shift mutations) have been described in *MC4R* genes; most of them were identified in extremely obese individuals. The prevalence of *MC4R* mutations varies from 0.2% to 6%. It is still not clear if the penetrance of the heterozygous condition is 100% because it has been described that some carriers of heterozygous mutation were not obese [31,32].

The analysis of 284 subjects including extreme obese children and their parents had shown that mutations in *MC4R* increased BMI by 14% ($4 \, kg/m^2$) and 36% ($9.5 \, kg/m^2$) in males and females, respectively, compared to their family members without mutations [33]. In vitro studies have suggested the potential benefit of two specific MC4R agonists activating the mutated receptor that depended on the functional mutation [34].

4. COMMON OBESITY

The most common form of obesity is caused by the contribution and interaction of both environmental factors and multiple genes. In this sense, the "obesogenic" environment (sedentary lifestyle and high caloric intake) is shared by many individuals; however, those subjects who have a genetic predisposition to gain weight will be more sensitive to the environmental factors and will probably become obese. The genetic predisposition to common obesity is complex and results from the simultaneous presence of DNA variations in multiple genes. The combined effect of multiple variants in different genes, some of them previously described in monogenic obesity, on a quantitative trait are known as polygenic effect. It is generally assumed that each allele associated with an obese phenotype has a small effect on the trait, and the combination of the variants is additive. Many variants are known to participate in body weight regulation, and it is estimated that the total number of genes with small effect exceeds 120 [14]. If an individual has many variants that increase body weight, obesity could develop. A polygenic basis of obesity implies that the specific set of polygenic variants relevant for obesity in one individual is unlikely to be the same in another subject. There is an individual difference in the vulnerability to obesity under the same environmental exposure [35].

The candidate gene approach and GWAS have identified at least 50 loci consistently associated with BMI and obesity. Some of the most frequent SNPs explaining higher BMI variance are shown in Table 2. Before GWAS, most candidate gene studies of common obesity looked at variants in genes already implicated in the monogenic forms of the disease discussed above with a limited success. Some examples of those genes associated with BMI analyzed through candidate gene approach are *LEP* and *LEPR*, cannabinoid receptor 1 (*CNR1*), and the Pro12Ala substitution in the peroxisome proliferator-activated receptor-gamma (*PPAR-γ*), among others [36]. Another interesting gene acting as an environmental sensor is the gene encoding G-protein coupled receptor 120 (*GPR120*). This protein is a trans membrane receptor for unsaturated long-chain free fatty acids with an important role regulating energy homeostasis. In human, it has been shown that GPR120 expression in adipose tissue is significantly higher in obese individuals than in lean controls and, on the other hand, the p.R270H variant in *GPR120* gene increases the risk of obesity in European populations [37].

To date, genetic variations of fat mass and obesity (*FTO*) gene (16q12.2), first described in 2007, are the strongest predictors of human polygenic obesity [38,39]. In the initial report of Frayling et al., the genetic variation rs9939609 in *FTO* gene was significantly associated with T2DM in Europeans (1924 T2DM subjects and 2938 population-based controls), but after the adjustment with BMI, this association disappeared, indicating that the effect of *FTO* on T2DM was dependent on its effect on BMI [38]. This association with BMI was subsequently replicated in several studies including more than 19,000 adults and more than 10,000 children,

TABLE 2   Some Single Nucleotide Polymorphisms Associated with BMI at GWAS Significance Level

Gene associated to SNP	Location	SNP	Frequency effect allele (%)	Explained variance in BMI (%)	Increase in BMI (kg/m$^{2)}$ for each additional risk allele
FTO	16q12.2	rs1558902	42	0.34	0.39
TMEM18	2p25.3	rs2867125	83	0.15	0.31
MC4R40	18q21.32	rs571312	24	0.10	0.23
GNPDA2	4p12	rs10938397	43	0.08	0.18
BDNF	11p14.1	rs10767664	78	0.07	0.19
SEC16B[a]	1q25.2	rs543874	19	0.07	0.22
RBJ[b]	2p23.3	Rs713586	47	0.06	0.14
SH2B1	16p11.2	Rs7359397	40	0.05	0.15
NEGR1	1p31.1	rs2815752	61	0.04	0.13

[a]Also known associated with waist and weight.
[b]New loci identified associated to BMI.
From Speliotes et al. [53].

suggesting that *FTO* variants affect T2DM through its effect on adiposity [40]. However, since then, different meta-analysis has suggested that the *FTO* gene variant influenced the risk of T2DM independently of BMI in both Europeans and East Asians [41,42].

The exact mechanisms by which *FTO* contributes to obesity are still not well known. *FTO* has been identified as an RNA-modifying enzyme, demethylating N6-methyladenosine on single-stranded RNA. Studies in animals showed that mice without *FTO* have reduced fat mass and on the other hand, overexpression of *FTO* in mice increases energy intake and adiposity [43].

Since the discovery of its association with variants in intron 1 of the *FTO* gene, many studies have been conducted to identify new variants in the gene, and to clarify the biological function of the protein and intend to explain its role in the origin and development of obesity [44,45]. The association of *FTO* variants and obesity has been observed in different populations [46—48] and among the identified variants, the rs9939609 is the one most strongly associated with common obesity [49,50]. Other SNPs located at intron 1 in the *FTO* gene (rs9935401, rs9928094, and rs9930333) have also been associated with obesity in children [50,51], and on the other hand, an SNP (rs8061518) located at intron 3 in *FTO* gene associated with a reduced risk of obesity and low plasma of leptin has been described [52]. Each minor allele of any commonly investigated variant in *FTO* increases BMI by 0.30–0.40 kg/m^2 and risk of obesity by 20% [21,53], with the effect estimate being smaller (0.26 kg/m^2) in East and South Asians [41].

A recent meta-analysis comprising more than 19,000 cases with cardiovascular disease and more than 103,000 control subjects also showed that the rs9939609 variant is significantly associated with cardiovascular disease risk (odds ratio 1.18, 95%, confidence interval = 1.07–1.30, $P = 0.001$) independent of BMI and other classical risk factors [52].

After the discovery of *FTO* gene variants, research focused on the detection of variants in other genes that contributed to adiposity [53]. A meta-analysis of data from four European population-based studies ($n = 11,012$) and three disease-case series ($n = 5864$) not only confirmed the role of *FTO* locus, but also an SNP near the *MC4R* gene, previously implicated in the development of early-onset obesity, was convincingly replicated in larger studies with different populations [54,55].

A previous study has shown that carriers for the missense variant V103I in the *MC4R* gene had a 31% reduction in risk for obesity [56]. A recent meta-analysis of 10,975 cases and 18,588 controls confirmed this result, although the risk reduction was lower (18%) [57]; the mechanisms underlying this association is not totally understood. Some studies suggest that this variant is not functional, and other functional studies showed that *MC4R* with this variant is less responsive to AGRP, an endogenous antagonist of the MC4R, leading to a weaker orexigenic signal compared to wild-type receptor [58].

With respect to PCSK1, an initial meta-analysis comprising a total of 13,659 individuals of European ancestry from seven case–control studies and one family study, showed a highly significant association between the nonsynonymous variants rs6232 and rs6235 in the *PCSK1* gene and the increased risk of obesity (1.32- and 1.22-fold, respectively) [59]. Later studies with other populations, showed no association in the general population or only a modest contribution with an effect that seems to be age dependent for the rs6232 variant [60,61].

With the main idea to increase the identification of new alleles associated with variations in BMI, obesity and other anthropometric traits, the Genetic Investigation of Anthropometric Traits (GIANT) consortium was created to facilitate large-scale meta-analysis of data from multiple GWA studies. In 2009, a meta-analysis including 32,387 subjects of European origin from 15 cohorts identified six new loci with genome-wide significance ($P < 5 \times 10^{-8}$) in or near transmembrane protein 18 (*TMEM18*), potassium channel tetramerization domain containing 15 (*KCTD15*), glucosamine-6-phosphate deaminase 2 (*GNPDA2*), SH2B adaptor protein 1 (*SH2B1*), mitochondrial carrier 2 (*MTCH2*), and neuronal growth regulator 1 (*NEGR1*) [62,63]. Many of these genes are expressed in the central nervous system, emphasizing the neuronal component in the predisposition to obesity [47]. These SNPs were replicated in another study including subjects from European (Icelandic and Dutch), European American and African-American origin, and other loci such as brain-derived neurotrophic factor (*BDNF*) and the transcription factor ETS variant gene (*ETV5*) were identified with strong evidence for association [63]. This is consistent with results in animal models and studies of monogenic forms of human obesity, where neuronal genes expressed in the hypothalamus are involved in regulation of appetite or energy balance. SH2B1 is a cytoplasmic adaptor protein that binds to a variety of protein tyrosine kinases (JAK 2 and insulin receptor), and it has been shown that SH2B1 is expressed in many tissues, mainly in the brain and hypothalamus. SH2B1-knockout mice have leptin resistance, hyperphagia, obesity, and insulin resistance, and the restoration of SH2B1 corrects all metabolic disorders and improves leptin signaling and leptin regulation of the orexigenic neuropeptide expression in the hypothalamus [64].

To find more loci associated with BMI, the GIANT Consortium was expanded using a two-stage GWA meta-analysis of 249,796 subjects of European origin. The association of 14 well-known obesity-susceptibility loci was confirmed, and 18 new loci associated with BMI were identified. The combined effect of these variants at the 32 loci was very modest, explaining only 1.45% of the variance in BMI, reaching 11% if other SNPs at various degrees of significance were included in the model [53]. Another interesting result of this meta-analysis was the finding of an SNP (rs1800437) in an incretin receptor (*GIPR*) locus encoding a receptor of gastric inhibitory polypeptide (GIP). This polypeptide is an incretin hormone that mediates insulin secretion in response to oral glucose intake. The association of a variation in *GIPR* gene with BMI suggests a link between insulin secretion and body weight regulation in humans.

5. THE MISSING HERITABILITY

Most of the studies on monogenic and polygenic forms of obesity have been focused on SNPs. The results from GWAS have identified loci explaining less than 10% of the heritability of obesity. Currently, there is still a great debate about the genetic factors responsible for this missing heritability. The main hypothesis is that some rare variants, copy number variants (CNVs), and epigenetic changes that are not detected by GWAS could explain part of the missing heritability [65].

The role of CNVs in obesity remains poorly understood. CNVs are genomic rearrangements resulting in deletions, insertions, and duplications in DNA sequences with more than 1 kb in length. The most known

CNV associated with obesity is a deletion of 593 kb at 16p11.2 [66,67], with a 43-fold increased risk for morbid obesity and also associated with cognitive disorders and autism. Deletions at this locus are responsible for 0.7% of morbid obesity, and it is not present in non-obese subjects. Other 17 CNVs loci associated with obesity have been described in 1080 children with European-American origin, and half of them were replicated in an African-American cohort of 1479 childhood obesity cases [68]. Although these CNVs are not very frequent, it has become quite clear that they have a role in the susceptibility to obesity and need further research on their effects.

6. GENE–ENVIRONMENT INTERACTION STUDIES

It is widely assumed that gene–environment interaction (GEI) must have an effect on adiposity and several studies have explored the relationship between lifestyle and obesity susceptibility genes, reporting significant interactions. However, due to the difficulties in obtaining accurate data on diet and physical activity, the sample size required to detect interaction should be very large [69].

It has been shown that physical activity can modify the genetic predisposition to obesity. An interaction between *FTO* variant and physical activity was demonstrated in a study with more than 6100 Danish subjects, showing that in subjects who reported more physical activity, the *FTO* effect was attenuated by 30% compared to sedentary subjects [70]. Physical inactivity was associated with an increase in BMI of 1.95 kg/m^2 in homozygous A-allele carriers of rs9939609 variant in *FTO* gene, and no effect was observed in non-carriers or in heterozygous subjects.

Another study analyzing near 12,000 subjects from the EPIC-Norfolk study demonstrated that the genetic predisposition to obesity could be reduced about 40% with active physical activity. Moreover, the benefit was higher in the people with genetic predisposition to obesity compared with the protected subjects [71]. A recent meta-analysis including data from 218,166 adults and 19,268 children and adolescents confirmed that the minor allele of the *FTO* rs9939609 variant reduced the risk of obesity among physically active individuals by 27% [72].

With respect to diet, different studies have confirmed the interaction between the genetic susceptibility to obesity and different dietary patterns and BMI. In a cross-sectional study that examined 4839 subjects in the population-based Malmö Diet and Cancer study with dietary data (from a modified diet history method) and information on the genetic variant *FTO* (rs9939609), it has been observed that an increase in BMI across *FTO* genotypes was restricted to those subjects who reported a high-fat diet, and was not associated with a higher BMI among subjects with lower fat intakes. In the same study, the increase in BMI across genotypes was mainly restricted to individuals who reported low leisure-time physical activity, indicating that high-fat diets and low physical activity levels may accentuate the susceptibility to obesity by the FTO variant [73].

Recently, in an analysis of 737 overweight adults from the Preventing Overweight Using Novel Dietary Strategies trial, individuals with the A allele in the *FTO* rs9939609 variant had a greater decrease in food cravings and appetite by choosing a high-protein weight-loss diet [74]. Most of the studies performed to analyze gene–diet interaction are limited because they only examined a single genetic variant explaining only a small fraction of the variation in BMI. The use of a more global genetic risk score (GRS) combining multiple variants with low effect on BMI will improve the identification of those subjects with a higher risk of developing obesity [75]. In this sense, a recent study including 9623 initially healthy women and 6379 initially healthy men of European ancestry with genotype data available, from the Nurses' Health Study and from The Health Professionals Follow-up Study, has demonstrated a significant interaction between fried food consumption and genetic predisposition to adiposity. The association between total fried food consumption and BMI was stronger in participants with a higher genetic risk score than in those with a lower genetic risk score. The findings were further replicated in a large independent cohort of 21,421 US women [75].

The interaction of the type of dietary fat and GRS using more than 50 well-established loci related to obesity has been analyzed in two US populations (783 subjects from GOLDEN study and 2035 subjects from MESA study) showing that saturated fatty acids (SFA) modulates significantly the relationship between GRS and BMI. In both, GOLDEN and MESA populations, those subjects with high SFA intake (about $\geq 11\%$ in both studies) and high GRS had higher BMI compared to subjects with low SFA intake [76]. The mechanisms of association are still not clear, but some evidence suggests that SFA reduces satiety and thermogenesis compared to unsaturated fatty acids [77].

The GEI has also been studied with weight loss interventions. It has been observed that individuals with the insulin receptor substrate 1 (*IRSI 1*) gene variant rs2943641 CC have greater weight loss and improvement in insulin resistance after 2 years on a high-carbohydrate/low-fat diet [78]. Another randomized study in 3234 obese subjects showed that some genes are involved in short-term weight loss, while others are related to long-term weight loss (*NEGR1* rs2815752 and *FTO* rs9939609) and others to regaining weight (*NEGR1* rs2815752, *BDNF* rs6265, and *PPARG* rs1801282), irrespective of treatment [79].

Taken together, all of these results provide further evidence that particular individuals who are genetically predisposed to obesity would benefit more from increased physical activity levels and specific dietary modifications than individuals who are genetically protected.

7. GENETIC TESTING IN OBESITY

A clinical application of the discovery of variants associated with obesity is that it enables the genetic testing of individuals to know their predisposition to develop the disorder. In this case, the ACCE (**A**nalytical and **C**linical validity, **C**linical utility, and **E**thical and legal implications) framework should be used for evaluating the test [80].

For common obesity, the analytical validity can be considered very high if the procedure is performed in an accredited laboratory using validated methods. As it has been discussed before, the ability of the identified single loci or genetic risk score to predict obesity is very limited, explaining less than 5% of the heritability and about 1.5% of the variability in BMI; therefore, additional factors like gene interactions and environmental exposure need to be considered in developing genetic methods to classify individuals at high risk. Currently, there are some companies offering genetic testing including not only the SNPs related to obesity, but also the SNPs related to response to exercise and dietary interventions. However, clinical utility of these tests should still be demonstrated.

It is possible that in the near future, with further evidence, genetic information will play a role in making decisions about the therapeutic options in obesity. Up to now, research in this field is still limited and evidence is not robust yet. Moreover, it is important to know if the knowledge of the genetic susceptibility to obesity will change the approach to management of or increase motivation for life-style changes in a subject.

8. CONCLUSIONS

In summary, obesity is a complex disease resulting from the interaction between hereditary and environmental factors. Genetic and molecular basis of obesity have been the focus of intensive research for a long time, and this has permitted a better understanding of this condition, although much still remains unknown. Family studies and animal models have helped us to identify numerous genes involved in the development of some severe forms of monogenic obesity. Subsequently, the genome-wide association studies permitted the identification of novel loci associated to more common forms of obesity across populations of diverse ethnicities and ages. To date, less than 5% of the heritability of body weight has been explained. Subsequent studies identified rare variants, CNVs, and epigenetic changes that contribute to the heritability of obesity. Although clinical applications of these findings are limited, a better knowledge of the interplay of genetic factors with environmental conditions will help clarify the etiology of obesity and its metabolic consequence, as well as identify at-risk individuals or groups in terms of their genetic profile with the goal of developing personalized prevention and treatment strategies.

References

[1] Haslam DW, James WP. Obesity. Lancet 2005;366(9492):1197−209.
[2] Genné-Bacon EA. Thinking evolutionarily about obesity. Yale J Bio Med 2014;87(2): 99−112.
[3] World Health Organization. Obesity and overweight. 2014. http://www.who.int/mediacentre/factsheets/fs311/en/ [accessed 16.08.14].
[4] Neel JV. The "thrifty genotype" in 1998. Nutr Rev 1999;57(5Pt2):S2−9.
[5] Farias MM, Cuevas AM, Rodriguez F. Set-point theory and obesity. Metab Syndr Relat Disord 2011;9(2):85−9.
[6] Speakman JR. Thrifty genes for obesity, an attractive but flawed idea, and an alternative perspective: the "drifty gene" hypothesis. Int J Obes Lond 2008;32(11):1611−7.
[7] Stunkard AJ, Foch TT, Hrubec Z. A twin study of human obesity. JAMA 1986;256(1): 51−4.
[8] Hjelmborg JV, Fagnani C, Silventoinen K, McGue M, Korkelia M, Christensen K, et al. Genetic influences on growth traits of BMI: a longitudinal study of adult twins. Obes (Silver Spring) 2008;16(4):847−52.
[9] Barsh GS, Farooqui IS, O'Rahilly S. Genetics of body weight regulation. Nature 2000; 404(6778):644−51.
[10] Hebebrand J, Hinney A. Environmental and genetic risk factors in obesity. Child Adolesc Psychiatr Clin N Am 2009;18(1):83−94.
[11] Stunkard AJ, Harris JR, Pedersen NL, Mc Clearn GE. The body-mass index of twins who have been reared apart. N Engl J Med 1990;322(21):1483−7.
[12] Tabor HK, Risch NJ, Myers RM. Candidate-gene approaches for studying complex genetic traits: practical considerations. Nat Rev Genet 2002;3(5):391−7.
[13] Bell CG, Walley AJ, Froguel P. The genetics of human obesity. Nat Rev Genet 2005;6(3): 221−34.
[14] Rankinen T, Zuberi A, Chagnon YC, Weisnagel SJ, Argyropoulos G, Walts B, et al. The human obesity gene map: the 2005 update. Obes (Silver Spring) 2006;14(4):529−644.
[15] Saunders CL, Chiodini BD, Sham P, Lewis CM, Abkevich V, Adeyemo AA, et al. Meta-analysis of genome-wide linkage studies in BMI and obesity. Obes (Silver Spring) 2007;15(9):2263−75.
[16] Mantzoros CS, Magkos F, Brinkoetter M, Sienkiewicz E, Dardeno TA, Kim SY, et al. Leptin in human physiology and pathophysiology. Am J Physiol Endocrinol Metab 2011;301(4):E567−84.
[17] Kim JH, Choi JH. Pathophysiology and clinical characteristics of hypothalamic obesity in children and adolescents. Ann Pediatr Endocrinol Metab 2013;18(4):161−7.
[18] Coll AP, Farooqi IS, Challis BG, Yeo GS, O'Rahilly S. Proopiomelanocortin and energy balance: insights from human and murine genetics. J Clin Endocrinol Metab 2004; 89(6):2557−62.
[18A] Kleinridders A, Ferris HA, Cai W, Kahn CR. Insulin action in brain regulates systemic metabolism and brain function. Diabetes 2014;63(7):2232−43.
[19] Valladares M, Domínguez-Vásquez P, Obregón AM, Weisstaub G, Burrows R, Maiz A, et al. Melanocortin-4 receptor gene variants in Chilean families: association with childhood obesity and eating behavior. Nutr Neurosci 2010;13(2):71−8.
[20] Farooqi IS, O'Rahilly S. Monogenic obesity in humans. Annu Rev Med 2005;56:443−58 [review].

[21] Zhang Y, Proenca R, Maffei M, Barone M, Leopold L, Friedman JM. Positional cloning of the mouse obese gene and its human homologue. Nature 1994;372(6505):425–32.
[22] Chua SC, Chung WK, Wu-Peng XS, Zhang Y, Liu SM, Tartaglia L, et al. Phenotypes of mouse diabetes and rat fatty due to mutations in the OB (leptin) receptor. Science 1996;271(5251):994–6.
[23] Montague CT, Farooqi IS, Whitehead JP, Soos MA, Rau H, Wareham NJ, et al. Congenital leptin deficiency is associated with severe early-onset obesity in humans. Nature 1997;387(6636):903–8.
[24] Clément K, Vaisse C, Lahlou N, Cabrol S, Pelloux V, Cassuto D, et al. A mutation in the human leptin receptor gene causes obesity and pituitary dysfunction. Nature 1998;392(6674):398–401.
[25] Farooqi IS, Wangensteen T, Collins S, Kimber W, Matarese G, Keogh JM, et al. Clinical and molecular genetic spectrum of congenital deficiency of the leptin receptor. N Engl J Med 2007;356(3):237–47.
[26] Farooqi IS, Matarese G, Lord GM, Keogh J, Laurence E, Agwu C, et al. Beneficial effects of leptin on obesity, T cell hyporesponsiveness, and neuroendocrine/metabolic dysfunction of human congenital leptin deficiency. J Clin Invest 2002;110(8):1093–103.
[27] Krude H, Biebermann H, Luck W, Horn R, Brabant G, Grüters A. Severe early-onset obesity, adrenal insufficiency and red hair pigmentation caused by POMC mutations in humans. Nat Genet 1998;19(2):155–7.
[28] Philippe J, Stijnen P, Meyre D, De Graeve F, Thuillier D, Delplanque J, et al. A nonsense loss-of-function mutation in PCSK1 contributes to dominantly inherited human obesity. Int J Obes (Lond) 2015;39(2):295–302.
[29] Yeo GS, Farooqi IS, Aminian S, Halsall DJ, Stanhope RG, O'Rahilly S. A frameshift mutation in MC4R associated with dominantly inherited human obesity. Nat Genet 1998;20(2):111–2.
[30] Vaisse C, Clement K, Guy-Grand B, Froguel P. A frameshift mutation in human MC4R is associated with a dominant form of obesity. Nat Genet 1998;20(2):113–4.
[31] Hinney A, Bettecken T, Tarnow P, Brumm H, Reichwald K, Lichtner P, et al. Prevalence, spectrum, and functional characterization of melanocortin-4 receptor gene mutations in a representative population-based simple and obese adults from Germany. J Clin Endocrinol Metab 2006;91(5):1761–9.
[32] Vaisse C, Clement K, Durand E, Hercberg S, Guy-Grand B, Froguel P. Melanocortin-4 receptor mutations are a frequent and heterogeneous cause of morbid obesity. J Clin Invest 2000;106(2):253–62.
[33] Dempfle A, Hinney A, Heinzel-Gutenbrunner M, Raab M, Geller F, Gudermann T, et al. Large quantitative effect of melanocortin-4 receptor gene mutations on body mass index. J Med Genet 2004;41(10):795–800.
[34] Roubert P, Dubern B, Plas P, Lubrano-Berthelier C, Alihi R, Auger F, et al. Novel pharmacological MC4R agonists can efficiently activate mutated MC4R from obese patient with impaired endogenous agonist response. J Endocrinol 2010;207(2):177–83.
[35] Hinney A, Vogel CI, Hebebrand J. From monogenic to polygenic obesity: recent advances. Eur Child Adolesc Psychiatry 2010;19(3):297–310.
[36] Xia Q, Grant SF. The genetics of human obesity. Ann N Y Acad Sci 2013;1281:178–90.
[37] Ichimura A, Hirasawa A, Poulain-Godefroy O, Bonnefond A, Hara T, Yengo L, et al. Dysfunction of lipid sensor GPR120 leads to obesity in both mouse and human. Nature 2012;483(7389):350–4.
[38] Frayling TM, Timpson NJ, Weedon MN, Zeggini E, Freathy RM, Lindgren CM, et al. A common variant in the FTO gene is associated with body mass index and predisposes to childhood and adult obesity. Science 2007;316(5826):889–94.

[39] Fawcett KA, Barroso I. The genetics of obesity; FTO leads the way. Trends Genet 2010; 26(6):266—74.

[40] Day FR, Loors RJ. Developments in obesity genetics in the era of genome-wide association studies. J Nutrigenet Nutrigenomics 2011;4(4):222—38.

[41] Li H, Kilpeläinen TO, Liu C, Zhu J, Liu Y, Hu C, et al. Association of genetic variation in FTO with risk of obesity and type 2 diabetes with data from 96,551 East and South Asians. Diabetologia 2012;55(4):981—95.

[42] Hertel JK, Johansson S, Sonestedt E, Jonsson A, Lie RT, Platou CG, et al. FTO, type 2 diabetes, and weight gain throughout adult life: a meta-analysis of 41,504 subjects from the Scandinavian HUNT, MDC, and MPP studies. Diabetes 2011;60(5):1637—44.

[43] Church C, Moir L, McMurray F, Girard C, Banks GT, Teboul L, et al. Overexpression of FTO leads to increased food intake and results in obesity. Nat Genet 2010;42(12): 1086—92.

[44] Scuteri A, Sanna S, Chen WM, Uda M, Albai G, Strait J, et al. Genome-wide association scan shows genetic variants in the FTO gene are associated with obesity-related traits. PLoS Genet 2007;3(7):e115.

[45] Peng S, Zhu Y, Xu F, Ren X, Li X, Lai M. FTO gene polymorphisms and obesity risk: a meta-analysis. BMC Med 2011;8:71.

[46] Labayen I, Ruiz JR, Ortega FB, Gottrand F, Huybrechts I, Dallongeville J, et al. Body size at birth modifies the effect of fat mass and obesity associated (FTO) rs9939609 polymorphism on adiposity in adolescents: the Healthy Lifestyle in Europe by Nutrition in Adolescence (HELENA) study. Br J Nutr 2012;107(10):1498—504.

[47] Xi B, Shen Y, Zhang M, Liu X, Zhao X, Wu L, et al. The common rs9939609 variant of the fat mass and obesity-associated gene is associated with obesity risk in children and adolescents of Beijing, China. BMC Med Genet 2010;11:107.

[48] Wing MR, Ziegler J, Langefeld CD, Ng MC, Haffner SM, Norris JM, et al. Analysis of FTO gene variants with measures of obesity and glucose homeostasis in the IRAS Family Study. Hum Genet 2009;125(5—6):615—26.

[49] Rutters F, Nieuwenhuizen AG, Bouwman F, Mariman E, Westerterp-Plantenga MS. Associations between a single nucleotide polymorphism of the FTO Gene (rs9939609) and obesity-related characteristics over time during puberty in a Dutch children cohort. J Clin Endocrinol Metab 2011;96(6):E939—42.

[50] Martínez-García F, Mansego ML, Rojo-Martínez G, De Marco-Solar G, Morcillo S, Soriguer F, et al. Impact of obesity-related genes in Spanish population. BMC Genet 2013;14:111.

[51] Olza J, Ruperez AI, Gil-Campos M, Leis R, Fernandez-Orth D, Tojo R, et al. Influence of FTO variants on obesity, inflammation and cardiovascuar disease risk biomarkers in Spanish children: a case-control multicentr study. BMC Med Genet 2013;14:123.

[52] Liu C, Mou S, Pan C. The FTO gene rs9939609 polymorphism predicts risk of cardiovascular disease: a systematic review and meta-analysis. PLoS One 2013;8(8):e71901.

[53] Speliotes EK, Willer CJ, Berndt SI, Monda KL, Thorleifsson G, Jackson AU, et al. Association analyses of 249,796 individuals reveal 18 new loci associated with body mass index. Nat Genet 2010;42(11):937—48.

[54] Loos RJF, Lindgren CM, Li S, Wheeler E, Zhao JH, Prokopenko I, et al. Common variants near MC4R are associated with fat mass, weight and risk of obesity. Nat Genet 2008;40(6):768—75.

[55] Chambers JC, Elliott P, Zabaneh D, Zhang W, Li Y, Froguel P, et al. Common genetic variation near MC4R is associated with waist circumference and insulin resistance. Nat Genet 2008;40(6):716—8.

[56] Geller F, Reichwald K, Dempfle A, Illig T, Vollmert C, Herpertz S, et al. Melanocortin-4 receptor gene variant I103 is negatively associated with obesity. Am J Hum Genet 2004;74(3):572—81.

[57] Young E, Wareham N, Farooqi S, Hinney A, Hebebrand J, Scherag A, et al. The V103I polymorphism of the MC4R gene and obesity: population based studies and meta-analaysis of 29563 individuals. Int J Obes (Lond) 2007;31(9):1437–41.

[58] Xiang Z, Litherland SA, Sorensen NB, Wood MS, Shaw AM, Millard WJ, et al. Pharmacological characterization of 40 human melanocortin-4 receptor polymorphisms with the endogenous proopiomelanocortin-derived agonists and the agouti-related protein (AGRP) antagonist. Biochemistry 2006;45(23):7277–88.

[59] Benzinou M, Creemers JW, Choquet H, Lobbens S, Dina C, Durand E, et al. Common nonsynonymous variants in PCSK1 confer risk of obesity. Nat Genet 2008;40(8):943–5.

[60] Kilpeläinen TO, Bingham SA, Khaw KT, Wareham NJ, Loos RJ. Association of variants in the PCSK1 gene with obesity in the EPOC-Norfolk study. Hum Mol Genet 2009; 18(18):3496–501.

[61] Choquet H, Kasberger J, Hamidovic A, Jorgenson E. Contribution of common PCSK1 genetic variants to obesity in 8,359 subjects from multi-ethnic American population. PLoS One 2013;8(2):e57857.

[62] Willer CJ, Speliotes E, Loos RJ, Li S, Lindrgren CM, Heid IM, et al. Six new loci associated with body mass index highlight a neuronal influence on body weight regulation. Nat Genet 2009;41(1):25–34.

[63] Thorleifsson G, Walters GB, Gudbjartsson DF, Steinthorsdottir V, Sulem P, Helgadottir A, et al. Genome-wide association yields new sequence variants at seven loci that associate with measures of obesity. Nat Genet 2009;41(1):18–24.

[64] Ren D, Zhou Y, Morris D, Li M, Li Z, Rui L. Neuronal SH2B1 is essential for controlling energy and glucose homeostasis. J Clin Invest 2007;117(2):397–406.

[65] Walley AJ, Asher JE, Froguel P. The genetic contribution to non-syndromic human obesity. Nat Rev Genet 2009;10(7):431–42.

[66] Walters RG, Jacquemont S, Valessia A, de Smith AJ, Martinet D, Andersson J, et al. A new highly penetrant form of obesity due to deletions on chromosome 16p11.2. Nature 2010;463(7281):671–5.

[67] Bochukova EG, Huang N, Keogh J, Henning E, Purmann C, Blaszczyk K, et al. Large, rare chromosomal deletions associated with severe early-onset obesity. Nature 2010; 463(7281):666–70.

[68] Glessner JT, Bradfield JP, Wang K, Takahashi N, Zhang H, Sleiman PM, et al. A genome-wide study reveals copy number variants exclusive to childhood obesity cases. Am J Hum Genet 2010;87(5):661–6.

[69] Wong M, Day N, Luan J, Chan K, Wareham N. The detection of gene-environment interaction for continuous traits: should we deal with measurement error by bigger studies or better measurement? Int J Epidemiol 2003;32(1):51–7.

[70] Andreasen CH, Stender-Petersen KL, Mogensen MS, Torekov SS, Wegner L, Andersen G, et al. Low physical activity accentuates the effect of the FTO rs9939609 polymorphism on body fat accumulation. Diabetes 2008;57(1):95–101.

[71] Li S, Zhao JH, Luan J, Ekelund U, Luben RN, Khaw KT, et al. Physical activity attenuates the genetic predisposition to obesity in 20,000 men and women from EPIC-Norfolk prospective population study. PLoS Med 2010;7(8):e1000332.

[72] Kilpelainen TO, Qi L, Brage S, Sharp SJ, Sonestedt E, Demerath E, et al. Physical activity attenuates the influence of FTO variants on obesity risk: a meta-analysis of 218,166 adults and 19,268 children. PLoS Med 2011;8(11):e1001116.

[73] Sonestedt E, Roos C, Cullberg B, Ericson U, Wirfält E, Orho-Melander M. Fat and carbohydrate intake modify the association between genetic variation in the FTO genotype and obesity. Am J Clin Nutr 2009;90(5):1418–25.

[74] Huang T, Qi Q, Li Y, Hu FB, Bray GA, Sacks FM, et al. FTO genotype, dietary protein, and change in appetite: the Preventing Overweight Using Novel Dietary Strategies trial. Am J Clin Nutr 2014;99(5):1126–30.

[75] Qi Q, Chu AY, Kang JH, Huang J, Rose LM, Jensen MK, et al. Fried food consumption, genetic risk, and body mass index: gene-diet interaction analysis in three US cohort studies. BMJ 2014;348:g1610.

[76] Casas-Agustench P, Arnett DK, Smith CE, Lai CQ, Parnell LD, Borecki IB, et al. Saturated fat intake modulates the association between an obesity genetic risk score ad Body Mass Index in two US populations. J Acad Nutr Diet 2014;114(12):1954—66.

[77] Casas-Agustench P, López-Uriarte P, Bulló M, Ros E, Gómez-Flores A, Salas-Salvadó J. Acute effects of three high-fat meals with different fat saturations on energy expenditure, substrate oxidation and satiety. Clin Nutr 2009;28(1):39—45.

[78] Qi Q, Bray GA, Smith SR, Hu FB, Sacks FM, Qi L. Insulin receptor substrate 1 gene variation modifies insulin resistance response to weight-loss diets in a 2-year randomized trial: the Preventing Overweight Using Novel Dietary Strategies (POUNDS LOST) trial. Circulation 2011;124(5):563—71.

[79] Delahanty LM, Pan Q, Jablonski KA, Watson KE, McCaffery JM, Shuldiner A, et al. Genetic predictors of weight loss and weight regain after intensive lifestyle modification, metformin treatment, or standard care in the diabetes prevention program. Diabetes Care 2012;35(2):363—6.

[80] CDC. Genomic testing: ACCE model process for evaluating genetic tests, http://www.cdc.gov/genomics/gtesting/ACCE/; [accessed 7.08.14].

The Epidemiology and Genetics of Vascular Dementia: Current Knowledge and Next Steps

Adam Naj

Department of Biostatistics and Epidemiology, Center for Clinical Epidemiology and Biostatistics, Perelman School of Medicine, University of Pennsylvania, Philadelphia, PA, USA

1. DEFINITION AND FEATURES OF VASCULAR DEMENTIA

Vascular dementia (VaD) is broadly defined as dementia, characterized by reduced cognitive function and memory loss, resulting from problems in the blood supply to the brain. More specifically, it is a form of dementia pathologically characterized by ischemic, ischemic—hypoxic, or hemorrhagic brain lesions that are a product of cerebrovascular disease and cardiovascular pathological alterations [1]. VaD often occurs as a result of conditions that block or reduce blood flow to the brain, leading brain cells to be deprived of oxygen and nutrients; typically, VaD arises from a series of non-fatal strokes. VaD is the second most common form of dementia after Alzheimer's disease (AD) and accounts for 15—20% of all cases of dementia [2].

Historically, the earliest identifications of VaD can be traced back to the work of Thomas Willis in 1672 who described cases of "dementia post-apoplexy" [3]. The eighteenth and early nineteenth centuries saw the frequent diagnosis of "brain congestion" for a variety of diseases and conditions currently well characterized, including anxiety, stroke, and cognitive decline symptomatology. The modern understanding of VaD derives from the work of Otto Binswanger and Alois Alzheimer, who as early as the 1890s identified VaD to be a distinct phenomenon from neurosyphilitic dementia. For much of the twentieth century, VaD was described as "cerebral arteriosclerosis" and "arteriosclerotic dementia," and remained so until seminal works by Roth, Blessed, and Tomlinson [4–8] established AD as the leading cause of dementia. Reclassified in the 1970s as "multi-infarct dementia" (MID) [9], the current definitions of VaD and "vascular cognitive impairment" (VCI) arose in the mid-1990s, respectively, from the need for consensus diagnostic criteria for VaD and from the need to prevent and diagnose the condition early in its course [10,11].

The term VCI arose from a need to identify and classify individuals for whom cerebrovascular abnormalities significantly affect the quality of life and the ability to perform activities of daily living, but whose impairments may not meet the criteria for classic dementia. Among individuals with dementia, postmortem pathological studies [12–17] have shown that 15–34% of dementia cases demonstrate significant vascular pathology, with a subset of these cases also showing AD pathology [18]. The inclusion of the term "cognitive impairment" parallels its usage as a descriptor for preclinical Alzheimer's symptomatology ("mild cognitive impairment" or MCI), and has been found to be especially appropriate as inflammation and its effects on cerebral vasculature produce a similar level of risk for AD to that of MCI [19,20].

Importantly, VaD and VCI are terms that describe a highly etiologically heterogeneous set of diseases and conditions. While there exist multiple syndromic forms of VaD, the ICD-10 [21] and NINDS-AIREN criteria [11] recognize multiple common etiologic subtypes, including subcortical VaD, Binswanger's disease, and thalamic dementia. VaD in all its varieties likely results from one or more types of cerebrovascular lesions and external risk factors. Among the types of cerebrovascular lesions contributing to VaD are (adapted from [11]): (1) MID, where multiple complete infarcts often from large-vessel occlusion lead to clinical dementia; (2) strategic single-infarct dementia, where small localized ischemic damage is present in functionally important cortical and subcortical areas; (3) small vessel disease with dementia, multiple lacunar strokes, and Binswanger's disease (BD); (4) hypoperfusion, where dementia results from global ischemia following cardiac arrest or hypotension; (5) hemorraghic dementia, resulting from lesions as they characterize chronic subdural hematoma; and (6) other mechanisms.

2. EPIDEMIOLOGY OF VaD

While VaD is the second most prevalent form of dementia in the USA and Europe (AD is most prevalent), it has been at times the most common form of dementia among some Asian groups. Prevalence of the VaD in the USA in 2007 was estimated to be 2.4% among persons of 71 years of age or older, compared with 9.7% prevalence of AD and 13.9% for all dementias [22]. Among race/ethnic groups in the USA, estimates suggest a higher prevalence of VaD among African-Americans than whites, due to the higher frequency of cardiovascular and cerebrovascular risk factors among African-Americans [23,24]. Studies in the UK and Italy have estimated the prevalence of VaD as 0.99 cases/1000 person-years (py) and 3.30/1000 py, respectively [25,26]. Early studies of Asian populations suggested that VaD accounted for a larger proportion of dementias in multiple groups, whereas AD was more common in most European studies [27,28]. More recent studies of Japanese populations have shown that, despite earlier studies showing a greater prevalence of VaD than AD [29,30], trends have shifted with decreases in prevalence of VaD and increases in AD prevalence [31,32].

Incidence and prevalence of dementia are both significantly elevated after a stroke, with 25–30% of stroke survivors developing dementia at some point, with incidence increasing from about 7% for 1 year after stroke to about 48% for 25 years after stroke [33].

Risk factors for VaD are similar to those for cardiovascular and cerebrovascular diseases, and share considerable overlap with risk factors for AD. Non-modifiable risk factors for VaD and VCI include (adapted from Ref. [34]):

- Age: The overall risk and prevalence of VCI increases with age. During 65–90 years of age, the incidence of dementia doubles, and may continue to increase after age 90 years [35]. In the European population-based studies, the prevalence of VaD was 1.6% in subjects of age 65 years and older [36,37].
- Sex: Men may have a higher incidence of VaD than women, although a combined analysis of incidence studies showed no significant difference, and the trends are inconsistent [2,38].
- Race: There may be a higher incidence of VaD in blacks than in whites, which is part of an overall higher incidence in blacks for vascular risk factors [39,40].
- Amyloid level: Cerebral amyloid angiopathy (CAA) in elderly patients can cause white matter lesions (WMLs), lobar cerebral micro bleeds, and VCI. Amyloid beta peptide is deposited in and weakens the walls of cerebral vessels, disrupting the blood–brain barrier and resulting in perivascular leakage of red blood cells [41–43]. For controlling the severity of AD-related pathology, cognitive performance is significantly impaired in patients with advanced CAA [44,45].
- Education level: Increased risk of VCI may be connected to lower level of educational attainment [46]; however, other factors may confound this association.

Among modifiable risk factors are:

- Diet: A Mediterranean-type dietary pattern may protect against cognitive decline [47]; however, there is minimal support for associations between dietary elements or supplements and VCI. Vitamin D studies have produced different results; one study showed an association between low levels of vitamin D and low cognitive function [48] while another showed no association [49]. No cognitive benefits have been shown for the consumption of antioxidants [50,51]. Additionally, vitamins B6, B12, and folic acid, and homocysteine have not been shown to impact cognitive function [40].
- Physical activity: Multiple studies have shown long-term physical activity may have a positive impact on cognitive functions and slow decline, as well as protecting both brain health and plasticity [40,52–55].
- Hypertension: Elevated blood pressure in midlife has been linked to a decline in cognitive function and outcome later in life [56–58]. In the elderly, it is less clear if treating high blood pressure will positively impact cognitive function [40,59]. Chronic hypertension can lead to impaired auto-regulation, subclinical and overt infarcts, small vessel ischemic disease (SVID), smaller brain volumes, and cognitive impairment [60,61]. Treating hypertension in persons of all ages can prevent stroke and ischemic heart disease. A study in progress, SPRINT-MIND, will determine if a systolic blood pressure goal of less than 120 mmHg will have lower rates of SVID and VCI compared to a pressure of 140 mmHg.
- Hyperglycemia and hypoglycemia: Chronic hyperglycemia, like chronic hypertension, can lead to impaired auto-regulation and neuronal damage from functional changes in cerebral blood vessels that impair cognitive function [62,63]. Saczynski et al. showed that this relationship between hyperglycemia and VCI may be dependent on both time and duration [64]. The ability to prevent VCI in diabetics with tight glycemic control is still inconclusive [65].
- Hyperlipidemia: The effect of cholesterol on VCI is still debated. Some studies have shown an increased risk of cognitive impairment in later life in individuals with elevated total cholesterol in midlife [66]. Statins may aid in the prevention of VCI in later life [67].
- Metabolic syndrome: The Honolulu–Asia Aging Study (HAAS) showed some evidence that metabolic syndrome in Japanese-American men increased the risk of VCI [68]. Additional studies may help solidify this finding.
- Smoking: Little evidence exists regarding preventing dementia with smoking cessation [40], but heavy smoking in middle age increases the long-term risk of VaD for both men and women in the later life [69].

- Alcohol consumption: While heavy drinking is associated with an increased risk of dementia and lowered cognitive function [70,71], some studies show that those who partake in moderate drinking may have better cognitive function than others who completely abstain.
- Carotid and intracranial atherosclerosis: The exact mechanism of VCI in patients with atherosclerotic pathology has yet to be determined, but carotid and intracranial atherosclerosis seems to be associated with VCI. Mechanisms could potentially be cerebral hypoperfusion, brain embolism, alteration of blood pressure, and impaired blood—brain barrier [72]. Many studies have seen an inverse relation between carotid intima media thickness (IMT) and cognitive function; however, the exact causal relation of carotid IMT and VCI is not clear [73,74]. These studies controlled for age, education, and symptoms related to depression.
- Atrial fibrillation (Afib): The Afib is a risk factor for both stroke and VCI [75,76]. Afib and VCI may be related by the occurrence of silent or subclinical ischemic damage from cardioembolism, hypercoagulable states associated with Afib, and shared cardiovascular risk factors [77—79]. Kalantarian et al. conducted a meta-analyses of 21 studies that showed a higher risk of both dementia and cognitive impairment in Afib patients with and without a history of stroke [80].
- Peripheral vascular disease: Ankle—brachial index (ABI) is one measure of peripheral vascular disease (PVD). It may help to identify those at risk of cognitive impairment [81]. A low ABI was connected to increased risk of VCI in the HAAS [82].
- Depression: Depression has been linked to both vascular disease and dementia [83], and may be a risk factor in the development of future dementia [84].

3. GENETICS OF VaD

Genetic factors for VaD have not been widely studied, largely due to variable definitions and classifications of VCI. Finding causal genetic risk factors for VaD is complicated further by frequent comorbidity with AD, of whom 20—25% show VCI [85]; across multiple autopsy series of AD cases, cerebrovascular lesions have been seen in 18—80% of AD brains [85—88]. However, a range of genetic studies have been performed to identify genetic associations with VaD and VCI, including candidate gene studies, studies of familial monogenic subtypes, and genome-wide association studies (GWAS).

3.1 Candidate Genes for VaD

Prior to the wide availability of cost-effective genome-wide single nucleotide variant genotyping and sequencing technologies, most genetic studies involved fine-mapping genomic regions containing linkage signals (examining positional candidate genes) or genotyping variation in genes likely to have a functional role in disease pathogenesis (examining biological candidate genes).

Genome-wide linkage studies have not been done on VaD as a phenotype specifically, but they have been done on closely related phenotypes or endophenotypes (intermediate traits in pathogenesis; that is, phenotypes "endogenous" to biological pathways) of VaD. For instance, genome-wide linkage studies have been done on imaging of WMLs [89–92], brain lesions that arise from ischemic or hemorrhagic damage to brain tissues and contribute to subcortical VaD. WMLs also represent a primary imaging biomarker for the presence of subcortical VaD. These studies identified a linkage at 4 cM on chromosome 4 [89,91], chromosome 1q24 in hypertensive sibships, with suggestive linkage signals on chromosomes 11, 17, 21, and 22 [92]. Bivariate linkage analyses looking at pleiotropic contribution to WML volumes and blood pressure identified candidate regions on chromosomes 1q24, 1q42, 10q22-q26, and 15q26. While these studies yielded a multitude of candidate genomic regions for fine-mapping, none yielded strong biological candidates for follow-up [90,92].

A multitude of biological candidate gene association studies have been performed for VaD. The candidate genes examined were selected for a variety of potential links with VaD: (1) some are known risk factors in dementia, specifically AD dementia (e.g., *APOE*); (2) some are genes implicated in stroke and risk factors of stroke (e.g., *ACE* and *AGTR1* in the renin–angiontensin system; *MTHFR*) because of the close relationship of cerebrovascular disease with dementia; (3) some are associated with phenotypes used to characterize VaD phenotypes (WML for subcortical VaD); and (4) some are contributors to highly genetic or monogenic forms of VaD (e.g., *NOTCH3*, discussed in more detail in the following section).

Few candidate genes for VaD have been found among genes contributing to AD dementia. A study of 29 single nucleotide polymorphisms (SNPs) in eight biological candidate genes, including *ACE* (encoding angiotensin I-converting enzyme), *APOE, BDNF* (encoding brain-derived neurotrophic factor), *DAPK1* (death-associated protein kinase 1), EIF2AK2 (eukaryotic translation initiation factor 2-alpha kinase 2), *GAB2* (GRB-associated binding protein 2), and *GOLM1* (Golgi membrane protein 1) identified no associations of these variants with VaD [93]. While this suggests a lack of biological candidates shared between AD and VaD, it should be noted that none of the above variants other than those in the *APOE* region have associations with AD that have been replicated in even extremely large GWAS analyses of AD [94]. Although the *APOE* was not associated with VaD ($P = 0.55$) in the aforementioned candidate gene study, the ε4 has demonstrated associations in multiple ethnic groups including Northern European white populations [95–99] and multiple Chinese cohorts [100–102], although this association has not been observed in all studies [103–106].

Genes likely to increase the risk of stroke and those associated with the increased risk of cardiovascular disease have been frequently explored as biological candidates for VaD and VCI, and include such genes as *ACE* (encoding angiotension converting enzyme) [107–109], *MTHFR* (encoding methylenetetrahydrofolate reductase), and *F5* (encoding coagulation factor V). The deletion (D) allele of an insertion/deletion (I/D) polymorphism in the *ACE* gene has been linked to higher levels of ACE, which correlates with diabetes, hypertension, and stroke, all well-established risk factors for VaD. *MTHFR* is involved in the folate–homocysteine pathway, with homocysteine levels a notable risk factor for stroke. Among schizophrenic patients, T/T homozygotes at the C677T variant of the *MTHFR* have demonstrated reduced executive function [110]. Severe deficiency of MTHFR (enzyme activity <20%) is associated with developmental delays and extreme neurological impairment [111]. The Arg → Glu mutation (R506Q) of the gene encoding coagulation factor V is associated with hypercoagulability [112] and thrombosis [113], and it has been shown that coagulation pathways may play a role in dementia, including AD [114,115]. In several studies investigating these candidate genes, support for a potential role in VaD risk has only been found for the C677T variant of *MTHFR*, which has demonstrated consistent associations with generalized vascular disease [116], WMLs [117], and VaD specifically [118]. Although *ACE* and *F5* have demonstrated associations with ischemic stroke and cerebrovascular disease, they have not been shown to contribute to VaD risk [119,120]. A number of other candidate genes have been explored, including *GSTO1*, *HSPA1A*, *MMP1/3/9*, and *PON1/2*, among others. These are described in detail in Table 1.

A final category of candidate genes for non-familial VaD include genes that are found to be highly associated with biomarkers of disease, like WMLs that are a marker for subcortical VaD [139], thoroughly reviewed genetic contributors to WMLs, implicating several known VaD candidates genes including *ACE*, *AGT*, and *APOE*, and implicated other genes including angiotensin receptor genes 1 and 2 (*AGTR1* and *AGTR2*), and *SORL1*, a known risk gene for AD. More detail on these candidate genes is presented in Table 2.

TABLE 1 Genomic Variants in Candidate Gene Studies Associated with Susceptibility to VaD

Gene	Symbol	Chromosome	MIM	Sequence variant				References	
				Variant ID	Location	Risk allele(s)	# Cases	Article	
Angiotensin I-converting enzyme	ACE	17q23.3	106180	rs4340 (288bp I/D)	Intron	Del	80	[121]	
Angiotensinogen	AGT	1q42–43	106150	rs699 (T235M)	Exon	M, MM + TM	207	[122]	
Apolipoprotein E	APOE	19q13.2	107741	e4 [rs7412,rs429358] (R112,R158)	Promoter	e4	89 144	[96] [100]	
Glutathione S-transferase o-1	GSTO1	10q25.1	605482	A140D	Exon	Asp/Asp	97	[123]	
Heat shock protein 70-1	HSPA1A	6p21.3	140550	A-110C	Promoter	CC	57	[124]	
Insulin-like growth-1 factor receptor	IGF1R	15q26.3	147370	G3179A	Exon	A, AA	75	[125]	
Interleukin-1a	IL1A	2q14	147760	C-889T	Promoter	CT	82	[126]	
Interleukin-1b	IL1B	2q14	147720	C-511T	Promoter	TT	104	[127]	
Intracellular adhesion molecule 1	ICAM1	19p13.3–13.2	147840	K469E	Exon	E, EE	107	[128]	
Interleukin-6	IL6	7p21	147620	G-174C	Promoter	GG	122	[129]	
Matrix metalloproteinase 1	MMP1	11q22–q23	120353	G-1607GG	Promoter	2G, 2G2G, 2G1G	193	[130]	
Matrix metalloproteinase 3	MMP3	11q23	185250	−1171 5A/6A	Promoter	5A, 5A5A	193	[130]	
Matrix metalloproteinase	MMP9	20q11.2–13.1	120361	C-1562T	Promoter	T, TT	193	[130]	

Matrix metalloproteinase 9								
5,10-Methylenetetrahydrofolate reductase	MTHFR	1p36.3	607093	A222V	Exon	TT	60	[120]
							130	[119]
Paraoxonase 1	PON1	7q21.3	168820	T-107C (a) / Q192R (b)	Promoter / Exon	$T_{(a)}T_{(a)}$, $Arg_{(b)}$ / $T_{(a)}Arg_{(b)}$	24	[131]
Paraoxonase 2	PON2	7q21.3	602447	C311S	Exon	Ser	55	[132]
Sterol regulatory element binding transcription factor 2	SREBF2	22q13	600481	G34995T	Intron	GT	207	[133]
Transforming growth factor B1	TGFB1	19q13.1	190180	P10L	Exon	Leu	207	[122]
					Exon	Leu/Leu	99	[134]
Tumor necrosis factor a	TNF	6p21.3	191160	T-1031C (a)	Promoter	$T_{(b)}C_{(b)}T_{(b)}$, $T_{(b)}T_{(b)}$	81	[135]
				C-857T (b)	Promoter	$C_{(a)}C_{(b)}$	57	[124]
Vascular endothelial growth factor	VEGFA	6p12	192240	G-1154A (a)	Promoter	$G_{(a)}T_{(b)}C_{(c)}$	207	[136]
				C-7T (b)	5' UTR	$G_{(a)}T_{(b)}$		
				C13553T (c)	3' UTR	$T_{(b)}C_{(c)}$		
Very low-density lipoprotein receptor	LRP1	9p24	107770	−19(CGG)n	5' UTR	$(CGG)_5$	47	[137]

Adapted from Ref. [138].

TABLE 2 Genetic Sequence Variant Associated with Susceptibility to VaD

Gene	Symbol	Chromosome	MIM	Phenotype	Variant ID	Location	Risk allele(s)	OR/P-value	Sample	Article
						Sequence variant			References	
Angiotensin I-converting enzyme	ACE	17q23.3	106180	WML burden	rs4340 (288bp I/D)	Intron	D/D	OR = 1.95 (95% CI: 1.09, 3.48)	Meta-analysis of 46 studies	[140]
Angiotensinogen	AGT	1q42–43	106150	WML presence	Promoter haplotype	Promoter	B/B or B+/A−	OR = 8.0, P = 0.003	Cohort study in elderly Caucasians	[141]
				WML burden	rs699 (T235M)	Exon	235T	NS	Meta-analysis of six studies	[140]
				WML burden				P = 0.01	GWAS meta-analyses in elderly Caucasians	[142]
Angiotensin receptor 1B	AGTR1	3q24	106165	WML progression	A1166C/ rs5186	3' UTR	1166C	P = 0.0002 in males	Depressed and non-depressed elderly Caucasians	[143]
Angiotensin II receptor, type 2	AGTR2	Xq23	300034	WML progression	C3123A/ rs2148582	Intron	3123A	P = 0.014 in males	Depressed and non-depressed elderly Caucasians	[143]
Apolipoprotein E	APOE	19q13.2	107741	WML burden	e2/e3/e4	Exon	e4 carrier	P = 0.016 in hypertensives	Cohort study of elderly Caucasians	[144]
					e2/e3/e4			P = 0.003	Cohort study of elderly Caucasians	[145]
				WML and lacunes	e2/e3/e4	Exon	e2/e3 genotype	OR = 3.0	Cohort study of elderly Caucasians	[146]
				WML presence	e2/e3/e4	Exon	e4 carrier	NS	Meta-analysis of 24 studies	[140]
				Lacunes	e2/e3/e4	Exon	e4+/e4+	OR = 4.7, P = 0.04	Genetic isolate, elderly hypertensives	[147]
				Micro bleeds	e2/e3/e4	Exon	e4 carrier	OR = 1.35 (95% CI: 1.10, 1.65)	Cohort study of elderly Caucasians	[148]

Gene	Symbol	Locus	OMIM	Phenotype	SNP	Region	Variant	Statistics	Study	Ref.
Notch (*Drosophila* homolog) 3	NOTCH3	19p13.12	600276	WML presence	rs10404382	Intron	1036 + 846G > T	OR = 1.8, P = 0.02 OR = 3.2, P = 0.002 in hypertensives	Cohort study of elderly Caucasians	[146]
				WML progression	rs10404382	Intron	1036 + 846G > T	β = 0.087, P = 0.05 β = 0.136, P = 0.013 in hypertensives		
Sortilin-related receptor 1	SORL1	11q24.1	602005	Micro bleeds	rs2282649	Intron	5239 + 73 TT	OR = 6.87, P = 0.005	Genetic isolate, elderly hypertensives	[147]
Tripartite motif-containing protein 47	TRIM47	17q25.1	611041	WML burden	rs3744017	Intron	A	$P = 7.3 \times 10^{-9}$	Meta-analyses of GWAS in elderly Caucasian	[142]
Tripartite motif-containing protein 65	TRIM65	17q25.1	201292	WML burden	rs3744028	Intron	C	$P = 4.0 \times 10^{-9}$	Meta-analyses of GWAS in elderly Caucasian	[142]
WW domain-binding protein 2	WBP2	17q25.1	606962	WML burden	rs11869977	Intergenic	G	$P = 5.7 \times 10^{-9}$	Meta-analyses of GWAS in elderly Caucasian	[142]

WML, white matter lesion; OR, odds ratio; 95% CI, 95% confidence interval; β, regression coefficient; GWAS, genome-wide association study; NS, not statistically significant.
Adapted from Ref. [139].

3.2 Common Polygenic Contributors to VaD from Genome-wide Studies

GWASs that utilize high-density genotyping arrays in thousands of samples are a cost-effective approach for interrogating the genome for common variants contributing to complex disease, and these studies have been highly successful in identifying genetic contributors to disease [149]. The utility of these studies is constrained by the statistical power to detect sample size [150], and as most GWAS have identified variants with modest effects, collecting sufficiently large numbers of cases and controls has proved challenging for even studies of highly prevalent disease. The modest prevalence of VaD among dementias (lower prevalence than AD); the diversity of disease subtypes, pathogenetic mechanisms, and case definitions; and the absence of well-established genetic risk factors of strong effect (odds ratio > 2) has made examining VaD using GWAS a challenging prospect.

Thus far, only two GWASs with very small numbers of cases have been performed for VaD. The first, a Korean case–control study of 84 VaD cases and 200 controls, was not able to detect any associations with genome-wide significance ($P < 5 \times 10^{-8}$) [151]; however, secondary analyses in this study provided suggestive evidence for a role of variation in the chr9q22.2 gene *SYK* (encoding spleen tyrosine kinase) in VaD. The second GWAS [152] examined 67 VaD cases and 5700 dementia-free individuals in the Rotterdam Study, a large, population-based prospective study from the Netherlands. One genomic variant, rs12007229 which is located 15.5 kb upstream of the chromosome Xq12 gene *AR* (encoding androgen receptor), demonstrated a statistically significant genome-wide association with increased risk of VaD (OR = 3.7 [95% CI: 2.3, 5.8]; $P = 1.3 \times 10^{-8}$). This association was replicated in a panel of 221 VaD cases and 213 controls recruited from German memory and gerontological clinics, although this association was with a smaller effect size (OR = 1.5 [95% CI: 1.0–2.4]) and a nominal significance ($P = 0.02$). A second variant demonstrating genome-wide significance in the discovery phase of the study (OR = 3.5 [95% CI: 2.2–5.4]; $P = 3.7 \times 10^{-8}$), rs10491487 in the chromosome 5q14.1 gene *RASGRF2* (which encodes Ras protein-specific guanine nucleotide-releasing factor 2), however, did not replicate ($P = 0.42$). Neither of these variants, nor the top 31 SNPs in the Dutch Study ($P < 10^{-5}$) replicated in a follow-up in the Korean sample of the first GWAS; however, this could be due to issues of statistical power or frequency of the variants of interest [153].

Several GWASs have been published that examined proxy phenotypes, including WMLs (proxy for subcortical VaD). A first GWAS on WML was performed in the CHARGE consortium [142], examining 9361 stroke-free individuals of European descent from seven community-based cohorts for white matter hyperintensity (WMH) burden, and following up significant associations in two additional cohorts with 3024 individuals. Six genomic variants in one locus on chromosome 17q25 demonstrated associations with genome-wide statistical significance, the most significant being rs3744028 ($P_{discovery} = 4.0 \times 10^{-9}$; $P_{replication} = 1.3 \times 10^{-7}$; $P_{combined} = 4.0 \times 10^{-15}$). These variants were distributed across six known genes which included *WBP2*, *TRIM65*, *TRIM47*, *MRPL38*, *FBF1*, and *ACOX1* (see Table 2 for GWAS and candidate gene associations), and their combined effects conferred a modest increase in WMH burden (4—8%). The study also reported three additional SNPs with suggestive P-values ($P < 5 \times 10^{-5}$) in the genes *PMF1*, *COL25A1*, and *MTHFD1*. Associations of the six variants with genome-wide statistical significance were also replicated in a separate wave of the Rotterdam Study, RS III, with all variants demonstrating significant associations after adjustment for multiple testing (e.g., $P = 1.1 \times 10^{-3}$ for rs3744028) [154].

While these studies provide some interesting loci for further fine-mapping and functional genomics studies, expanded GWASs with larger sample sizes will have more statistical power to identify loci with more modest effects (OR < 2) on both VaD and proxy phenotypes, providing further opportunities to identify additional VaD loci.

3.3 Monogenic Forms of VaD

While candidate gene studies have typically focused on identifying common variants underlying VaD, several rare subtypes of VaD have been shown to be highly genetic (Table 3).

One such form of VaD is cerebral autosomal dominant arteriopathy with subcortical infarcts and leukoencephalopathy (CADASIL) [156]. CADASIL (MIM: 125310) is a highly heritable type of cerebrovascular disease that occurs when the thickening of blood vessel walls blocks the flow of blood to the brain, causing infarcts in the brain which often lead to stroke. Among individuals with CADASIL, stroke often occurs repeatedly and can occur any time from childhood to late adulthood. Subcortical infarcts affect regions associated with memory and cognitive function, contributing to dementia. VaD is one of the most common symptoms of CADASIL, which include seizures, vision problems, and a broad spectrum of psychiatric disturbances including depression and changes in personality and mood. CADASIL is caused by various missense mutations or deletions in the *NOTCH3* gene [156,157]. About 95% of these mutations affect amino acid residues in the tandem EGF (epidermal growth factor)-like repeat region of the Notch3 transmembrane receptor and result in either a gain or loss of one (or sometimes three) cysteine residue(s) in the amino-terminal region of the receptor [158]. Notch signaling regulates cellular differentiation and is critical during

TABLE 3 Monogenic Causes of VaD and VaD-Precipitating Diseases

Disease	Affected protein	Gene	Chromosome	Inheritance	Notes
CADASIL (cerebral autosomal dominant arteriopathy with subcortical infarcts and leukoencephalopathy)	Transmembrane receptors of vascular smooth muscle cell	NOTCH3	19p13.12	Dominant	• Non-hypertensive young and middle-aged adults affected
CARASIL (cerebral autosomal recessive arteriopathy with subcortical infarcts and leukoencephalopathy, Maeda syndrome)	HtrA serine peptidase 1	HTRA1	10q 26.13	Recessive	• May be late-onset. • Alopecia and disco-vertebral degeneration common • Observed in Japanese and Chinese populations
FCAAS					
• Hereditary cerebral hemorrhage with amyloidosis, Dutch type (HCHWA-D) ("Dutch" variant)	APP	APP (E693G, D694N)	21q21.3	Dominant	• AD-like dementia • Cortical calcifications • Leukoencephalopathy
• APP-related amyloid angiopathy ("Italian," "Iowa," "Flemish," and "Arctic" variants)	APP	APP (E693K, L705V, N694D)	21q21.3	Dominant	

• Cystatin C amyloid angiopathy ("Icelandic" variant) (HCHWA-I)	Cystatin C	CST3	20p11.21	Dominant	
• ITM2B amyloid angiopathy ("British" and "Danish" variants)	"ABri," "ADan"	ITM2B	13q14.2	Dominant	
HERNS (hereditary endotheliopathy, retinopathy, nephropathy, and stroke)	3'–5' repair exonuclease 1	TREX1 (suspected)	3p21.31	Dominant	• Retinal vasculopathy and cerebral leukodystrophies, encompassing three conditions with a common etiology-cerebroretinal vasculopathy, hereditary vascular retinopathy and HERNS
HEREDITARY HEMORRHAGIC TELANGIECTASIA (OSLER–WEBER–RENDU SYNDROME)					
• Type 1	ENG (endoglin)	ENG	9q34.1	Dominant	• Type 1 is earlier onset; blood vessel malformations in brain more common
• Type 2	Alk-1 (activin receptor-like kinase 1)	ACVRL1	12q13.13	Dominant	• Type 2 and 3 more likely to affect liver
• Type 3	SMAD4	SMAD4	18q21.2	Dominant	• Brain AVMs (10% of patients) may increase stroke risk; overall stroke risk similar to population

Continued

TABLE 3 Monogenic Causes of VaD and VaD-Precipitating Diseases—cont'd

Disease	Affected protein	Gene	Chromosome	Inheritance	Notes
Hyperhomocysteinaemia	Methyltetrahydrofolate reductase (MTHFR)	*MTHFR*	1p36.22	Recessive	• *MTHFR C677T* carriers may not respond to folic acid supplements; instead use 5-MTHF
Fabry disease	α-galactosidase A	*GLA*	Xq22.1	X-linked (recessive)	• Lysosomal storage disease; stroke common
MELAS (mitochondrial encephalopathy with lactic acidosis and seizures)	tRNAs and multiple mitochondrial proteins	*MT–ND1, MT–ND5, MT-TH, MT-TL1, MT-TV*	Mitochondrial genome	Maternal inheritance	• Repeated stroke-like episodes before age 40 may lead to dementia
Moyamoya disease	Various smooth muscle cell proteins involved in proliferation	Type 2- *RNF213*; type 5- *ACTA2*; type 6- *GUCY1A3*; linked loci for type 1 (3p), type 3 (8q23); type 4 (X-linked syndrome)	Type 1- 3p; Type 2- 17q25; Type 3- 8q23; Type 4- X; Type 5- 10q23.3; Type 6- 4q32	Different transmission modes	• Identified in Japanese populations, but observed in multiple racial ethnic groups
Sickle-cell disease	Hemoglobin S	*HBB* (E6V)	11p15.4	Recessive	• Risk of stroke

AVM, arteriovenous malformation.
Adapted from Ref. [155].

development; however, in adults Notch3 is expressed only in vascular smooth muscle cells (VSMCs). The odd number of residues causes defective vessel homeostasis leading to apoptosis of VSMCs.

Another highly heritable form of VaD results from familial cerebral amyloid angiopathy (FCAA). The FCAA is characterized by the accumulation of amyloid in the walls of multiple types of blood vessels (arteries, arterioles, capillaries, and veins) which causes ischemic or hemorrhagic strokes and subsequent VaD. A variety of mutations contribute to different subtypes of FCAA: (1) the "Dutch" variant (MIM: 605714), also known as hereditary cerebral hemorrhage with amyloidosis-Dutch type (HCHWA-D), is caused by mutations in the *APP* gene (a known early-onset AD risk gene, encoding amyloid precursor protein); (2) the "Icelandic" variant (MIM: 105150) is caused by mutations in the gene *CST3* (which encodes Cystatin C); and (3) the "British" (MIM: 176500) or "Danish" (MIM: 117300) variant are caused by mutations in the gene *ITM2B* (encoding integral membrane protein 2B). "Dutch" variant FCAA is caused by an E693Q mutation in *APP* leading to vascular deposition of irregular beta-amyloid [159]. Different *APP* mutations have led to the manifestation of similar FCAA subtypes in several other populations, including "Italian" (E693K, L705V) [160,161] and "Iowa" (N694D) [162] subtypes. "Icelandic" FCAA, often characterized as cystatin C amyloid angiopathy, is caused by an amino acid substitution (L68Q) in *CST3*, which encodes the cystatin C protein that is known to affect amyloid deposition in the brain [163]. FCAA contributing to British familial dementia is the result of a stop-loss mutation (X267R) in the *ITM2B* gene resulting in the production of an abnormal peptide ("ABri"), which is deposited as amyloid fibrils causing neuronal dysfunction and dementia [164]. FCAA causing Danish familial dementia is caused by a 10-bp duplication (795−796insTTTAATTTGT) between codons 265 and 266 [165], one codon ahead of the stop-loss mutation occurring in British familial dementia, which results in the formation of an amyloid peptide ("ADan") having effects similar to ABri in British familial dementia on cerebral vascular amyloid deposition.

Monogenic diseases with cerebrovascular pathologies are also known to precipitate cognitive impairment (Table 3); however, the mutations causing these diseases may have a wide spectrum of neurological effects (e.g., seizures, headaches, and deafness) in addition to cognitive impairment that they are infrequently classified as genes causing VaD. These diseases include moyamoya (MIM: 252350, 607151, 608796, 300845, 614042, and 615750), a disease caused by blocked arteries in the basal ganglia often characterized by transient ischemic attacks; Fabry disease (MIM: 301500), a lysosomal storage disease; and hereditary hemorrhagic telangiectasia (MIM: 187300 and 600376), which presents as cerebral vasculopathy as well as retinal vasculopathy and cerebral leukodystrophy. However, the frequent co-occurrence of VCI with other traits in these syndromes suggests that these genes may be classified as genetic risk factors for VaD with pleiotropic effects on other traits.

4. FUTURE OF GENETIC STUDIES

4.1 Limitations

Identifying genetic risk factors for VaD remains challenging for several reasons, not the least because of the heterogeneity in cerebrovascular pathologies underlying it. Individuals demonstrating VaD may present with acute large vessel infarcts, subcortical small vessel disease, or the more distinctive CAA and CADASIL phenotypes. While recruitment of individuals with VaD may be facilitated by following patients with acute ischemic stroke and examining cognitive function, differences in features of lesions such as volume, location, and laterality may themselves be driven by genetic variation, and thus extreme large sample sizes may be necessary to collect sufficient numbers of individuals with strokes in similar brain regions. Furthermore, there is reason to be concerned with the recruitment of individuals with preexisting dementia—two studies using the Informant Questionnaire on Cognitive Decline in the Elderly (IQCODE) found that between 12% and 16% of stroke patients demonstrated prestroke dementia [166,167], suggesting that a large portion of poststroke dementia may be attributable to pathological factors partially or wholly independent of the occurrence of stroke. Degree of cognitive impairment, and rate of loss of cognitive function, may reflect multiple genetic and environmental factors. However, the length of follow-up and other factors may also influence VaD case definitions—between 6 and 12 months after stroke has occurred, as many as half of patients presenting with poststroke dementia may recover considerably, although the degree of recovery and subsequent impairment may both be functions of neural tissue responses to injury which may vary across individuals. Long-term deterioration of cognitive function in stroke sufferers is likely to manifest itself on a timescale of months or years [168–170]. The occurrence of additional ischemic events may modify the rate of progression of cognitive loss, and the recurrence of these events may be a function of genetic risk factors as well as quality of treatment of initial stroke events. The selection of an aging population for genetic studies also carries with it the risk of identifying individuals with some mixed dementias, including AD pathology. Among these individuals, the prevalence of common genetic risk factors for AD may be sufficiently high as to confound the relationships of these and other genes with VaD.

Study designs may be modified to account for some of these characteristics of poststroke dementia, especially the acute recovery phase, by having longer follow-up periods and repeated follow-up time points to clearly characterize VaD development and progression; however, these features represent multiple costly challenges to study design. For some forms of VaD, such as subcortical VaD, expensive imaging studies using magnetic resonance imaging (MRI) and computed tomography (CT) may be necessary to adequately identify VaD pathology [18], and even with variation across lesions including which regions of the brain are affected, identifying common genetic risk factors for VaD may be possible. However, adequate sample sizes are necessary for identifying modest effects, and this may require collecting data on thousands if not tens of thousands of cases. Phenotyping many thousands of individuals may prove prohibitively expensive, especially if detailed characterizations of the dementia may identify distinctive pathological subgroups to be studied separately (thereby reducing sample size and power to detect associations).

4.2 Opportunities

While characterizing the genetic architecture of complex diseases in general and VaD in particular has proved to be an extremely complex challenge, the advent and advancement of genome-wide studies present opportunities for finding the genetic causes of VaD. GWAS, whole exome sequencing (WES), and whole genome sequencing (WGS) provide ways to agnostically identify genetic risk factors, allowing for the identification of unexpected genetic contributions to disease. These types of studies are not subject to the high degree of false-positive findings and publication bias that have affected candidate gene studies. They have proven useful in confirming known associations as well as identifying novel genetic susceptibility loci, and have facilitated conditional association analyses and studies of gene—environment interaction genome-wide; however, the availability of data on environmental factors and limitations of statistical power represents important challenges to be confronted in designing these studies.

After genome-wide studies, functional studies will be necessary to breakdown and identify what pathological mechanisms these causal variants may affect the most [138]. Examining expression through luciferase assays and performing knockouts of genes in animal models to determine gene function are just two types of studies that will be necessary to characterize the role of truly causal genetic variants for stroke and subsequent VaD. For the foreseeable future, characterizing the genetic architecture of VaD will remain a challenging but hopeful effort.

5. SUMMARY

VaD remains a highly heterogeneous phenotype with a broad spectrum of known genetic and environmental contributors. Studies focusing on biological candidate genes, highly homogeneous and genetically driven subtypes, and genome-wide association analyses have done much to identify genetic risk factors for disease; however, more studies are needed to dissect the complex genetic underpinnings of this heritable and highly variable form of dementia.

ABBREVIATIONS

ABI Ankle—brachial index
AD Alzheimer's disease
Afib Atrial fibrillation
CAA Cerebral amyloid angiopathy
CADASIL Cerebral autosomal dominant arteriopathy with subcortical infarcts and leukoencephalopathy
GWAS Genome-wide association study
MCI Mild cognitive impairment
MID Multi-infarct dementia
NINDS-AIREN National Institute of Neurological Disorders and Stroke and Association Internationale pour la Recherche et l'Enseignement en Neurosciences
PVD Peripheral vascular disease
VaD Vascular dementia
VCI Vascular cognitive impairment
WES Whole exome sequencing
WGS Whole genome sequencing
WMH White matter hyperintensity
WML White matter lesions

References

[1] Roman GC. Vascular dementia revisited: diagnosis, pathogenesis, treatment, and prevention. Med Clin N Am 2002;86:477—99.
[2] Ruitenberg A, Ott A, van Swieten JC, Hofman A, Breteler MM. Incidence of dementia: does gender make a difference? Neurobiol Aging 2001;22:575—80.
[3] Roman G. Vascular dementia: a historical background. Int Psychogeriatr 2003; 15(Suppl. 1):11—3.
[4] Blessed G, Tomlinson BE, Roth M. The association between quantitative measures of dementia and of senile change in the cerebral grey matter of elderly subjects. Br J Psychiatry J Ment Sci 1968;114:797—811.
[5] Roth M, Tomlinson BE, Blessed G. Correlation between scores for dementia and counts of 'senile plaques' in cerebral grey matter of elderly subjects. Nature 1966;209:109—10.
[6] Roth M, Tomlinson BE, Blessed G. The relationship between quantitative measures of dementia and of degenerative changes in the cerebral grey matter of elderly subjects. Proc R Soc Med 1967;60:254—60.

[7] Tomlinson BE, Blessed G, Roth M. Observations on the brains of non-demented old people. J Neurol Sci 1968;7:331—56.

[8] Tomlinson BE, Blessed G, Roth M. Observations on the brains of demented old people. J Neurol Sci 1970;11:205—42.

[9] Hachinski VC, Lassen NA, Marshall J. Multi-infarct dementia. A cause of mental deterioration in the elderly. Lancet 1974;2:207—10.

[10] Bowler JV, Hachinski V. Vascular cognitive impairment: a new approach to vascular dementia. Baillieres Clin Neurol 1995;4:357—76.

[11] Roman GC, Tatemichi TK, Erkinjuntti T, Cummings JL, Masdeu JC, Garcia JH, et al. Vascular dementia: diagnostic criteria for research studies. Report of the NINDS-AIREN International Workshop. Neurology 1993;43:250—60.

[12] Bowler JV, Steenhuis R, Hachinski V. Conceptual background to vascular cognitive impairment. Alzheimer Dis Assoc Disord 1999;13(Suppl. 3):S30—7.

[13] Feldman H, Levy AR, Hsiung GY, Peters KR, Donald A, Black SE, et al. A Canadian cohort study of cognitive impairment and related dementias (ACCORD): study methods and baseline results. Neuroepidemiology 2003;22:265—74.

[14] Hachinski V. Vascular dementia: a radical redefinition. Dementia 1994;5:130—2.

[15] O'Brien JT, Erkinjuntti T, Reisberg B, Roman G, Sawada T, Pantoni L, et al. Vascular cognitive impairment. Lancet Neurol 2003;2:89—98.

[16] Rockwood K, Davis H, MacKnight C, Vandorpe R, Gauthier S, Guzman A, et al. The consortium to investigate vascular impairment of cognition: methods and first findings. Can J Neurol Sci Le Journal Canadien des Sciences Neurologiques 2003;30:237—43.

[17] Rosand J, Altshuler D. Human genome sequence variation and the search for genes influencing stroke. Stroke J Cereb Circ 2003;34:2512—6.

[18] Leblanc GG, Meschia JF, Stuss DT, Hachinski V. Genetics of vascular cognitive impairment: the opportunity and the challenges. Stroke J Cereb Circ 2006;37:248—55.

[19] Candore G, Balistreri CR, Grimaldi MP, Vasto S, Listi F, Chiappelli M, et al. Age-related inflammatory diseases: role of genetics and gender in the pathophysiology of Alzheimer's disease. Ann N Y Acad Sci 2006;1089:472—86.

[20] Iemolo F, Duro G, Rizzo C, Castiglia L, Hachinski V, Caruso C. Pathophysiology of vascular dementia. Immun Ageing 2009;6:13.

[21] World Health Organization. The ICD-10 classification of mental and behavioural disorders : diagnostic criteria for research. Geneva: World Health Organization; 1993.

[22] Plassman BL, Langa KM, Fisher GG, Heeringa SG, Weir DR, Ofstedal MB, et al. Prevalence of dementia in the United States: the aging, demographics, and memory study. Neuroepidemiology 2007;29:125—32.

[23] Froehlich TE, Bogardus Jr ST, Inouye SK. Dementia and race: are there differences between African Americans and Caucasians? J Am Geriatr Soc 2001;49:477—84.

[24] Still CN, Jackson KL, Brandes DA, Abramson RK, Macera CA. Distribution of major dementias by race and sex in South Carolina. J S C Med Assoc 1990;86:453—6.

[25] Imfeld P, Brauchli Pernus YB, Jick SS, Meier CR. Epidemiology, co-morbidities, and medication use of patients with Alzheimer's disease or vascular dementia in the UK. J Alzheimer's Dis 2013;35:565—73.

[26] Ravaglia G, Forti P, Maioli F, Martelli M, Servadei L, Brunetti N, et al. Incidence and etiology of dementia in a large elderly Italian population. Neurology 2005;64:1525—30.

[27] Jorm AF. Cross-national comparisons of the occurrence of Alzheimer's and vascular dementias. Eur Arch Psychiatry Clin Neurosci 1991;240:218—22.

[28] Yeo G, Gallagher-Thompson D, Lieberman M. Variations in dementia characteristics by ethnic category. In: Yeo G, Gallagher-Thompson D, editors. Ethnicity and the dementias. Washington, DC: Taylor and Francis; 1996. p. 21—30.

[29] Ikeda M, Hokoishi K, Maki N, Nebu A, Tachibana N, Komori K, et al. Increased prevalence of vascular dementia in Japan: a community-based epidemiological study. Neurology 2001;57:839−44.

[30] Shadlen MF, Larson EB, Yukawa M. The epidemiology of Alzheimer's disease and vascular dementia in Japanese and African-American populations: the search for etiological clues. Neurobiol Aging 2000;21:171−81.

[31] Yamada M, Mimori Y, Kasagi F, Miyachi T, Ohshita T, Sudoh S, et al. Incidence of dementia, Alzheimer disease, and vascular dementia in a Japanese population: Radiation Effects Research Foundation adult health study. Neuroepidemiology 2008;30: 152−60.

[32] Yamada M, Sasaki H, Mimori Y, Kasagi F, Sudoh S, Ikeda J, et al. Prevalence and risks of dementia in the Japanese population: RERF's adult health study Hiroshima subjects. Radiation Effects Research Foundation. J Am Geriatr Soc 1999;47:189−95.

[33] Leys D, Henon H, Mackowiak-Cordoliani MA, Pasquier F. Poststroke dementia. Lancet Neurol 2005;4:752−9.

[34] Farooq MU, Gorelick PB. Vascular cognitive impairment. Curr Atheroscler Rep 2013; 15:330.

[35] Corrada MM, Brookmeyer R, Paganini-Hill A, Berlau D, Kawas CH. Dementia incidence continues to increase with age in the oldest old: the 90+ study. Ann Neurol 2010;67:114−21.

[36] Fratiglioni L, Launer LJ, Andersen K, Breteler MM, Copeland JR, Dartigues JF, et al. Incidence of dementia and major subtypes in Europe: a collaborative study of population-based cohorts. Neurologic Diseases in the Elderly Research Group. Neurology 2000;54:S10−5.

[37] Lobo A, Launer LJ, Fratiglioni L, Andersen K, Di Carlo A, Breteler MM, et al. Prevalence of dementia and major subtypes in Europe: a collaborative study of population-based cohorts. Neurologic Diseases in the Elderly Research Group. Neurology 2000; 54:S4−9.

[38] Andersen K, Launer LJ, Dewey ME, Letenneur L, Ott A, Copeland JR, et al. Gender differences in the incidence of AD and vascular dementia: the EURODEM Studies. EURODEM Incidence Research Group. Neurology 1999;53:1992−7.

[39] Fitzpatrick AL, Kuller LH, Ives DG, Lopez OL, Jagust W, Breitner JC, et al. Incidence and prevalence of dementia in the Cardiovascular Health Study. J Am Geriatr Soc 2004;52:195−204.

[40] Gorelick PB, Scuteri A, Black SE, Decarli C, Greenberg SM, Iadecola C, et al. Vascular contributions to cognitive impairment and dementia: a statement for healthcare professionals from the american heart association/american stroke association. Stroke J Cereb Circ 2011;42:2672−713.

[41] del Valle J, Duran-Vilaregut J, Manich G, Pallas M, Camins A, Vilaplana J, et al. Cerebral amyloid angiopathy, blood−brain barrier disruption and amyloid accumulation in SAMP8 mice. Neurodegener Dis 2011;8:421−9.

[42] Grysiewicz R, Gorelick PB. Key neuroanatomical structures for post-stroke cognitive impairment. Curr Neurol Neurosci Rep 2012;12:703−8.

[43] Viswanathan A, Patel P, Rahman R, Nandigam RN, Kinnecom C, Bracoud L, et al. Tissue microstructural changes are independently associated with cognitive impairment in cerebral amyloid angiopathy. Stroke J Cereb Circ 2008;39:1988−92.

[44] Arvanitakis Z, Leurgans SE, Wang Z, Wilson RS, Bennett DA, Schneider JA. Cerebral amyloid angiopathy pathology and cognitive domains in older persons. Ann Neurol 2011;69:320−7.

[45] Greenberg SM, Gurol ME, Rosand J, Smith EE. Amyloid angiopathy-related vascular cognitive impairment. Stroke J Cereb Circ 2004;35:2616−9.

REFERENCES

201

[46] Ott A, Breteler MM, van Harskamp F, Claus JJ, van der Cammen TJ, Grobbee DE, et al. Prevalence of Alzheimer's disease and vascular dementia: association with education. The Rotterdam study. BMJ 1995;310:970–3.

[47] Scarmeas N, Stern Y, Mayeux R, Manly JJ, Schupf N, Luchsinger JA. Mediterranean diet and mild cognitive impairment. Arch Neurol 2009;66:216–25.

[48] Buell JS, Scott TM, Dawson-Hughes B, Dallal GE, Rosenberg IH, Folstein MF, et al. Vitamin D is associated with cognitive function in elders receiving home health services. J Gerontol A Biol Sci Med Sci 2009;64:888–95.

[49] Slinin Y, Paudel ML, Taylor BC, Fink HA, Ishani A, Canales MT, et al. 25-Hydroxyvitamin D levels and cognitive performance and decline in elderly men. Neurology 2010;74:33–41.

[50] Kang JH, Cook NR, Manson JE, Buring JE, Albert CM, Grodstein F. Vitamin E, vitamin C, beta carotene, and cognitive function among women with or at risk of cardiovascular disease: the Women's Antioxidant and Cardiovascular Study. Circulation 2009;119: 2772–80.

[51] Kang JH, Grodstein F. Plasma carotenoids and tocopherols and cognitive function: a prospective study. Neurobiol Aging 2008;29:1394–403.

[52] Cotman CW, Berchtold NC. Exercise: a behavioral intervention to enhance brain health and plasticity. Trends Neurosci 2002;25:295–301.

[53] Soumare A, Tavernier B, Alperovitch A, Tzourio C, Elbaz A. A cross-sectional and longitudinal study of the relationship between walking speed and cognitive function in community-dwelling elderly people. J Gerontol A Biol Sci Med Sci 2009;64:1058–65.

[54] Sturman MT, Morris MC, Mendes de Leon CF, Bienias JL, Wilson RS, Evans DA. Physical activity, cognitive activity, and cognitive decline in a biracial community population. Arch Neurol 2005;62:1750–4.

[55] Weuve J, Kang JH, Manson JE, Breteler MM, Ware JH, Grodstein F. Physical activity, including walking, and cognitive function in older women. JAMA 2004;292:1454–61.

[56] Gorelick PB. Can we save the brain from the ravages of midlife cardiovascular risk factors? Neurology 1999;52:1114–5.

[57] Kilander L, Nyman H, Boberg M, Hansson L, Lithell H. Hypertension is related to cognitive impairment: a 20-year follow-up of 999 men. Hypertension 1998;31:780–6.

[58] Kivipelto M, Helkala EL, Hanninen T, Laakso MP, Hallikainen M, Alhainen K, et al. Midlife vascular risk factors and late-life mild cognitive impairment: a population-based study. Neurology 2001;56:1683–9.

[59] Qiu C, Winblad B, Fratiglioni L. The age-dependent relation of blood pressure to cognitive function and dementia. Lancet Neurol 2005;4:487–99.

[60] Gorelick PB. Risk factors for vascular dementia and Alzheimer disease. Stroke J Cereb Circ 2004;35:2620–2.

[61] Rincon F, Wright CB. Vascular cognitive impairment. Curr Opin Neurol 2013;26: 29–36.

[62] Cox DJ, Kovatchev BP, Gonder-Frederick LA, Summers KH, McCall A, Grimm KJ, et al. Relationships between hyperglycemia and cognitive performance among adults with type 1 and type 2 diabetes. Diabetes Care 2005;28:71–7.

[63] McNay EC. The impact of recurrent hypoglycemia on cognitive function in aging. Neurobiol Aging 2005;26(Suppl. 1):76–9.

[64] Saczynski JS, Jonsdottir MK, Garcia ME, Jonsson PV, Peila R, Eiriksdottir G, et al. Cognitive impairment: an increasingly important complication of type 2 diabetes: the age, gene/environment susceptibility—Reykjavik study. Am J Epidemiol 2008; 168:1132–9.

[65] Evans JG, Areosa SA. Effect of the treatment of type II diabetes mellitus on the development of cognitive impairment and dementia. Cochrane Database Syst Rev 2003;1: CD003804. http://dx.doi.org/10.1002/14651858.CD003804.

[66] Solomon A, Kareholt I, Ngandu T, Wolozin B, Macdonald SW, Winblad B, et al. Serum total cholesterol, statins and cognition in non-demented elderly. Neurobiol Aging 2009;30:1006—9.

[67] Trompet S, van Vliet P, de Craen AJ, Jolles J, Buckley BM, Murphy MB, et al. Pravastatin and cognitive function in the elderly. Results of the PROSPER study. J Neurol 2010;257:85—90.

[68] Kalmijn S, Foley D, White L, Burchfiel CM, Curb JD, Petrovitch H, et al. Metabolic cardiovascular syndrome and risk of dementia in Japanese-American elderly men. The Honolulu-Asia aging study. Arterioscler Thromb Vasc Biol 2000;20:2255—60.

[69] Rusanen M, Kivipelto M, Quesenberry Jr CP, Zhou J, Whitmer RA. Heavy smoking in midlife and long-term risk of Alzheimer disease and vascular dementia. Arch Intern Med 2011;171:333—9.

[70] Elias PK, Elias MF, D'Agostino RB, Silbershatz H, Wolf PA. Alcohol consumption and cognitive performance in the Framingham Heart Study. Am J Epidemiol 1999;150: 580—9.

[71] Wright CB, Elkind MS, Rundek T, Boden-Albala B, Paik MC, Sacco RL. Alcohol intake, carotid plaque, and cognition: the Northern Manhattan Study. Stroke J Cereb Circ 2006;37:1160—4.

[72] Demarin V, Zavoreo I, Kes VB. Carotid artery disease and cognitive impairment. J Neurol Sci 2012;322:107—11.

[73] Komulainen P, Kivipelto M, Lakka TA, Hassinen M, Helkala EL, Patja K, et al. Carotid intima-media thickness and cognitive function in elderly women: a population-based study. Neuroepidemiology 2007;28:207—13.

[74] Sander K, Bickel H, Forstl H, Etgen T, Briesenick C, Poppert H, et al. Carotid- intima media thickness is independently associated with cognitive decline. The INVADE study. Int J Geriatr Psychiatry 2010;25:389—94.

[75] Bunch TJ, Weiss JP, Crandall BG, May HT, Bair TL, Osborn JS, et al. Atrial fibrillation is independently associated with senile, vascular, and Alzheimer's dementia. Heart Rhythm Off J Heart Rhythm Soc 2010;7:433—7.

[76] Dublin S, Anderson ML, Haneuse SJ, Heckbert SR, Crane PK, Breitner JC, et al. Atrial fibrillation and risk of dementia: a prospective cohort study. J Am Geriatr Soc 2011;59: 1369—75.

[77] Barber M, Tait RC, Scott J, Rumley A, Lowe GD, Stott DJ. Dementia in subjects with atrial fibrillation: hemostatic function and the role of anticoagulation. J Thromb Haemost 2004;2:1873—8.

[78] Duron E, Hanon O. Vascular risk factors, cognitive decline, and dementia. Vasc Health Risk Manag 2008;4:363—81.

[79] Ezekowitz MD, James KE, Nazarian SM, Davenport J, Broderick JP, Gupta SR, et al. Silent cerebral infarction in patients with nonrheumatic atrial fibrillation. The veterans affairs stroke prevention in nonrheumatic atrial fibrillation investigators. Circulation 1995;92:2178—82.

[80] Kalantarian S, Stern TA, Mansour M, Ruskin JN. Cognitive impairment associated with atrial fibrillation: a meta-analysis. Ann Intern Med 2013;158:338—46.

[81] Price JF, McDowell S, Whiteman MC, Deary IJ, Stewart MC, Fowkes FG. Ankle brachial index as a predictor of cognitive impairment in the general population: ten-year follow-up of the Edinburgh Artery Study. J Am Geriatr Soc 2006;54:763—9.

[82] Laurin D, Masaki KH, White LR, Launer LJ. Ankle-to-brachial index and dementia: the Honolulu-Asia Aging Study. Circulation 2007;116:2269—74.

[83] Korczyn AD, Halperin I. Depression and dementia. J Neurol Sci 2009;283:139—42.

[84] Pohjasvaara T, Erkinjuntti T, Ylikoski R, Hietanen M, Vataja R, Kaste M. Clinical determinants of poststroke dementia. Stroke J Cereb Circ 1998;29:75—81.

[85] Jellinger KA. The enigma of mixed dementia. Alzheimers Dement J Alzheimers Assoc 2007;3:40−53.

[86] Heyman A, Fillenbaum GG, Welsh-Bohmer KA, Gearing M, Mirra SS, Mohs RC, et al. Cerebral infarcts in patients with autopsy-proven Alzheimer's disease: CERAD, part XVIII. Consortium to Establish a Registry for Alzheimer's Disease. Neurology 1998;51: 159−62.

[87] Jellinger KA, Attems J. Prevalence and pathogenic role of cerebrovascular lesions in Alzheimer disease. J Neurol Sci 2005;229-230:37−41.

[88] Jellinger KA, Attems J. Prevalence and impact of cerebrovascular pathology in Alzheimer's disease and parkinsonism. Acta Neurol Scand 2006;114:38−46.

[89] DeStefano AL, Atwood LD, Massaro JM, Heard-Costa N, Beiser A, Au R, et al. Genome-wide scan for white matter hyperintensity: the Framingham Heart Study. Stroke J Cereb Circ 2006;37:77−81.

[90] Kochunov P, Glahn D, Lancaster J, Winkler A, Kent Jr JW, Olvera RL, et al. Whole brain and regional hyperintense white matter volume and blood pressure: overlap of genetic loci produced by bivariate, whole-genome linkage analyses. Stroke J Cereb Circ 2010;41:2137−42.

[91] Seshadri S, DeStefano AL, Au R, Massaro JM, Beiser AS, Kelly-Hayes M, et al. Genetic correlates of brain aging on MRI and cognitive test measures: a genome-wide association and linkage analysis in the Framingham Study. BMC Med Genet 2007;8(Suppl. 1):S15.

[92] Turner ST, Fornage M, Jack Jr CR, Mosley TH, Knopman DS, Kardia SL, et al. Genomic susceptibility Loci for brain atrophy, ventricular volume, and leukoaraiosis in hypertensive sibships. Arch Neurol 2009;66:847−57.

[93] Kim Y, Kong M, Lee C. Lack of common genetic factors for susceptibility to vascular dementia and Alzheimer's disease. Gene 2012;497:298−300.

[94] Lambert JC, Ibrahim-Verbaas CA, Harold D, Naj AC, Sims R, Bellenguez C, et al. Meta-analysis of 74,046 individuals identifies 11 new susceptibility loci for Alzheimer's disease. Nat Genet 2013;45:1452−8.

[95] Chuang YF, Hayden KM, Norton MC, Tschanz J, Breitner JC, Welsh-Bohmer KA, et al. Association between APOE epsilon4 allele and vascular dementia: the Cache County study. Dement Geriatr Cogn Disord 2010;29:248−53.

[96] Davidson Y, Gibbons L, Purandare N, Byrne J, Hardicre J, Wren J, et al. Apolipoprotein E epsilon4 allele frequency in vascular dementia. Dement Geriatr Cogn Disord 2006; 22:15−9.

[97] Frisoni GB, Calabresi L, Geroldi C, Bianchetti A, D'Acquarica AL, Govoni S, et al. Apolipoprotein E epsilon 4 allele in Alzheimer's disease and vascular dementia. Dementia 1994;5:240−2.

[98] Kalman J, Juhasz A, Csaszar A, Kanka A, Rimanoczy A, Janka Z, et al. Increased apolipoprotein E4 allele frequency is associated with vascular dementia in the Hungarian population. Acta Neurol Scand 1998;98:166−8.

[99] Slooter AJ, Tang MX, van Duijn CM, Stern Y, Ott A, Bell K, et al. Apolipoprotein E epsilon4 and the risk of dementia with stroke. A population-based investigation. JAMA 1997;277:818−21.

[100] Baum L, Lam LC, Kwok T, Lee J, Chiu HF, Mok VC, et al. Apolipoprotein E epsilon4 allele is associated with vascular dementia. Dement Geriatr Cogn Disord 2006;22: 301−5.

[101] Katzman R, Zhang MY, Chen PJ, Gu N, Jiang S, Saitoh T, et al. Effects of apolipoprotein E on dementia and aging in the Shanghai Survey of Dementia. Neurology 1997;49: 779−85.

[102] Yang J, Feng G, Zhang J, Hui Z, Breen G, St Clair D, et al. Is ApoE gene a risk factor for vascular dementia in Han Chinese? Int J Mol Med 2001;7:217−9.

[103] Engelborghs S, Dermaut B, Goeman J, Saerens J, Marien P, Pickut BA, et al. Prospective Belgian study of neurodegenerative and vascular dementia: APOE genotype effects. J Neurol Neurosurg Psychiatry 2003;74:1148—51.

[104] Kim KW, Youn JC, Han MK, Paik NJ, Lee TJ, Park JH, et al. Lack of association between apolipoprotein E polymorphism and vascular dementia in Koreans. J Geriatr Psychiatry Neurol 2008;21:12—7.

[105] Lin HF, Lai CL, Tai CT, Lin RT, Liu CK. Apolipoprotein E polymorphism in ischemic cerebrovascular diseases and vascular dementia patients in Taiwan. Neuroepidemiology 2004;23:129—34.

[106] Nakayama S, Kuzuhara S. Apolipoprotein E phenotypes in healthy normal controls and demented subjects with Alzheimer's disease and vascular dementia in Mie Prefecture of Japan. Psychiatry Clin Neurosci 1999;53:643—8.

[107] Cambien F, Poirier O, Lecerf L, Evans A, Cambou JP, Arveiler D, et al. Deletion polymorphism in the gene for angiotensin-converting enzyme is a potent risk factor for myocardial infarction. Nature 1992;359:641—4.

[108] Catto A, Carter AM, Barrett JH, Stickland M, Bamford J, Davies JA, et al. Angiotensin-converting enzyme insertion/deletion polymorphism and cerebrovascular disease. Stroke J Cereb Circ 1996;27:435—40.

[109] Markus HS, Barley J, Lunt R, Bland JM, Jeffery S, Carter ND, et al. Angiotensin-converting enzyme gene deletion polymorphism. A new risk factor for lacunar stroke but not carotid atheroma. Stroke J Cereb Circ 1995;26:1329—33.

[110] Roffman JL, Weiss AP, Deckersbach T, Freudenreich O, Henderson DC, Purcell S, et al. Effects of the methylenetetrahydrofolate reductase (MTHFR) C677T polymorphism on executive function in schizophrenia. Schizophr Res 2007;92:181—8.

[111] Sibani S, Christensen B, O'Ferrall E, Saadi I, Hiou-Tim F, Rosenblatt DS, et al. Characterization of six novel mutations in the methylenetetrahydrofolate reductase (MTHFR) gene in patients with homocystinuria. Hum Mutat 2000;15:280—7.

[112] Bertina RM, Koeleman BP, Koster T, Rosendaal FR, Dirven RJ, de Ronde H, et al. Mutation in blood coagulation factor V associated with resistance to activated protein C. Nature 1994;369:64—7.

[113] Ridker PM, Hennekens CH, Lindpaintner K, Stampfer MJ, Eisenberg PR, Miletich JP. Mutation in the gene coding for coagulation factor V and the risk of myocardial infarction, stroke, and venous thrombosis in apparently healthy men. N Engl J Med 1995;332:912—7.

[114] Borroni B, Akkawi N, Martini G, Colciaghi F, Prometti P, Rozzini L, et al. Microvascular damage and platelet abnormalities in early Alzheimer's disease. J Neurol Sci 2002;203-204:189—93.

[115] Gupta A, Pansari K. The association between blood coagulation markers, atherothrombosis and dementia. Int J Clin Pract 2003;57:107—11.

[116] Frosst P, Blom HJ, Milos R, Goyette P, Sheppard CA, Matthews RG, et al. A candidate genetic risk factor for vascular disease: a common mutation in methylenetetrahydrofolate reductase. Nat Genet 1995;10:111—3.

[117] Kohara K, Fujisawa M, Ando F, Tabara Y, Niino N, Miki T, et al. MTHFR gene polymorphism as a risk factor for silent brain infarcts and white matter lesions in the Japanese general population: the NILS-LSA Study. Stroke J Cereb Circ 2003;34:1130—5.

[118] Yoo JH, Choi GD, Kang SS. Pathogenicity of thermolabile methylenetetrahydrofolate reductase for vascular dementia. Arterioscler Thromb Vasc Biol 2000;20:1921—5.

[119] Chapman J, Wang N, Treves TA, Korczyn AD, Bornstein NM. ACE, MTHFR, factor V Leiden, and APOE polymorphisms in patients with vascular and Alzheimer's dementia. Stroke J Cereb Circ 1998;29:1401—4.

[120] Zuliani G, Ble A, Zanca R, Munari MR, Zurlo A, Vavalle C, et al. Genetic polymorphisms in older subjects with vascular or Alzheimer's dementia. Acta Neurol Scand 2001;103:304−8.

[121] Rigat B, Hubert C, Alhenc-Gelas F, Cambien F, Corvol P, Soubrier F. An insertion/ deletion polymorphism in the angiotensin I-converting enzyme gene accounting for half the variance of serum enzyme levels. J Clin Invest 1990;86:1343−6.

[122] Kim Y, Kim JH, Nam YJ, Kim YJ, Yu KH, Lee BC, et al. Sequence variants of ACE, AGT, AT1R, and PAI-1 as genetic risk factors for vascular dementia. Neurosci Lett 2006;401: 276−9.

[123] Kolsch H, Linnebank M, Lutjohann D, Jessen F, Wullner U, Harbrecht U, et al. Polymorphisms in glutathione S-transferase omega-1 and AD, vascular dementia, and stroke. Neurology 2004;63:2255−60.

[124] Fung HC, Chen CM, Wu YR, Hsu WC, Ro LS, Lin JC, et al. Heat shock protein 70 and tumor necrosis factor alpha in Taiwanese patients with dementia. Dement Geriatr Cogn Disord 2005;20:1−7.

[125] Garcia J, Ahmadi A, Wonnacott A, Sutcliffe W, Nagga K, Soderkvist P, et al. Association of insulin-like growth factor-1 receptor polymorphism in dementia. Dement Geriatr Cogn Disord 2006;22:439−44.

[126] Wang HK, Hsu WC, Fung HC, Lin JC, Hsu HP, Wu YR, et al. Interleukin-1alpha and -1beta promoter polymorphisms in Taiwanese patients with dementia. Dement Geriatr Cogn Disord 2007;24:104−10.

[127] Yucesoy B, Peila R, White LR, Wu KM, Johnson VJ, Kashon ML, et al. Association of interleukin-1 gene polymorphisms with dementia in a community-based sample: the Honolulu-Asia Aging Study. Neurobiol Aging 2006;27:211−7.

[128] Pola R, Flex A, Gaetani E, Papaleo P, De Martini D, Gerardino L, et al. Association between intercellular adhesion molecule-1 E/K gene polymorphism and probable vascular dementia in humans. Neurosci Lett 2002;326:171−4.

[129] Pola R, Gaetani E, Flex A, Aloi F, Papaleo P, Gerardino L, et al. -174 G/C interleukin-6 gene polymorphism and increased risk of multi-infarct dementia: a case-control study. Exp Gerontol 2002;37:949−55.

[130] Flex A, Gaetani E, Proia AS, Pecorini G, Straface G, Biscetti F, et al. Analysis of functional polymorphisms of metalloproteinase genes in persons with vascular dementia and Alzheimer's disease. J Gerontol A Biol Sci Med Sci 2006;61:1065−9.

[131] Helbecque N, Cottel D, Codron V, Berr C, Amouyel P. Paraoxonase 1 gene polymorphisms and dementia in humans. Neurosci Lett 2004;358:41−4.

[132] Janka Z, Juhasz A, Rimanoczy AA, Boda K, Marki-Zay J, Kalman J. Codon 311 (Cys−>Ser) polymorphism of paraoxonase-2 gene is associated with apolipoprotein E4 allele in both Alzheimer's and vascular dementias. Mol Psychiatry 2002;7:110−2.

[133] Kim Y, Nam YJ, Lee C. Analysis of the SREBF2 gene as a genetic risk factor for vascular dementia. Am J Med Genet B Neuropsychiatr Genet Off Publ Int Soc Psychiatric Genet 2005;139B:19−22.

[134] Peila R, Yucesoy B, White LR, Johnson V, Kashon ML, Wu K, et al. A TGF-beta1 polymorphism association with dementia and neuropathologies: the HAAS. Neurobiol Aging 2007;28:1367−73.

[135] McCusker SM, Curran MD, Dynan KB, McCullagh CD, Urquhart DD, Middleton D, et al. Association between polymorphism in regulatory region of gene encoding tumour necrosis factor alpha and risk of Alzheimer's disease and vascular dementia: a case-control study. Lancet 2001;357:436−9.

[136] Kim Y, Nam YJ, Lee C. Haplotype analysis of single nucleotide polymorphisms in VEGF gene for vascular dementia. Am J Med Genet B Neuropsychiatr Genet Off Publ Int Soc Psychiatric Genet 2006b;141B:332−5.

[137] Helbecque N, Berr C, Cottel D, Fromentin-David I, Sazdovitch V, Ricolfi F, et al. VLDL receptor polymorphism, cognitive impairment, and dementia. Neurology 2001;56: 1183–8.

[138] Kim Y, Kong M, An J, Ryu J, Lee C. Genetic dissection of susceptibility to vascular dementia. Psychiatr Genet 2011;21:69–76.

[139] Schmidt H, Freudenberger P, Seiler S, Schmidt R. Genetics of subcortical vascular dementia. Exp Gerontol 2012;47:873–7.

[140] Paternoster L, Chen W, Sudlow CL. Genetic determinants of white matter hyperintensities on brain scans: a systematic assessment of 19 candidate gene polymorphisms in 46 studies in 19,000 subjects. Stroke J Cereb Circ 2009;40:2020–6.

[141] Schmidt H, Fazekas F, Kostner GM, van Duijn CM, Schmidt R. Angiotensinogen gene promoter haplotype and microangiopathy-related cerebral damage: results of the Austrian Stroke Prevention Study. Stroke J Cereb Circ 2001;32:405–12.

[142] Fornage M, Debette S, Bis JC, Schmidt H, Ikram MA, Dufouil C, et al. Genome-wide association studies of cerebral white matter lesion burden: the CHARGE consortium. Ann Neurol 2011;69:928–39.

[143] Taylor WD, Steffens DC, Ashley-Koch A, Payne ME, MacFall JR, Potocky CF, et al. Angiotensin receptor gene polymorphisms and 2-year change in hyperintense lesion volume in men. Mol Psychiatry 2010;15:816–22.

[144] de Leeuw FE, Richard F, de Groot JC, van Duijn CM, Hofman A, Van Gijn J, et al. Interaction between hypertension, apoE, and cerebral white matter lesions. Stroke J Cereb Circ 2004;35:1057–60.

[145] Godin O, Tzourio C, Maillard P, Alperovitch A, Mazoyer B, Dufouil C. Apolipoprotein E genotype is related to progression of white matter lesion load. Stroke J Cereb Circ 2009;40:3186–90.

[146] Schmidt H, Zeginigg M, Wiltgen M, Freudenberger P, Petrovic K, Cavalieri M, et al. Genetic variants of the NOTCH3 gene in the elderly and magnetic resonance imaging correlates of age-related cerebral small vessel disease. Brain J Neurol 2011;134: 3384–97.

[147] Schuur M, van Swieten JC, Schol-Gelok S, Ikram MA, Vernooij MW, Liu F, et al. Genetic risk factors for cerebral small-vessel disease in hypertensive patients from a genetically isolated population. J Neurol Neurosurg Psychiatry 2011;82:41–4.

[148] Poels MM, Vernooij MW, Ikram MA, Hofman A, Krestin GP, van der Lugt A, et al. Prevalence and risk factors of cerebral microbleeds: an update of the Rotterdam scan study. Stroke J Cereb Circ 2010;41:S103–6.

[149] Manolio TA, Collins FS, Cox NJ, Goldstein DB, Hindorff LA, Hunter DJ, et al. Finding the missing heritability of complex diseases. Nature 2009;461:747–53.

[150] Spencer CC, Su Z, Donnelly P, Marchini J. Designing genome-wide association studies: sample size, power, imputation, and the choice of genotyping chip. PLoS Genet 2009;5:e1000477.

[151] Kim Y, Kong M, Lee C. Association of intronic sequence variant in the gene encoding spleen tyrosine kinase with susceptibility to vascular dementia. World J Biol Psychiatry Off J World Fed Soc Biol Psychiatry 2013;14:220–6.

[152] Schrijvers EM, Schurmann B, Koudstaal PJ, van den Bussche H, Van Duijn CM, Hentschel F, et al. Genome-wide association study of vascular dementia. Stroke J Cereb Circ 2012;43:315–9.

[153] Lee C, Kim Y. Complex genetic susceptibility to vascular dementia and an evidence for its underlying genetic factors associated with memory and associative learning. Gene 2013;516:152–7.

[154] Verhaaren BF, de Boer R, Vernooij MW, Rivadeneira F, Uitterlinden AG, Hofman A, et al. Replication study of chr17q25 with cerebral white matter lesion volume. Stroke J Cereb Circ 2011;42:3297–9.

[155] Korczyn AD, Vakhapova V, Grinberg LT. Vascular dementia. J Neurol Sci 2012;322: 2—10.

[156] Joutel A, Corpechot C, Ducros A, Vahedi K, Chabriat H, Mouton P, et al. Notch3 mutations in CADASIL, a hereditary adult-onset condition causing stroke and dementia. Nature 1996;383:707—10.

[157] Joutel A, Corpechot C, Ducros A, Vahedi K, Chabriat H, Mouton P, et al. Notch3 mutations in cerebral autosomal dominant arteriopathy with subcortical infarcts and leukoencephalopathy (CADASIL), a mendelian condition causing stroke and vascular dementia. Ann N Y Acad Sci 1997;826:213—7.

[158] Ungaro C, Mazzei R, Conforti FL, Sprovieri T, Servillo P, Liguori M, et al. CADASIL: extended polymorphisms and mutational analysis of the NOTCH3 gene. J Neurosci Res 2009;87:1162—7.

[159] Herzig MC, Eisele YS, Staufenbiel M, Jucker M. E22Q-mutant Abeta peptide (Abeta-Dutch) increases vascular but reduces parenchymal Abeta deposition. Am J Pathol 2009;174:722—6.

[160] Bugiani O, Giaccone G, Rossi G, Mangieri M, Capobianco R, Morbin M, et al. Hereditary cerebral hemorrhage with amyloidosis associated with the E693K mutation of APP. Arch Neurol 2010;67:987—95.

[161] Obici L, Demarchi A, de Rosa G, Bellotti V, Marciano S, Donadei S, et al. A novel Abe-taPP mutation exclusively associated with cerebral amyloid angiopathy. Ann Neurol 2005;58:639—44.

[162] Grabowski TJ, Cho HS, Vonsattel JP, Rebeck GW, Greenberg SM. Novel amyloid precursor mutation in an Iowa family with dementia and severe cerebral amyloid angiopathy. Ann Neurol 2001;49:697—705.

[163] Abrahamson M, Jonsdottir S, Olafsson I, Jensson O, Grubb A. Hereditary cystatin C amyloid angiopathy: identification of the disease-causing mutation and specific diagnosis by polymerase chain reaction based analysis. Hum Genet 1992;89:377—80.

[164] Vidal R, Frangione B, Rostagno A, Mead S, Revesz T, Plant G, et al. A stop-codon mutation in the BRI gene associated with familial British dementia. Nature 1999; 399:776—81.

[165] Vidal R, Revesz T, Rostagno A, Kim E, Holton JL, Bek T, et al. A decamer duplication in the 3′ region of the BRI gene originates an amyloid peptide that is associated with dementia in a Danish kindred. Proc Natl Acad Sci USA 2000;97:4920—5.

[166] Klimkowicz A, Dziedzic T, Polczyk R, Pera J, Slowik A, Szczudlik A. Factors associated with pre-stroke dementia: the cracow stroke database. J Neurol 2004;251: 599—603.

[167] Tang WK, Chan SS, Chiu HF, Ungvari GS, Wong KS, Kwok TC, et al. Frequency and determinants of prestroke dementia in a Chinese cohort. J Neurol 2004;251:604—8.

[168] Desmond DW, Moroney JT, Sano M, Stern Y. Incidence of dementia after ischemic stroke: results of a longitudinal study. Stroke J Cereb Circ 2002;33:2254—60.

[169] Ivan CS, Seshadri S, Beiser A, Au R, Kase CS, Kelly-Hayes M, et al. Dementia after stroke: the Framingham Study. Stroke J Cereb Circ 2004;35:1264—8.

[170] Tatemichi TK, Paik M, Bagiella E, Desmond DW, Stern Y, Sano M, et al. Risk of dementia after stroke in a hospitalized cohort: results of a longitudinal study. Neurology 1994;44:1885—91.

Genomic Medicine and Ethnic Differences in Cardiovascular Disease Risk

Alexis C. Frazier-Wood [1], *Stephen S. Rich* [2]

[1] Department of Pediatrics, USDA/ARS Children's Nutrition Research Center, Baylor College of Medicine, Houston, TX, USA; [2] Center for Public Health Genomics, University of Virginia, Charlottesville, VA, USA

1. INTRODUCTION

Despite advances in the ability to prevent, predict and treat disease in the developed world, the incidence of cardiovascular disease (CVD) and its risk factors are increasing [1]. Alongside this overall increase in CVD prevalence, ethnic differences in health status are widening [2]. A key reason for this is economics; ethnic minorities (especially Hispanic, African-American, and Native Americans) have, on average, completed less years of high school, earn less, and have a lower poverty to income ratio [3]. Socioeconomic status (SES) is negatively correlated with access to health care, and the poorest groups have seen the least benefits from medical advancement over the past decade [4]. Thus, as the economic gap continues to widen between non-Hispanic whites (NHW) and ethnic minorities in America, the gap in health status is expected to widen. Yet, despite the contribution of economics to health differences, the health status of non-white minorities still ranks lower than that of NHW when controlling for income and related economic variables [2]. A large body of research attributes the origins of any remaining differences in health disparities to a variety of environmental factors, including cultural factors such as diet, environmental exposures to home-based toxins and hazardous waste sites, and acculturation issues related to social marginalization, discrimination, and stress [4–9]. Yet, while evidence suggests that these factors do contribute to ethnic differences in health across a variety

Translational Cardiometabolic Genomic Medicine
http://dx.doi.org/10.1016/B978-0-12-799961-6.00009-3 **209**

of conditions, the role of ancestry-dependent genetic variation is potentially a critical etiological factor and the focal point of the discussion on health differences here.

2. METHODOLOGICAL ISSUES WITH CROSS-CULTURAL RESEARCH

In an ethnic-specific risk analysis, several methodological aspects need evaluation. First, there are few direct ethnic comparisons within a single study and this necessitates cross-study comparisons. Drawing cross-study comparisons is difficult because this may necessitate the comparison of different measures of the same construct across studies. Here measure-specific effects may be misattributed to population differences between the studies. Second, within-study cross-ethnic comparisons need equal validity of measures across races. Dietary assessment methods are an example of measures that, when validated for one ethnic group, may be inadequate to capture the dietary intake in other populations. Third, cross-study comparisons need to control for the correlates of risk factors that may vary between study populations (i.e., urbanization and dietary intake), but remain unmeasured. Fourth, sample size representing ethnic minorities often are a smaller proportion of the overall study population than NHW, reducing the power to detect associations in the minority sample. This can lead to non-significant findings of true associations in the minority group, but not in the NHW group, and spurious conclusions about ethnic differences. Finally, ethnic categories, when not derived from genetic data, can be crude and potentially misleading. It has been noted that the terms "race" and "ethnicity" do not have a generally agreed upon definition, making it hard to derive group membership for an ethnicity [5]. The problems this may cause are evident in the use of the term "Hispanic"; there are two large Hispanic populations in the United States; those in the southwest are thought to represent an admixture background from Spanish and Native American ancestry, while those in the East are predominantly Spanish, with smaller contributions from African ancestry [10,11]. Conversely Mexican-American and Hispanic-American have been used to refer to both the same genetic ancestry across studies, and different ethnic group in a study-dependent manner. Throughout this chapter, we will define ancestry groups as in the original research, although we highlight again that this can make cross-study comparisons difficult where different definitions of ancestry have been subsumed under the same ethnic category. The genomic impact on ethnic differences in CVD risk must consider the issues and limitations inherent in prior observational studies.

3. THE IMPACT OF ETHNICITY ON THE EPIDEMIOLOGY OF CARDIOVASCULAR DISEASE

CVD is used to describe variables related to atherosclerosis, a clinical outcome that occurs when plaque accumulates in the artery walls, causing a narrowing of the arterial diameter that restricts blood flow. Within the United States, there are major ethnic differences in CVD incidence, CVD events, and mortality attributed to CVD. There are also ethnic differences in the type of CVD. For example, NHW males had a higher incidence of coronary heart disease (CHD) in 2010 (8.2%) than African-American males (6.8%); however, African-American males had a higher risk of stroke (4.3%) than NHW males (2.4%) [1].

In 2009, the total direct and indirect costs of CVD to the United States were estimated to be $312.6 billion, up from $228 billion in 2008 [1]. By 2030, the American Heart Association (AHA) estimates that CVD will cost the United States approximately $1.48 trillion [1]. NHW ancestry forms a substantial part of the US population; data from the US Census Bureau indicates that the population increase of 27.3 million people (9.7%) between 2000 and 2010 was mostly due to people who reported their race as something other than white alone [12]. In 2010, 13.6% of the US population reported themselves as "Black or African-American," 1.7% as "American Indian or Alaska Native," 5.6% as "Native Hawaiian and other Pacific Islander," and 7.0% as "some other race" [12]. Furthermore, the prevalence of those reporting themselves as a minority ethnicity is increasing; 17% of the US population is Hispanic or Latino (almost 50 million individuals), which is projected to rise to over one-fourth of the US population (103 million individuals) by 2050 [3]. Given the increasing burden of CVD within minority populations, its economic and personal cost, it is important to consider factors (non-genomic and genomic) that lead to ethnic differences in CVD and the role of environment on an at-risk genomic architecture.

3.1 All-Cause Cardiovascular Disease

Greater than one in three American adults have at least one form of CVD, and the prevalence is expected to rise to almost 41% by 2030 [1]. Current ethnic differences in CVD are expected to widen, indicating that minority groups are going to see the greatest increases in all forms of CVD [1]. Within the United States, the prevalence of all CVD in 2010 was highest among African-Americans, where it is the leading cause of death (44.4% for males; 48.9% for females). Hispanic ancestry has historically been considered to convey the lowest prevalence and mortality of CVD [1,13—17]; however, the estimated prevalence was controversial given the higher prevalence of risk factors such as hypertriglyceridemia, obesity, and type-2 diabetes (T2D) [15,18—23]. Although health records in the

1960s indicated that Native Americans had a low risk of CVD compared to other US populations [24], CVD is increasing the fastest in this population, and now is the leading cause of death [25].

3.2 Stroke

The incidence of stroke among Mexican-Americans has been reported to be higher (16.8/1000) than in NHW (13.6/1000) in rural Texas [26]. In contrast, the age-adjusted incidence of first ischemic stroke was highest in African-Americans (1.91/1000), the lowest in NHW (0.88/1000) with Hispanic-Americans falling intermediate (1.49/1000) [27], and the Native American rate was at 6.79/1000 person-years [28]. Stroke not only occurs earlier in minorities but also kills earlier; death certificate data suggested that the mean age of death in NHW is 80.7 years and 72.6 years in African-Americans, 71.4 years in Native Americans, 75.4 years in Asian/Pacific islanders, and 72.6 years in Hispanic-Americans. These rates indicate a substantially increased loss of person-years in minority populations [29]. This is further reflected in a greater stroke mortality in African-Americans in the United States (men: 80/100,000; women: 76/100,000) compared to NHW (men: 62/100,000; women 58/100,000) [30].

3.3 Coronary Heart Disease

In the United States, CHD mortality is similar among African-American men (224/100,000) and NHW men (236/100,000); however, a higher mortality is found among African-American women (160/100,000) compared to NHW women (140/100,000) [30]. Native Americans stand out as disproportionately affected by CHD; in cross-study comparisons, the rate of CHD in Native Americans is nearly double as compared to that among African-American and NHW populations [25]. An exception is the rate of CHD in Pima Indians that is the lowest of all US ethnic groups; however, it is known that with the recent adoption of a more Westernized lifestyle the rates of CHD, like other forms for CVD, are climbing [25,31].

Despite small differences by CVD type, African-Americans, Native Americans, and Hispanic-Americans show a dramatically increased risk of CVD compared to NHW, and data indicate that these differences will become greater with time. Such differences in CVD incidence, morbidity, and mortality correlate highly, and likely reflect, the ethnic differences in CVD risk factors and biomarkers.

4. ETHNIC DIFFERENCES IN CVD RISK FACTORS

The health conditions most associated with CVD include hypertension, dyslipidemia (high low-density lipoprotein (LDL) cholesterol and triglycerides), and chronic kidney disease. These risk factors often occur with T2D and obesity; for example, 75–80% of those with T2D have hypertension and 70–80% have elevated LDL [32]. Factors implicated in the risk of complications of diabetes, whether type 1 (T1D) or T2D, include endothelial dysfunction, oxidation, inflammation, and vascular remodeling. These factors also contribute to the deposition of atherosclerotic plaque, independent of other risk conditions such as obesity [33]. Obesity is similarly associated with increased hypertension and hyperlipidemia, and an independent risk factor for T2D [34–37]. Obesity and T2D act to increase CVD risk both synergistically and independently [1]. Excess adipose tissue increases CVD risk through a number of pathways. Adipose tissue synthesizes and secretes molecules that are associated with CVD risk (e.g., adiponectin, PAI-1, TNF-α (http://europepmc. org/abstract/MED/14678864/?whatizit_url_gene_protein=http://www. uniprot.org/uniprot/?query=tumor%20necrosis%20factor&sort=score), and IL-6 [38]) and can be insulin resistant, which in adults is associated with a twofold increased 7-year hazard ratio for CVD [39,40]. Deposition of adipose tissue centrally, as opposed to peripherally, is particularly problematic and associated with morbidity and mortality from stroke, congestive heart failure, myocardial infarction, and cardiovascular death independent of the association between obesity and other CVD risk factors [41,42].

4.1 Obesity

The National Health and Nutrition Examination Survey (NHANES) is a large and ongoing national US survey conducted by the National Center for Health Statistics of the centers for Disease Control and Prevention. NHANES employed a complex multistage sampling approach based on the selection of counties, blocks, households, and the number of people within a household, designed to accurately represent the US population across socioeconomic status and ethnicity (achieved via sampling weights in analysis) which, therefore, enhances our understanding of ethnicity-driven epidemiology. NHANES data suggest that prevalence of excess adiposity varies with ethnicity. Age-adjusted overweight status (a body mass index; BMI $\geq$ 25) was 68.5% across the US adult population [43]. The lowest prevalence of overweight status in the United States was found in non-Hispanic-Asians (38.6%), with higher rates in NHW (67.2%), non-Hispanic blacks (76.2%), and Hispanics (77.9%) [43]. The ethnic distribution of obesity (BMI $\geq$ 30) exhibited similar patterns. Over one-third (34.9%) of US adults are obese, with non-Hispanic-Asians having the lowest rates of obesity (10.8%) while higher rates were seen in NHW (32.6%) [43]. In contrast to the similar prevalence of overweight, Hispanics had a lower prevalence of obesity (42.5%) than non-Hispanic blacks (47.8%) [43]. National

Health Interview Survey (NHIS) data report Native Americans to have a more healthy weight (BMI < 25) than "black-Americans", but less likely to have a healthy weight than "white-Americans"; in contrast, Native Americans were also more likely to be obese (prevalence of 40.8%) than whites (27.2%) or Asians (9.3%) [1].

The direction of association between body fat and CVD risk factor profile appears to be similar across ethnicities, with higher adiposity (especially central adiposity) conveying higher CVD risk [44]. However, the threshold for waist circumference associated with clinically significant CVD risk may be ethnic-specific for men, but not for women [44]. This ethnic-specific threshold for equal CVD risk and CVD mortality may relate to the differing prevalence of overweight and obesity between US ethnic groups; however, as the risk effects of obesity on CVD are likely mediated by other risk factors relating to other etiologic pathways, other explanations, such as biomarker levels and the etiology of adiposity, must be considered [45,46].

4.2 Type 2 Diabetes

The overall prevalence of diabetes (the majority T2D) is estimated as 8.9% in the US adult population (25—70 years old). Like obesity, T2D prevalence is higher in Hispanic (11.1%) and African (10.8%) ancestry populations than NHW (8.0%) [47]. Ethnic differences in T2D have widened in recent years with increasing rates in minority populations, and are expected to continue, contributing to the increasing CVD differences by ethnicity. From 1994 to 2004, the age-adjusted prevalence of T2D increased the most among Native Americans and Alaska Natives (101%) [48]. Anticipated increases in T2D to 2050 are projected to impact Hispanics (127% projected increase) and non-Hispanic blacks (107% projected increase) more than NHW (although there is a projected 99% increase) [49]. In addition to overall prevalence differences, the health outcomes associated with T2D differ by ethnic group; T2D is less likely to be diagnosed among African (24.3% of those with T2D) than Hispanic (21.4%) or NHW (21.2%) populations, and so may lead to worse outcomes for these populations [47].

As T2D and obesity are strong risk factors for CVD, the higher prevalence of these conditions among minority populations clearly contributes to their higher incidence of CVD. Like obesity, T2D does not account for all CVD risk as there are complex etiologic pathways that contribute through other biological and environmental mechanisms or ethnic differences [50]. Factors that may operate independently to these T2D and obesity include CVD biomarkers such as inflammatory cytokines, lipids, and fatty acids, all that may contribute to CVD and often exhibit ethnic differences in their distribution.

5. ETHNIC DIFFERENCES IN BIOMARKERS OF CVD

Recent advances in CVD research have facilitated the identification of several biomarkers of cardiometabolic disease risk. As with the prevalence of T2D and obesity levels of these biological indicators, which are known to associate with CVD, differ between ethnic populations and so provide a conceptual link to differences in CVD prevalence.

5.1 Markers of Inflammation

Atherosclerosis is an inflammatory process, and inflammation may be part of the mechanistic link between obesity and T2D and CVD [51,52]. C-reactive protein (CRP) is among the most studied markers of inflammation, due to the availability of inexpensive, standardized, high-sensitivity assays, and its ability to provide additive predictive information to both the lipid profile and the Framingham Risk Score, independently of other markers of inflammation [53–57]. The Women's Health Initiative (WHI) compared CRP levels across 24,455 self-reported white, 254 self-reported black, and 357 self-reported Asian female participants [58]. Median CRP levels were the lowest among Asian women (1.12 mg/L) and highest among black women (2.96 mg/L), with NHW women (2.02 mg/L) and Hispanic women (2.06 mg/L) intermediate [58]. The differences between black and Asian, between white and Asian, and between black and white women remained statistically significant even after adjusting for age and estrogen use. The positive relationship between BMI and CRP was in the same direction between women of all ethnicities and not attenuated when controlling for BMI, but was stronger in the black women than in other ethnicities. These data implicate obesity as one of the multiple pathways through which CRP could account for ethnic-based CVD differences [58].

Ethnic-specific aspects of CVD risk may also be sex-specific. In the Dallas Heart Study, being female and being black were both associated with increased CRP levels compared to white males [59]. These differences remained after controlling for oral statin and estrogen usage and subsequently excluding users, but the differences by ethnicity disappeared in the males after adjusting for traditional cardiovascular risk factors, BMI, and statin use (odds ratios when compared to white men: black women 1.7; white women: 1.6; black men: 1.3) [59]. NHANES data reported higher mean CRP concentrations for African-American women and Mexican-American women compared to NHW women. These differences were attributed to differences in education, smoking status, total cholesterol, systolic blood pressure, waist circumference, alcohol use, and hormone replacement therapy, thus highlighting the role of the environment as well as ancestry in health differences [60]. In this respect,

NHANES and WHI disagree, with environmental control negating ethnic differences in CRP in NHANES but not in WHI. The reasons why NHANES and WHI differ remain unclear; however, there are socioeconomic differences between the two groups. WHI is largely (>80%) composed of professional women, while NHANES is designed to be more representative of the US population [60]. Socioeconomic status may be a surrogate for additional factors that represent ethnic differences in CVD risk factors. Of note, there are limited data on CRP levels in Native American women. One report found Native American women to be similar in CVD risk to Caucasian women [61], yet this remains an important direction for future research.

Other markers of inflammation that are not the focus of public health screening or intervention efforts nevertheless are associated with CVD and exhibit ethnic differences. The Multi-Ethnic Study of Atherosclerosis (MESA) reports the highest levels of CRP and IL-6 for African ancestry in the United States, and the lowest for Asian ancestry, with Caucasian and Hispanic ancestry intermediate [62]. Plasma fibrinogen levels among adults living in the United Kingdom were lower in Blacks than in Whites in men (2.73 g/L vs 2.91 g/L) and in women (2.83 g/L vs 2.92 g/L) [63]. Data from the Study of Health Assessment and Risk in Ethnic Groups show that South Asians have higher levels for markers of inflammation compared to Europeans (e.g., PAI-1 activity: 16.6 ng/mL vs 10.1 ng/mL; fibrinogen: 3.22 mg/dL vs 2.69 mg/dL) [64], with these results supported from other studies. Research that combines many inflammatory markers and ethnic groups into a single, longitudinal study should clarify the role these markers play in CVD development and form a baseline for the impact of other markers of inflammation on patterns of CVD risk and ethnic differences.

An understudied question is the mechanism by which markers of inflammation differ in their association with CVD by ethnicity. In the Atherosclerosis Risk in Communities (ARIC) study a composite "low-grade inflammation score" was developed, in which the score was composed of CRP, IL-6, sialic acid, and orosomuciod. The score was predictive of incident T2D in "whites," but not in "blacks," over a 9-year period [65], a result consistent with "inflammation" determined by multiple ethnic-specific pathways.

5.2 Lipids

Increased cholesterol and triglyceride (TG) levels are among the most recognized indicators of CVD risk. In general, lipids are the target of therapeutic interventions and the focus of medical prevention strategies. A recent meta-analysis of 17 population-based prospective studies estimated that risk ratios between TGs and CVD events were 1.32 for men and 1.76 for women [66]. When controlling for high-density lipoprotein (HDL) cholesterol and other risk factors, the risk ratios for TGs and CVD events were reduced but remained statistically significant (1.14 for men and 1.37 for women) [66]. Thus, hypertriglyceridemia remained a significant predictor of CVD events, independent of cholesterol levels [66]. Increased low density lipoprotein (LDL) cholesterol levels appear to be causally related to incident CVD and adverse CVD events from prospective epidemiological studies, Mendelian randomization designs, and intervention trials [67–69]. Increased levels of HDL have been considered protective, but clinical trial data suggest that increasing HDL does not significantly reduce CVD risk. Thus, the causal role of HDL in the pathophysiology of CVD is uncertain [68,69].

African-Americans and non-Hispanic blacks have the most favorable lipid profiles with respect to CVD risk [70]. In NHANES 1999–2002, HDL levels were significantly higher in non-Hispanic black men (51.0 mg/dL) than NHW (45.5 mg/dL) or Mexican-American (45.0 mg/dL) men [70]. Among women, similarly high HDL levels were seen in non-Hispanic black (57.3 mg/dL) and NHW (56.6 mg/dL), with lower HDL levels seen in Mexican-American women (52.9 mg/dL) [70]. These ethnic differences have also been reported in aging populations [71]. TG levels are also more favorable for African-Americans. Mexican-American (142 mg/dL) and NHW (134 mg/dL) men have significantly higher TGs than non-Hispanic black (99 mg/dL) men. In women all three ethnicities had significantly different TG levels with those in Mexican-American women (135 mg/dL) higher than NHW women (117 mg/dL), both being higher than those in non-Hispanic black women (90 mg/dL) [70]. LDL-C was not significantly different between NHW, non-Hispanic blacks, and Mexican-Americans [70]. Total cholesterol differed only among the men with Mexican-Americans (202 mg/dL) and NHWs (202 mg/dL) having higher levels than non-Hispanic black men (195 mg/dL) [70]. Trends for cholesterol mirror those of CVD. Overall, Mexican-American and NHW cholesterol levels did not change significantly between the two collection periods, but those for non-Hispanic black decreased significantly [70]. These differences may have been largely driven by men, as cholesterol levels in women dropped significantly for all races, but only non-Hispanic black men had a drop in overall cholesterol [70]. Data comparing serum lipids other than cholesterol were only available in NHANES from 1976 to 1980. When compared to the data taken during 1999–2002, NHWs and non-Hispanic blacks had decreases in their overall LDL [70]. However, in the Mexican-Americans, only women saw a significant decrease in LDL.

5.3 Lipoproteins

LDL and HDL levels do not account for all lipid-based CVD risks [72]. Within each lipoprotein fraction, the constituent particles are heterogeneous in their size and composition. Specific constellations of lipoprotein sizes have been associated with insulin resistance (IR) and atherosclerosis [73–79], and attention has turned to perturbations in the size and distribution of the HDL subclass, as a causal factor in incident CVD. Very few studies have systematically compared lipoprotein subclasses by ethnicity. In MESA, mean VLDL diameter significantly differed between all ethnicities, with mean size increasing from Chinese-Americans (46.39 nm) through African-Americans (47.26 nm) and Caucasian-Americans (49.10 nm) to Hispanic-Americans (50.98 nm) [80]. African (20.79 nm) and Caucasian (20.81 nm) ancestry showed similar mean diameters for LDL diameter, which were larger than the diameters from participants of Chinese ancestry (20.61 nm); Hispanic ancestry had the largest size of LDL particles (20.64 nm) [80]. HDL diameter also differed significantly across all ethnicities, with size increasing from Hispanic (9.19 nm), Caucasian- (9.25 nm), and Chinese- (9.29 nm) to African (9.30 nm) ancestry. Conclusions on the role of absolute differences in lipoprotein size by ethnicity on CVD risk is complicated, especially given that risk profiles associated with lipoprotein diameters may be ethnic-specific and diverse in correlated CVD risk factors. For example, smaller LDL particle size is associated with hypertension in Caucasians but not in African-Americans [81], while total and LDL cholesterol levels are more strongly associated with CVD risk in Caucasian than in African-Americans [82].

The differences in CVD risk profiles comprised risk factors and biomarkers are compelling, yet the paucity of research in ethnic minorities leaves several questions to be resolved. First, data are limited on the relationships between baseline levels of risk factors across ethnic groups (such as Native Americans) and their longitudinal prediction of CVD events. Second, there are little data on the mechanisms that would result in ethnic differences observed at specific ages (as well as sex differences across ethnic groups) in the association between CVD biomarkers and disease risk. Third, the causal pathways between CVD biomarkers and disease mechanism may vary with ethnicity due to multiple factors, most of which exhibit strong correlations between the biomarkers but not consistently across ethnic groups. These questions highlight difficulties in the design of screening and prevention programs that are effective across all populations that will enable translation of population-based research to the clinical setting.

6. THE ROLE OF THE GENOME

Complex traits arise from combinations of, and interactions between, environmental and genetic factors. Just as ethnic health disparities arise from a combination of environmental factors, including environmental toxins, acculturation and cultural differences, so is the genome likely to contribute to health disparities. Research on this has been scant, but needed [5]. The underlying reasons for the lack of data in this research area are complex, and include the difficulty of accruing large samples of ethnic minorities, as well as the concern about the potential for stereotyping and marginalizing minority ethnic groups [5]. Nonetheless, the presence of data that the human genome is important in the development of CVD and related traits motivates the examination of whether features of the genome (or epigenome) contributes to ethnic health disparities in CVD.

Most quantitative traits associated with CVD, or CVD risk, are moderately to strongly heritable. Family studies show that CVD and most CVD risk factors aggregate more frequently among relatives than unrelated individuals. However, it is difficult to separate the contribution of genetic factors from shared environmental factors between family members. Comparisons of the correlation in CVD and CVD risk factors between monozygotic twin pairs (identical twins; MZ) and dizygotic twin pairs (non-identical twins; DZ) have provided estimates of the role of genetics in these traits. As MZ twins are identical genetically, while DZ pairs share (on average) one-half of their genome, twin studies can partition the variance in disease risk into genetic factors, shared environmental factors (from the environment which make members of a family more similar), and non-shared, or unique, environmental factors (factors from the environment which make members of a family less similar on a given trait). However, twin studies on disease traits with a late age-of-onset can be challenging, given the difficulty of recruiting enough complete twin pairs to achieve robust estimation of genetic and non-genetic components to the trait [83]. Further, twin and family studies on non-Caucasian samples are rare; however, these data are highlighted as they permit a comparison of the relative role genetic variance plays in estimation of the "heritability" of a trait between ethnic groups.

The search for genes underlying CVD and CVD risk factors has been moderately successful, primarily through the use of genome-wide genotyping arrays applied to large cohort studies in a meta-analytic framework. Yet, despite a number of replicated and validated CVD and CVD risk loci, much of the heritable variance remains "missing" [84]. There are numerous hypotheses for the limited discovery of genes and variants on

CVD risk, including a focus on common (rather than infrequent or rare) variants, incomplete coverage of the genome with earlier molecular technologies, the limited inference of genetic background on environmental risk, and heterogeneous genetic architecture that may be enhanced across ethnically diverse populations. The paucity of genetic research in minority populations makes it hard to draw firm, evidence-based conclusion on the role of genome in ethnic differences in CVD, but it is clear that as individual differences across the genome are factors in the development of CVD and cluster by ancestry, that genomics must be incorporated into both our understanding of ethnic differences in CVD, and our methods to minimize these differences (personalized medicine).

6.1 Heritability of CVD and Its Risk Factors

A common measure of the impact of genetic (or familial) factors on a trait is heritability. There are numerous definitions of heritability for diseases (heritability of liability) or a measured trait, with the majority of estimates coming from twin and family studies. The general concept is that heritability represents the proportion of variation in a trait that is attributed to genetic factors (either only those that can be tracked from parent to child, defined by the additive effects of genes (narrow sense heritability) or all effects of genes including allelic and gene interactions).

Family studies indicate that a parental history of premature CHD (defined as myocardial infarction or angina pectoris) before the age of 60 years is associated with an increased risk ratio for acute myocardial infarction or coronary death (1.61 in men and 1.85 in women) independent of known risk factors, including smoking, high serum cholesterol, blood pressure, diabetes, obesity, and socioeconomic status [85]. These risk ratios increase for early offspring outcomes, when CHD is assessed at earlier than 55 years of age, moderately in men, and significantly in women [85]. Twin studies estimated the heritability of death from CHD, finding higher heritability in male twins (57%), than among female twins (38%) [86]. As the heritability is less than 100%, the residual effect appears to be due to non-shared environmental factors [83]. It has been noted that the heritability of CVD/CHD decreases over the life span, yet the decrease is small and nonsignificant, emphasizing the consistent importance of genetic factors [86]. Similarly, family history is a strong and consistent independent predictor of coronary artery disease [87–89].

Numerous studies have reported that stroke aggregates in families, with a positive family history of any stroke associated with an odds ratio of 1.76 in a meta-analysis [90]. The odds ratio is lower (1.28) when including studies that meet several methodological criteria (e.g., separated the type of stroke (ischemic vs hemorrhagic), recruited patients consecutively to minimize the survival bias, accounted for the number of relatives with a history, and considered subtypes ischemic stroke (large vessel, small vessel, cardioembolic) as an indicator of presumed etiology). Nonetheless, familial aggregation of stroke remains significant compared to the population rate, thereby supporting the role of familial factors in the development of stroke [90]. Twin studies, as expected, demonstrate that MZ twins are more likely to be concordant for stroke than DZ twins (combined odds ratio 1.65), suggesting that, of the familial factors involved in stroke risk, the majority are genetic [90].

Obesity and T2D are heritable to a similar extent and each may contribute to the familial aggregation of CVD. Twin studies have shown that the majority of liability in T2D (up to 90%) can be attributed to genetic factors [91–95]. Fasting glucose and fasting insulin have been shown to exhibit a range of heritability estimates, depending on study design. In twins aged 55–74 years, all phenotypes related to glucose were heritable, ranging from 38% to 50% [96]. These estimates are similar in younger twin pairs [97]. The heritability of BMI, a common measure used as a surrogate of obesity, is consistently high based upon the eight countries containing Caucasian participants participating in the GenomeEUtwin project. Each of these countries estimated the heritability of BMI between 64% and 84%, with remainder thought largely due to unique environmental factors [98].

Given the heritability of CVD, obesity, and T2D, it is not surprising that biomarkers of CVD also show evidence of a significant contribution of genetics. Heritability estimates are high for inflammatory biomarkers such as CRP (61%), IL-6 (31%), and soluble IL-6 receptor (49%), as well as those for fibrinogen (52%) [99], lipids (48%), and apolipoproteins (86%) [100].

The majority of familial and twin studies have been conducted in populations of Caucasian ancestry. Methodological differences make it difficult to compare the relative contribution of genetic factors (heritability) across studies. In the small number of studies conducted in ethnic minorities, overall heritability estimates remain similar between different ethnicities. For heart rate variability, a correlate of CVD, there were no significant differences found in heritability estimates between African- and Caucasian-American twin pairs [101]. For cardiac index heart rate, blood pressure, and stroke, the differences in heritability estimates between African- and European-Americans were small and non-significant, and virtually disappeared after an adjustment for obesity [102]. No significant differences between whites, African-Americans, Hispanics, and Asians were found for the heritability of BMI in the National Longitudinal Study of Adolescent Health [103]. The heritability of BMI is reported as in a similar range to Caucasians for Chinese (50%), Korean (82–87%), Hispanic (62–82%), and African-American (64–83%) cohorts [104–108]. Similarly, the heritability of T2D was estimated at 60% in African-Americans [109]. The overwhelming evidence is that the proportion of variance in CVD and related traits accounted for by genetic factors does not differ by ethnicity. It is important to realize that although broad-sense estimates of heritability are similar across groups, the individual genes, variants, and their interactions with other genes and/or environment can differ and, therefore, contribute to group differences.

6.2 Genetic Associations with CVD and Differences between Populations

Recent advances in DNA technology have allowed more systematic and deep examinations of whether specific genetic variants can account for population differences in trait levels by ancestry. Nearly 10 million single nucleotide polymorphisms (SNPs) occurring in at least 1% of the population are thought to account for the majority of human genetic variation [110,111]. At the same time, the rapid expansion of the human population has generated an abundance of new variants, often in single copies, that significantly increases genetic diversity [112].

Examining genetic variation at the DNA sequence level is helping redefine our concept of ancestry. Ethnic groups, defined by country of origin, are more similar genetically than originally thought. All humans share a set of common African ancestors, with migration out of Africa occurring as recently as 200,000 years ago. Genetic sampling across individuals has revealed differential population growth and increased migration, and has overturned the notion of distinct ethnic groups that can be defined by differing genetic ancestry. There is more continuity in the genetic code between ethnic groups than expected, leading some to theorize that ancestry may not be seen by our current socio-political definitions, but through a more nuanced genetic definition which transcends and crosses current racial boundaries [113]. For example, about 15% of the NHW are closely related at the genetic level to a different racial group and, overall, about 20% of Americans are genetically related to more than one racial group [114].

Differing genetic backgrounds and the problems of identifying causal variants for disease (CVD) and biomarkers of disease in human populations further complicate the examination of ethnic-specific differences. It has been shown that *NOS1AP* SNPs are associated with sudden cardiac death in European-, Hispanic-, and Chinese-Americans (although not African-Americans); however, the NOS1AP SNP with strongest association with sudden cardiac death differs by ethnicity [115]. Interpretation of these findings is not clear. The same gene can be implicated and a specific variant (or variants) may differ by ethnicity in their disease associations within the gene (Figure 1; Panel A). Alternatively, the true causal variant has not been identified, and differing genomic backgrounds give rise to different correlational structures among SNPs, leading to differential

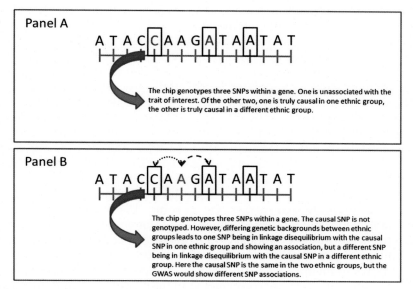

FIGURE 1   Patterns of gene variant—disease associations.

associations (Figure 1; Panel B). Thus, the gene (and biological pathway) may be consistent across ethnic groups with respect to the outcome, with the specific variant differing due to chance mutational events that affect the gene product, or the pathway may be the same and techniques such as finer mapping will help resolve this.

Reflecting an increased interest in population genetics, the International HapMap consortium has provided an open access database of how genetic background differs between many populations across the world. These collections are typically small in number, but include those with African ancestry in Southwest United States (ASW); Utah residents with Northern and Western European ancestry from the Centre d'Etude du Polymorphism Humain collection (CEU); the Han Chinese population in Beijing, China (CHB); the Chinese population in Metropolitan Denver, Colorado (CHD); the Gujarati Indians in Houston, Texas (GIH); Japanese in Tokyo, Japan (JPT); Luhya in Webuye, Kenya (LWK); populations with Mexican ancestry in Los Angeles, California (MXL); the Maasai in Kinyawa, Kenya (MKK); the Toscani in Italia (TSI); and the Yoruba in Ibadan, Nigeria (YRI) [116]. HapMap has focused on the common variation (greater or equal to 1% frequency in the population) [111]. The data that emerged from the HapMap permit questions to be addressed that focus on how genetic variation contributes to ethnic differences in a range of health conditions, through association of specific SNPs with outcomes, contrasting the SNP effects by population, and comparing allele frequencies of SNPs across populations.

Evidence from numerous data sources suggests that associations between CVD and the genes containing CVD risk variants are consistent across ethnic groups; however, there may be different SNPs within a gene that are most strongly associated across ethnic groups or, when the same SNP shows the strongest association, the frequencies of the SNP minor allele may be different across groups. Different SNPs showing associations within the same gene between ethnic groups likely reflects differing genetic backgrounds associated with ancestry. Patterns of genetic background occur when number of SNPs are correlated at rates greater than expected by chance. SNPs occurring together more frequently than expected are said to be in "linkage disequilibrium" and form haplotypes blocks of differing size across the genome. Populations defined by ancestry often show more similarities than differences in their haplotype patterns. Differing associations within a gene often occur where the causal variant is not genotyped, but variants in high linkage disequilibrium with the causal variant are. Due to differing haplotype patterns between ethnic groups, different SNPs are correlated with the causal SNPs, and therefore different SNPs show the strongest association with CVD risk factors. In addition, the frequency of haplotypes, like the frequency of alleles, differs which may partially explain ethnic differences in disease risk [117,118]. Genetic drift, defined by the variation in allele frequencies through restricted population size, is thought to be more extensive in African than in European ancestries, and similarly, greater in East Asian populations compared to European populations [119,120]. It should be noted, however, that while genetic variation is the focus of this chapter, other biological processes (e.g., gene expression) have been shown to contribute to population differences [121].

It is clear that no one racial/ethnic group is genetically homogeneous. Just as there are allelic differences across the spectrum of European populations that form the foundation of NHWs in the United States, there are allelic differences in frequency among African, Hispanic, Native, and Asian ancestry populations. In a recent report, patterns of Native American population substructure were identified in the genomes of Mexican mestizos from populations throughout Mexico [122]. Sonora and neighboring northern states had the highest average proportions of the northern native component (15%), compared with near absence in Oaxaca and the Yucatan peninsula. The southern native component was the most prevalent across states, reaching maximum values in Oaxaca and decreasing northward. Samples from the Yucatan peninsula were dominated by the Mayan component that diminished with northward spread. Mexican-Americans sampled in Los Angeles (MXL) also do not show a homogeneous pattern, consistent with their diverse origins within

Mexico. These results suggest that fine-scale patterns of native ancestry alone could have significant impacts on disease (e.g., CVD) and risk factors whose variations are defined, in part, by genetic factors and show how differences in genetic architecture across populations can be used to better understand genetic factors and ethnic differences contributing to CVD.

Other factors, such as population migration, could result in allelic differences across populations and resulting differential risk for CVD. Using the Human Genome Diversity Panel, the genetic risk of 102 diseases in 1043 unrelated individuals across 51 populations were determined [123]. Using over 650,000 SNPs as a basis, filtering identified ~700 SNPs associated with disease and correlations between genetic risk and migration trajectories were determined. It was demonstrated that genetic risk for T2D, an important risk factor for CVD, decreased as humans migrated from Africa toward East Asia. This differentiation could arise from the exposure to new environments through migration, so that the adaptation would result in allele frequency differences due to selection. Other, more complex patterns of adaptation or chance combinations of alleles could result in this differentiation of allele frequencies across populations.

6.3 Genetic Associations with Cardiovascular Disease Risk Factors

Multiple lines of evidence suggest that the same genes and, on occasion, the same variants contribute to disease traits and risk factors between populations. At the same time, comparisons of NHW associations with specific genes and variants in non-NHW cohorts can identify novel associations. Most comparisons have been made between NHW cohorts with those of African ancestry; however, the number of participants in the NHW studies often is 10-fold larger. This difference in sample size reflects differential power to detect variants contributing to CVD or CVD risk factor. With this differential in population size, often meta-analytic approaches are used, in which the effects within the smaller ethnic groups are obscured by combining with the larger NHW sample. Newer methods are emerging which meta-analyze both within and across ancestry-informative groups, and include information on sample size and allele frequency to generate estimates of heterogeneity in genetic risk [124]. These advances will only enhance our understanding of the role genetic heterogeneity plays in ancestry-based risk differences.

An important aspect of genetic association with CVD and CVD risk factors is the type of variants being interrogated. In previous years, candidate genes and primary, putative functional variants were interrogated. This design assumed that a common, functional variant in a gene with an association in one ethnic group should be important in all ethnic groups. Given the evolutionary arguments noted above with respect to genetic drift, selection and migration patterns that may affect the allele frequency spectrum across populations, it is not surprising that many of these candidate gene/variant studies supported genetic differences in association with primary phenotypes. An example of consistency of variants is fibrinogen level, which is associated with the G/A polymorphism at nucleotide 455 and the C/T at nucleotide 148 in the promoter of the β-fibrinogen gene (*FGB*) in blacks, South Asians Indians, and whites living in London [63]. In contrast, variants in the hepatic lipase (*LIPC*) and apolipoprotein B (*APOB*) genes associate with VLDL and HDL diameter in Caucasian and in Hispanic participants, but not in African-American or Chinese-American participants, in the Multi-Ethnic Study of Atherosclerosis [80]. Similar results showing differential variant association have been observed for *FTO* and BMI [125–130], and *MTHFR* with arterial thrombotic disease, coronary artery disease, and venous thrombosis [131–136].

Large individual studies of populations from African, Hispanic, and Asian ancestry are lacking; however, there are many studies of modest size in these populations that can now be investigated using meta-analytic methods. Recently, a large GWAS meta-analysis of BMI in 36 studies of African ancestry participants was conducted, totally over 39,000 participants [137]. Eleven SNPs at five loci achieved genome-wide significance for association with BMI. Four of the five loci were known as associated in NHW or Asian cohorts (*SEC16B, GNPDA2, FTO,* and *MC4R*) and one (*GALNT10*) was novel. Of 36 previously established BMI loci in NHW and Asian populations, the most associated SNP was directionally consistent with the original report in 32 of 36 loci, suggesting commonality of genes and pathways (if not specific variants) across ethnic groups contributing to variation in BMI. Similar efforts in T2D have shown a significant excess in the direction of SNPs associated with T2D across European, east Asian, south Asian, Mexican, and Mexican-American ancestry groups [138]. Heterogeneity in allelic effects, based upon discordance in the direction and/or magnitude of odds ratios, was tested at 69 established autosomal T2D susceptibility loci across the ancestry groups. Heterogeneity was observed at only three loci — *TCF7L2* (the odds ratio is largest in European ancestry), *PEPD*, and *KLF14*. Notably, 34 loci showed the same direction of effect across all ancestry groups.

With the increase in DNA sequencing technology, there has been a remarkable increase in the number of variants identified (many new, rare variants, with the rapid expansion of the human population) and, as a result, in the ability to better identify contributions of rare variants in genes as a burden of risk. One of the first large-scale applications of sequencing the coding regions of the human genome (the "exome") was the NHLBI Exome Sequencing Project (ESP), that conducted exome sequencing on ~4500 NHW and ~3000 African-American participants whose samples were obtained from cohorts focused on heart, lung, and blood disorders. In an analysis of LDL cholesterol, ~2000 participants were sequenced (554 from the extremes of the LDL distribution) with follow-up sequencing in 1302 and genotyping in over 50,000 subjects [139]. Association of LDL level was obtained with the burden of rare variants in three known genes (PCSK9, LDLR, and APOB) and one novel gene (PNPLA5). There was no statistical evidence of heterogeneity or burden-frequency difference between the NHW and African-American participants with respect to coding variants associated with LDL level, although the APOB loss-of-function variants were primarily found in the NHWs.

In evaluation of rare coding variants in the ESP on variation in triglyceride and HDL levels, a total of 3374 participants were including in exome sequencing [140]. Gene burden tests identified APOC3 as the gene most strongly associated with plasma triglyceride levels in NHW and African ancestry participants. The aggregate effect of the rare alleles in APOC3 was to reduce triglyceride levels by 39%, increase HDL levels by 22%, and decrease LDL levels by 16%. Following replication by genotyping an additional ~34,000 NHW and ~7000 African ancestry participants for four variants in APOC3, those carrying any of the four variants had lower TG levels in both NHW (42% lower) and African (32% lower) ancestry. Further, the risk of CHD for those having one of the four APOC3 variants was 40% lower across all studies/ethnic groups.

7. IMPACT OF PERSONALIZED GENOMICS FOR CVD

The advent of next-generation sequencing and high-throughput genotyping provides the ability to completely characterize an individual's genome, leading to the possibility of personalized genomic diagnosis of risk. This genetic diagnosis, however, is based upon relatively small contribution to the total disease risk (or variation in biomarkers of risk) by the known genetic variants for the vast majority of diseases that affect humans. Sequencing studies have revealed the incredible diversity in the human population at the fundamental genomic level, due to the contribution of rare variants from the expansion of the human population. Thus, although there are historical ethnic differences in disease and risk factors, the individual genes and variants associated with the diseases and risk factors may vary as much within a population as between populations. Critically, the impact of those differing genes and variants may be the same on a biological pathway; thus, the outcome of genomic medicine may be the identification of critical pathways and biomarkers of pathways that affect all individuals, regardless of ancestry, for targeted therapeutics or interventions.

With the availability of personalized genomic information, there are many additional issues that are tightly coupled with the generation of genomic data. These issues include the manner in which the genomic data are collected and integrated with existing health information, the manner in which primary care physicians and specialists are trained to communicate information and knowledge to the individual patient, as well as the impact of genomic information on family members. The inclusion of genomic data in standard medical care will require fundamental shifts in a number of research, health-care, and societal practices. New legal and regulatory approaches will need to be implemented based upon the advances in genomic applications to health care.

8. SYNTHESIS

Methodological difficulties make it difficult to draw conclusions on the extent to which the genome contributes to ethnic differences in CVD risk and to variation in CVD risk factors. These include, in addition to the broader problems with ethnic-sensitive research, difficulties peculiar to genetic analysis, including the difficulty of identifying causal variants, disparate sample sizes in available studies that lead to differences in power, and the difficulty of fully accounting for gene−environment interactions given the different dietary and environmental backgrounds associated with ethnicity. Taken together, current data suggest that the contribution of variation in genes associated with CVD traits is small, and the genes themselves are consistent across populations, with specific variants and their allele frequencies differing, and so contributing to risk and disease prevalence differences. This is consistent with the results of a meta-analysis on racial differences across all disease traits [141], that analyzed 43 gene−disease associations where significant SNP−trait associations in at least one ethnicity were attempted in another racial group (the small number of studies included highlighting the paucity of research in this badly needed area). In pair-wise comparisons statistically significant differences were seen in 15% of comparisons between European versus East Asian ancestry; 8% of comparisons between European versus African ancestry; and none of the comparisons of east Asian versus European ancestry, a rate described as "not much difference from what would be expected by chance" [141].

This consistent genetic effect across racial groups implies a common, consistent biological effect of SNPs on individuals, regardless of ethnicity. In other words, it is the prevalence of the risk allele, or those that are correlated with it in location or in effect, and not its individual effect that differs between groups of differing ancestry. This is promising as we look forward to a reduction in CVD risk as a whole, as similar biological pathways suggest treatments may be effective in an ethnicity-independent manner. However, it should be clear that more research on the contribution of the genome to ethnic differences is CVD risk should be an immediate research priority to aid prediction and understanding, and to reduce the economic and personal cost of CVD which disproportionally affects the NHW population.

References

[1] Go AS, et al. Heart disease and stroke statistics—2013 update: a report from the American Heart Association. Circulation 2013;127:e6—245.

[2] Nelson AR. Unequal treatment: report of the Institute of Medicine on racial and ethnic disparities in healthcare. Ann Thorac Surg 2003;76:S1377—81.

[3] U.S. Census Bureau. 1970, 1980, 1990, and 2000 decennial censuses; population projections, July 1, 2010 to July 1, 2050. 2008. at, http://www.census.gov/population/hispanic/publications/hispanics_2006.html.

[4] Sankar P, et al. Genetic research and health disparities. JAMA 2004;291:2985—9.

[5] Collins FS. What we do and don't know about "race", "ethnicity", genetics and health at the dawn of the genome era. Nat Genet 2004;36:13—15.

[6] Berkowitz GS, et al. Exposure to indoor pesticides during pregnancy in a multiethnic, urban cohort. Environ Health Perspect 2003;111:79—84.

[7] Istre GR, McCoy MA, Osborn L, Barnard JJ, Bolton A. Deaths and injuries from house fires. N Engl J Med 2001;344:1911—6.

[8] Hispanic environmental health: ambient and indoor air pollution. Otolaryngol Head Neck Surg 1996;114:256—64.

[9] Soliman MR, Derosa CT, Mielke HW, Bota K. Hazardous wastes, hazardous materials and environmental health inequity. Toxicol Ind Health 1993;9:901—12.

[10] Moore J, Pachon H. Hispanics in the United States. Prentice-Hall Inc; 1985.

[11] McLemore SD, Romo R. In: De la Garza RO, Bean FD, Bonjean CM, Romo R, Alvarez R, editors. Mex. Am. Exp. An interdiscip. anthol. University of Texas Press; 1985.

[12] Humes KR, Jones NA, Ramirez RR. Overview of race and hispanic origin: 2010. 2011.

[13] Stern MP, Rosenthal M, Haffner SM, Hazuda HP, Franco LJ. Sex difference in the effects of sociocultural status on diabetes and cardiovascular risk factors in Mexican Americans. The San Antonio Heart Study. Am J Epidemiol 1984;120:834—51.

[14] Diehl AK, Stern MP. Special health problems of Mexican-Americans: obesity, gallbladder disease, diabetes mellitus, and cardiovascular disease. Adv Intern Med 1989;34:73—96.

[15] Stern MP, et al. Secular decline in death rates due to ischemic heart disease in Mexican Americans and non-Hispanic whites in Texas, 1970—1980. Circulation 1987;76:1245—50.

[16] Stern MP, et al. Affluence and cardiovascular risk factors in Mexican-Americans and other whites in three northern California communities. J Chronic Dis 1975;28:623—36.

[17] Mitchell BD, Hazuda HP, Haffner SM, Patterson JK, Stern MP. Myocardial infarction in Mexican-Americans and non-Hispanic whites. The San Antonio heart study. Circulation 1991;83:45—51.

[18] Schoen R, Nelson VE. Mortality by cause among Spanish surnamed Californians, 1969—71. Soc Sci Q 1981;62:259—74.

[19] Frerichs RR, Chapman JM, Maes EF. Mortality due to all causes and to cardiovascular diseases among seven race-ethnic populations in Los Angeles County, 1980. Int J Epidemiol 1984;13:291–8.

[20] Buechley RW, Key CR, Morris DL, Morton WE, Morgan MV. Altitude and ischemic heart disease in tricultural New Mexico: an example of confounding. Am J Epidemiol 1979;109:663–6.

[21] Becker TM, Wiggins C, Key CR, Samet JM. Ischemic heart disease mortality in Hispanics, American Indians, and non-Hispanic whites in New Mexico, 1958–1982. Circulation 1988;78:302–9.

[22] Kautz JA, Bradshaw BS, Fonner E. Trends in cardiovascular mortality in Spanish-surnamed, other white, and black persons in Texas, 1970–1975. Circulation 1981;64:730–5.

[23] Stern MP, Gaskill SP. Secular trends in ischemic heart disease and stroke mortality from 1970 to 1976 in Spanish-surnamed and other white individuals in Bexar County, Texas. Circulation 1978;58:537–43.

[24] Howard BV, et al. Coronary heart disease prevalence and its relation to risk factors in American Indians. The Strong Heart Study. Am J Epidemiol 1995;142:254–68.

[25] Howard BV, et al. Rising tide of cardiovascular disease in American Indians: the Strong heart study. Circulation 1999;99:2389–95.

[26] Morgenstern LB, et al. Excess stroke in Mexican Americans compared with non-Hispanic whites: the Brain Attack Surveillance in Corpus Christi project. Am J Epidemiol 2004;160:376–83.

[27] White H, et al. Ischemic stroke subtype incidence among whites, blacks, and Hispanics: the Northern Manhattan Study. Circulation 2005;111:1327–31.

[28] Zhang Y, et al. Incidence and risk factors for stroke in American Indians: the Strong Heart Study. Circulation 2008;118:1577–84.

[29] Disparities in deaths from stroke among persons aged <75 years—United States, 2002. MMWR Morb Mortal Wkly Rep 2005;54:477–81.

[30] Forouhi NG, Sattar N. CVD risk factors and ethnicity—a homogeneous relationship? Atheroscler Suppl 2006;7:11–9.

[31] Nelson RG, et al. Low incidence of fatal coronary heart disease in Pima Indians despite high prevalence of non-insulin-dependent diabetes. Circulation 1990;81:987–95.

[32] Preis SR, et al. Trends in cardiovascular disease risk factors in individuals with and without diabetes mellitus in the Framingham Heart Study. Circulation 2009;120:212–20.

[33] Fisher M. Diabetes and atherogenesis. Heart 2004;90:336–40.

[34] Daniels SR, Jacobson MS, McCrindle BW, Eckel RH, Sanner BM. American Heart Association Childhood Obesity Research Summit: executive summary. Circulation 2009;119:2114–23.

[35] Magnussen CG, et al. Pediatric metabolic syndrome predicts adulthood metabolic syndrome, subclinical atherosclerosis, and type 2 diabetes mellitus but is no better than body mass index alone: the Bogalusa Heart Study and the Cardiovascular Risk in Young Finns Study. Circulation 2010;122:1604–11.

[36] Fox CS, et al. Trends in the incidence of type 2 diabetes mellitus from the 1970s to the 1990s: the Framingham Heart Study. Circulation 2006;113:2914–8.

[37] Hu FB, et al. Diet, lifestyle, and the risk of type 2 diabetes mellitus in women. N Engl J Med 2001;345:790–7.

[38] Sowers JR. Obesity as a cardiovascular risk factor. Am J Med 2003;115:37–41.

[39] Malik S, et al. Impact of the metabolic syndrome on mortality from coronary heart disease, cardiovascular disease, and all causes in United States adults. Circulation 2004;110:1245–50.

[40] Rutter MK, Meigs JB, Sullivan LM, D'Agostino RB, Wilson PW. Insulin resistance, the metabolic syndrome, and incident cardiovascular events in the Framingham Offspring Study. Diabetes 2005;54:3252–7.

[41] Kenchaiah S, et al. Obesity and the risk of heart failure. N Engl J Med 2002;347:305—13.

[42] Lakka TA, Lakka HM, Salonen R, Kaplan GA, Salonen JT. Abdominal obesity is associated with accelerated progression of carotid atherosclerosis in men. Atherosclerosis 2001;154:497—504.

[43] Ogden CL, Carroll MD, Kit BK, Flegal KM. Prevalence of childhood and adult obesity in the United States, 2011—2012. JAMA 2014;311:806.

[44] Zhu S, et al. Race-ethnicity-specific waist circumference cutoffs for identifying cardiovascular disease risk factors. Am J Clin Nutr 2005;81:409—15.

[45] Gower BA, Nagy TR, Goran MI. Visceral fat, insulin sensitivity, and lipids in prepubertal children. Diabetes 1999;48:1515—21.

[46] Ku CY, Gower BA, Nagy TR, Goran MI. Relationships between dietary fat, body fat, and serum lipid profile in prepubertal children. Obes Res 1998;6:400—7.

[47] Smith JP. Nature and causes of trends in male diabetes prevalence, undiagnosed diabetes, and the socioeconomic status health gradient. Proc Natl Acad Sci USA 2007;104: 13225—31.

[48] Diagnosed diabetes among American Indians and Alaska natives aged <35 years—United States, 1994—2004. MMWR Morb Mortal Wkly Rep 2006;55:1201—3.

[49] Narayan KMV, Boyle JP, Geiss LS, Saaddine JB, Thompson TJ. Impact of recent increase in incidence on future diabetes burden: U.S., 2005—2050. Diabetes Care 2006; 29:2114—6.

[50] Lindquist CH, Gower BA, Goran MI. Role of dietary factors in ethnic differences in early risk of cardiovascular disease and type 2 diabetes. Am J Clin Nutr 2000;71: 725—32.

[51] Libby P, Ridker PM, Maseri A. Inflammation and atherosclerosis. Circulation 2002;105: 1135—43.

[52] Ross R. Atherosclerosis—an inflammatory disease. N Engl J Med 1999;340:115—26.

[53] Albert Ma, Ridker PM. C-reactive protein as a risk predictor: do race/ethnicity and gender make a difference? Circulation 2006;114:e67—74.

[54] Ledue TB, Rifai N. Preanalytic and analytic sources of variations in C-reactive protein measurement: implications for cardiovascular disease risk assessment. Clin Chem 2003;49:1258—71.

[55] Ridker PM, Rifai N, Rose L, Buring JE, Cook NR. Comparison of C-reactive protein and low-density lipoprotein cholesterol levels in the prediction of first cardiovascular events. N Engl J Med 2002;347:1557—65.

[56] Pai JK, et al. Inflammatory markers and the risk of coronary heart disease in men and women. N Engl J Med 2004;351:2599—610.

[57] Koenig W, Löwel H, Baumert J, Meisinger C. C-reactive protein modulates risk prediction based on the Framingham Score: implications for future risk assessment: results from a large cohort study in southern Germany. Circulation 2004;109:1349—53.

[58] Albert MA, Glynn RJ, Buring J, Ridker PM. C-reactive protein levels among women of various ethnic groups living in the United States (from the Women's Health Study). Am J Cardiol 2004;93:1238—42.

[59] Khera A, et al. Race and gender differences in C-reactive protein levels. J Am Coll Cardiol 2005;46:464—9.

[60] Ford ES, Giles WH, Mokdad AH, Myers GL. Distribution and correlates of C-reactive protein concentrations among adult US women. Clin Chem 2004;50:574—81.

[61] LaMonte MJ. Cardiorespiratory fitness and C-reactive protein among a tri-ethnic sample of women. Circulation 2002;106:403—6.

[62] Ranjit N, et al. Socioeconomic position, race/ethnicity, and inflammation in the multi-ethnic study of atherosclerosis. Circulation 2007;116:2383—90.

[63] Cook DG, et al. Ethnic differences in fibrinogen levels: the role of environmental factors and the beta-fibrinogen gene. Am J Epidemiol 2001;153:799—806.

[64] Kain K, Catto AJ, Grant PJ. Impaired fibrinolysis and increased fibrinogen levels in South Asian subjects. Atherosclerosis 2001;156:457–61.

[65] Duncan BB, et al. Low-grade systemic inflammation and the development of type 2 diabetes: the atherosclerosis risk in communities study. Diabetes 2003;52: 1799–805.

[66] Hokanson JE, Austin MA. Plasma triglyceride level is a risk factor for cardiovascular disease independent of high-density lipoprotein cholesterol level: a meta-analysis of population-based prospective studies. J Cardiovasc Risk 1996;3:213–9.

[67] Juha Pekkanen M, et al. Ten-year mortality from cardiovascular disease in relation to cholesterol level among men with and without preexisting cardiovascular disease. N Engl J Med 1990;322:1700–7.

[68] Voight BF, et al. Plasma HDL cholesterol and risk of myocardial infarction: a mendelian randomisation study. Lancet 2012;380:572–80.

[69] Ridker PM, et al. Reduction in C-reactive protein and LDL cholesterol and cardiovascular event rates after initiation of rosuvastatin: a prospective study of the JUPITER trial. Lancet 2009;373:1175–82.

[70] Carroll MD, et al. Trends in serum lipids and lipoproteins of adults, 1960–2002. JAMA 2005;294:1773–81.

[71] Rodriguez C, Pablos-me A. Comparison of modifiable determinants of lipids and lipoprotein levels among Hispanic Caucasians > 65 years of age living in New York City. Am J Cardiol 2002;89:178–83.

[72] Executive summary of the third report of the National Cholesterol Education Program (NCEP) expert panel on detection, evaluation, and treatment of high blood cholesterol in adults (adult treatment panel III). JAMA 2001;285:2486–97.

[73] Campos H, Moye LA, Glasser SP, Stampfer MJ, Sacks FM. Low-density lipoprotein size, pravastatin treatment, and coronary events. JAMA 2001;286:1468–74.

[74] Gray RS, et al. Relation of LDL size to the insulin resistance syndrome and coronary heart disease in American Indians. The Strong Heart Study. Arterioscler Thromb Vasc Biol 1997;17:2713–20.

[75] Festa A, et al. Nuclear magnetic resonance lipoprotein abnormalities in prediabetic subjects in the Insulin Resistance Atherosclerosis Study. Circulation 2005;111: 3465–72.

[76] Frazier-Wood AC, et al. A clustering analysis of lipoprotein diameters in the metabolic syndrome. Lipids Health Dis 2011;10:237.

[77] Garvey WT, et al. Effects of insulin resistance and type 2 diabetes on lipoprotein subclass particle size and concentration determined by nuclear magnetic resonance. Diabetes 2003;52:453–62.

[78] Goff Jr DC, D'Agostino Jr RB, Haffner SM, Otvos JD. Insulin resistance and adiposity influence lipoprotein size and subclass concentrations. Results from the Insulin Resistance Atherosclerosis Study. Metabolism 2005;54:264–70.

[79] Syvanne M, et al. High density lipoprotein subfractions in non-insulin-dependent diabetes mellitus and coronary artery disease. J Lipid Res 1995;36:573–82.

[80] Frazier-Wood AC, et al. Genetic variants associated with VLDL, LDL and HDL particle size differ with race/ethnicity. Hum Genet 2013;132:405–13.

[81] Kullo IJ, Jan MF, Bailey KR, Mosley TH, Turner ST. Ethnic differences in low-density lipoprotein particle size in hypertensive adults. J Clin Lipidol 2007;1:218–24.

[82] Kral BG, et al. Racial differences in low-density lipoprotein particle size in families at high risk for premature coronary heart disease. Ethn Dis 2001;11:325–37.

[83] Neale MC, Cardon L. Methodology for genetic studies of twins and families. Kluwer; 1994.

[84] Manolio TA, et al. Finding the missing heritability of complex diseases. Nature 2009; 461:747–53.

[85] Jousilahti P, Puska P, Vartiainen E, Pekkanen J, Tuomilehto J. Parental history of premature coronary heart disease: an independent risk factor of myocardial infarction. J Clin Epidemiol 1996;49:497—503.

[86] Zdravkovic S, et al. Heritability of death from coronary heart disease: a 36-year follow-up of 20 966 Swedish twins. J Intern. Med 2002;252:247—54.

[87] Shea S, Ottman R, Gabrieli C, Stein Z, Nichols A. Family history as an independent risk factor for coronary artery disease. J Am Coll Cardiol 1984;4:793—801.

[88] Hopkins PN, et al. Family history as an independent risk factor for incident coronary artery disease in a high-risk cohort in Utah. Am J Cardiol 1988;62:703—7.

[89] Grech ED, Ramsdale DR, Bray CL, Faragher EB. Family history as an independent risk factor of coronary artery disease. Eur Heart J 1992;13:1311—5.

[90] Flossmann E, Schulz UGR, Rothwell PM. Systematic review of methods and results of studies of the genetic epidemiology of ischemic stroke. Stroke 2004;35:212—27.

[91] Kaprio J, et al. Concordance for type 1 (insulin-dependent) and type 2 (non-insulin-dependent) diabetes mellitus in a population-based cohort of twins in Finland. Diabetologia 1992;35:1060—7.

[92] Newman B, et al. Concordance for type 2 (non-insulin-dependent) diabetes mellitus in male twins. Diabetologia 1987;30:763—8.

[93] Barnett AH, Eff C, Leslie RD, Pyke DA. Diabetes in identical twins. A study of 200 pairs. Diabetologia 1981;20:87—93.

[94] Tattersall RB, Pyke DA. Diabetes in identical twins. Lancet 1972;2:1120—5.

[95] Gottlieb MS, Root HF. Diabetes mellitus in twins. Diabetes 1968;17:693—704.

[96] Poulsen P, Kyvik KO, Vaag a, Beck-Nielsen H. Heritability of type II (non-insulin-dependent) diabetes mellitus and abnormal glucose tolerance—a population-based twin study. Diabetologia 1999;42:139—45.

[97] Poulsen P, et al. Heritability of insulin secretion, peripheral and hepatic insulin action, and intracellular glucose partitioning in young and old Danish twins. Diabetes 2005; 54:275—83.

[98] Karoline S, Gonneke W, Kyvik Kirsten O, Mortensen J, Boomsma DI, Cornes BK, et al. Sex differences in heritability of BMI: a comparative study of results from twin studies in eight countries. Twin Res 2003;6:409—21.

[99] Su S, et al. Genetic and environmental influences on systemic markers of inflammation in middle-aged male twins. Atherosclerosis 2008;200:213—20.

[100] Beekman M, et al. Heritabilities of apolipoprotein and lipid levels in three countries. Twin Res 2002;5:87—97.

[101] Wang X, Thayer JF, Treiber F, Snieder H. Ethnic differences and heritability of heart rate variability in African- and European American youth. Am J Cardiol 2005;96: 1166—72.

[102] Snieder H, Harshfield Ga, Treiber FA. Heritability of blood pressure and hemodynamics in African- and European-American youth. Hypertension 2003;41:1196—201.

[103] North KE, et al. Genetic epidemiology of BMI and body mass change from adolescence to young adulthood. Obes (Silver Spring) 2010;18:1474—6.

[104] Deng F-Y, et al. Genetic determination and correlation of body mass index and bone mineral density at the spine and hip in Chinese Han ethnicity. Osteoporos Int 2006; 17:119—24.

[105] Hur Y-M. Sex difference in heritability of BMI in South Korean adolescent twins. Obes (Silver Spring) 2007;15:2908—11.

[106] Nelson TL, Brandon DT, Wiggins Sa, Whitfield KE. Genetic and environmental influences on body-fat measures among African-American twins. Obes Res 2002;10: 733—9.

[107] Hsu F-C, et al. Heritability of body composition measured by DXA in the diabetes heart study. Obes Res 2005;13:312—9.

[108] Norris JM, et al. Quantitative trait loci for abdominal fat and BMI in Hispanic-Americans and African-Americans: the IRAS Family study. Int J Obes (Lond) 2005; 29:67–77.

[109] Elbein SC, Das SK, Hallman DM, Hanis CL, Hasstedt SJ. Genome-wide linkage and admixture mapping of type 2 diabetes in African American families from the American Diabetes Association GENNID (Genetics of NIDDM) Study Cohort. Diabetes 2009;58:268–74.

[110] Reich DE, Gabriel SB, Altshuler D. Quality and completeness of SNP databases. Nat Genet 2003;33:457–8.

[111] Kruglyak L, Nickerson DA. Variation is the spice of life. Nat Genet 2001;27:234–6.

[112] Tennessen JA, et al. Evolution and functional impact of rare coding variation from deep sequencing of human exomes. Science 2012;337:64–9.

[113] Chakravarti A. Being human: kinship: race relations. Nature 2009;457:380–1.

[114] Goldstein JR. Kinship networks that cross racial lines: the exception or the rule? Demography 1999;36:399–407.

[115] Shah SA, et al. Associations between NOS1AP single nucleotide polymorphisms (SNPs) and QT interval duration in four racial/ethnic groups in the Multi-Ethnic Study of Atherosclerosis (MESA). Ann Noninvasive Electrocardiol 2013;18:29–40.

[116] Altshuler DM, et al. Integrating common and rare genetic variation in diverse human populations. Nature 2010;467:52–8.

[117] Consortium, T. I. H.. The international HapMap project. Nature 2005;63(Suppl. 1): 29–34.

[118] Gabriel SB, et al. The structure of haplotype blocks in the human genome. Science 2002;296:2225–9.

[119] Keinan A, Mullikin JC, Patterson N, Reich D. Measurement of the human allele frequency spectrum demonstrates greater genetic drift in East Asians than in Europeans. Nat Genet 2007;39:1251–5.

[120] Consortium, T. I. H.. A second generation human haplotype map of over 3. 1 million SNPs. Nature 2009;449:851–61.

[121] Spielman RS, et al. Common genetic variants account for differences in gene expression among ethnic groups. Nat Genet 2007;39:226–31.

[122] Moreno-Estrada A, et al. Human genetics. The genetics of Mexico recapitulates native American substructure and affects biomedical traits. Science 2014;344:1280–5.

[123] Corona E, et al. Analysis of the genetic basis of disease in the context of worldwide human relationships and migration. PLoS Genet 2013;9:e1003447.

[124] Morris AP. Transethnic meta-analysis of genomewide association studies. Genet Epidemiol 2011;35:809–22.

[125] Speliotes EK, et al. Association analyses of 249,796 individuals reveal 18 new loci associated with body mass index. Nat Genet 2010;42:937–48.

[126] Wing MR, et al. Analysis of FTO gene variants with measures of obesity and glucose homeostasis in the IRAS Family Study. Hum Genet 2009;125:615–26.

[127] Scuteri A, et al. Genome-wide association scan shows genetic variants in the FTO gene are associated with obesity-related traits. PLoS Genet 2007;3:e115.

[128] Hennig BJ, et al. FTO gene variation and measures of body mass in an African population. BMC Med Genet 2009;10:21.

[129] Villalobos-Comparán M, et al. The FTO gene is associated with adulthood obesity in the Mexican population. Obes (Silver Spring) 2008;16:2296–301.

[130] Peng S, et al. FTO gene polymorphisms and obesity risk: a meta-analysis. BMC Med 2011;9:71.

[131] Frosst P, et al. A candidate genetic risk factor for vascular disease: a common mutation in methylenetetrahydrofolate reductase. Nat Genet 1995;10:111–3.

[132] Kang SS, Passen EL, Ruggie N, Wong PW, Sora H. Thermolabile defect of methylene-tetrahydrofolate reductase in coronary artery disease. Circulation 1993;88:1463—9.

[133] Kang SS, et al. Thermolabile methylenetetrahydrofolate reductase: an inherited risk factor for coronary artery disease. Am J Hum Genet 1991;48:536—45.

[134] Kluijtmans LA, et al. Molecular genetic analysis in mild hyperhomocysteinemia: a common mutation in the methylenetetrahydrofolate reductase gene is a genetic risk factor for cardiovascular disease. Am J Hum Genet 1996;58:35—41.

[135] Arruda VR, von Zuben PM, Chiaparini LC, Annichino-Bizzacchi JM, Costa FF. The mutation Ala677—>Val in the methylene tetrahydrofolate reductase gene: a risk factor for arterial disease and venous thrombosis. Thromb Haemost 1997;77:818—21.

[136] Franco RF, Araújo AG, Guerreiro JF, Elion J, Zago MA. Analysis of the 677 C—>T mutation of the methylenetetrahydrofolate reductase gene in different ethnic groups. Thromb Haemost 1998;79:119—21.

[137] Monda KL, et al. A meta-analysis identifies new loci associated with body mass index in individuals of African ancestry. Nat Genet 2013;45:690—6.

[138] Mahajan A, et al. Genome-wide trans-ancestry meta-analysis provides insight into the genetic architecture of type 2 diabetes susceptibility. Nat Genet 2014;46:234—44.

[139] Lange LA, et al. Whole-exome sequencing identifies rare and low-frequency coding variants associated with LDL cholesterol. Am J Hum Genet 2014;94:233—45.

[140] Crosby J, et al. Loss-of-function mutations in APOC3, triglycerides, and coronary disease. N Engl J Med 2014;371:22—31.

[141] Ioannidis JPA, Ntzani EE, Trikalinos TA. "Racial" differences in genetic effects for complex diseases. Nat Genet 2004;36:1312—8.

Genomics-Guided Immunotherapy of Human Epithelial Ovarian Cancer

Sahar Al Seesi[1,2], *Fei Duan*[2], *Ion I. Mandoiu*[1], *Pramod K. Srivastava*[2,3], *Angela Kueck*[3]

[1] Department of Computer Science & Engineering, University of Connecticut, Storrs, CT, USA; [2] Department of Immunology, Carole and Ray Neag Comprehensive Cancer Center, University of Connecticut School of Medicine, Farmington, CT, USA; [3] Carole and Ray Neag Comprehensive Cancer Center, University of Connecticut School of Medicine, Farmington, CT, USA

1. EPITHELIAL OVARIAN CANCER FROM A CLINICAL PERSPECTIVE: CURRENT TREATMENTS AND CHALLENGES

Epithelial ovarian cancer (EOC) is the leading cause of death in women with gynecologic cancer in the United States. It is the fifth most common cause of cancer death overall and in 2014, it is estimated that there will be over 21,000 new diagnoses and over 14,000 deaths from EOC [1]. The incidence of EOC increases with age and is most prevalent in women in their 60s and 70s. The majority of these women (70%) will present with advanced stage disease (Stage III or IV) resulting in the overall poor prognosis with a 5-year survival rate of only 27% [2,3]. Although treatment advances over the past two decades have extended the median survival to 65.6 months, most women will spend this time on multiple chemotherapy regimens and will succumb to their disease [4].

Risk factors for developing EOC have been identified through epidemiologic studies and link increased ovulatory cycles with increased risk (e.g., nulliparity, older age (>35) at first pregnancy) [5,6]. The use of oral contraceptive pills, breastfeeding, and younger age (<25) at first

Translational Cardiometabolic Genomic Medicine
http://dx.doi.org/10.1016/B978-0-12-799961-6.00010-X **237**

pregnancy can be protective with a 30—60% decreased risk of EOC [6]. While the majority of cases are sporadic, 5—10% may be identified through the family history and linkage with breast and ovarian cancers in families with BRCA1 and BRCA2 mutations [7,8]. In addition, families with Lynch syndrome Hereditary nonpolyposis colorectal cancer (HNPCC) not only have an increased risk of colon cancer, they also have higher risks of endometrial and ovarian cancer [9]. Prophylactic removal of the ovaries and fallopian tubes are recommended in all of these high-risk women as a risk-reducing strategy [10].

Early detection of EOC in a more curable early stage has been difficult for many reasons. The ovaries are not easily accessible for evaluation or study without invasive procedures. In addition, the heterogeneity of ovarian cancer is another challenge in identifying a novel biomarker and currently routine screening is not recommended by any professional society [11]. The presenting symptoms of ovarian cancer include bloating, pelvic or abdominal pain, difficulty eating or feeling full quickly, and urinary symptoms (urgency or frequency), especially if these symptoms are new and frequent (>12 days/month) [12]. Women with the above symptoms should have ovarian pathology ruled out by ultrasound imaging and a gynecologic examination. Ongoing trials in the United Kingdom and the United States are evaluating serial cancer antigen 125 (CA-125) testing with yearly ultrasounds in addition to other serum markers, but to date, randomized data do not yet support screening in the general population [13,14].

The primary treatment for patients presenting with advanced ovarian cancer is surgical cytoreductive therapy (debulking) followed by systemic chemotherapy [15]. It is recommended that surgery be performed by a subspecialist in gynecologic oncology, based on studies demonstrating improved outcomes [16,17]. EOC arises on the surface of the ovaries and disseminates throughout the peritoneal cavity and commonly lines the serosal surfaces of the bowel, mesentery, diaphragms, and the omentum and may also produce extensive ascites. The goal of the surgical debulking procedure is to explore the entire abdomen and pelvis and remove all gross visible disease when possible [18]. Following surgery, patients are categorized into optimal or suboptimal groups with optimal being defined as the residual tumor nodules measuring less than 1 cm in diameter [19]. There is a significant correlation to the amount of residual disease and overall survival and therefore, extensive resection of metastases should be attempted in patients who can tolerate surgery [20]. Ovarian cancer is one of the most chemosensitive tumors at initial diagnosis. The most active chemotherapy drugs are platinum based and usually given as a doublet with paclitaxel for 6—8 cycles. Intravenous, dose-dense, and intraperitoneal therapies are all approved regimens and obtain up to 80% remission rates following optimal cytoreduction [20].

However, up to 80% of patients will recur with a median 18 months of progression-free survival and then the challenges of chemoresistance arise [21]. Recurrent ovarian cancer may be treated again with surgery followed by chemotherapy with the most important prognostic factor being the time to recurrence [22]. Those who recur greater than 6 months from their initial treatment are considered "platinum sensitive" and have a better prognosis. Platinum resistance eventually occurs in most patients and second and third line options provide dismal response rates (<20%) and clinical trials are generally recommended [20]. New agents that target tumor vasculature or inhibit DNA repair processes are now being used along with standard chemotherapy but with no impact on overall survival. Clearly, new strategies are needed.

2. IMMUNE RESPONSE TO OVARIAN CANCER

The critical role of immune response in EOC has been demonstrated by a significant correlation of survival with the presence of tumor-infiltrating lymphocytes; also, a down-regulatory CD4 response subverts cancer immunity [23–25]. However, these solid observations have not yet led to any substantial tangible progress in immunotherapy of EOC. Most clinical studies are phase 1 and phase 2 trials, which are primarily designed to evaluate safety, determine a safe dosage range, and identify side effects. After reviewing 55 studies, including 3051 women with EOC, from 1996 to October 2013, Leffers et al. concluded that "no clinically effective antigen-specific active immunotherapy is yet available for ovarian cancer" [26].

A key lacuna in progress is a lack of understanding of the antigens on ovarian cancer cells that elicit the CD8 T cells that are much beneficial. A small number of antigens (e.g., MAGE and NY-ESO1) have been tested. However, targeting of single tumor antigens in EOC has led to only modest clinical results [27]. It is our belief that the inability to generate an immune response against the broader array of antigens, including the mutational antigens many of which may be unique to individual patients, is a major obstacle in immunotherapy of EOC. In a genomic analysis of 316 ovarian cancer samples, about 61 somatic mutations were identified in each tumor, and the vast majority (>95%) mutations are passenger mutations [28]. The mutational antigens have been traditionally the most difficult to examine because they may be individually specific to each patient's tumor. A revolution in high-throughput DNA-sequencing technologies now makes it possible to overcome this challenge to therapeutically target this most promising class of tumor antigens [29–35]. In the next section, we briefly describe the methodological approaches that are being used to divine the immunome of ovarian cancers, starting from their genomes or exomes.

3. NEXT-GENERATION SEQUENCING AND ITS APPLICATION TO CANCER GENOMICS

There have been many advances in the study of cancer genomics in the past decades using Sanger sequencing and microarrays for gene quantification and single nucleotide variation (SNV) detection. However, next-generation sequencing made it possible to identify the complete genome and transcriptome at single base resolution. Next-generation sequencing comes with its own challenges, including higher cost and computational challenges, resulting from short read lengths and coverage depth. However, as the continuous technological advances lead to longer reads, higher sequencing depth and lower cost, these new technologies, such as SOLiD [36], 454 [37], Illumina [38], or ION Torrent [39] are becoming the method of choice in studying cancer genomics.

Some of the commonly used next-generation sequencing protocols are transcriptome sequencing, usually referred to as RNA-Seq, whole-genome sequencing, and exome sequencing. Whole-genome sequencing has many applications, including calling single point and structural variants and determining copy number variations in tumor samples. The same questions can be answered using exome sequencing, at a much lower cost, since the size of the human exome is only ~1% of the genome. However, exome sequencing cannot be used to find variations in regulatory regions outside of the exons. Transcriptome sequencing is used to study the expression levels of genes, and identify genes with expression patterns specific to certain tumor types.

3.1 RNA-Seq

RNA-Seq uses next-generation sequencing technologies, such as SOLiD, 454, Illumina, or ION Torrent [36–39]. Figure 1 depicts the main steps in an RNA-Seq experiment, ending with the first step of analysis, which is typically annotating or mapping the data to a reference. The mRNA extracted from a sample is converted to cDNA using reverse transcription and sheared into fragments. Fragments within lengths of a certain range are selected and ligated with sequencing adapters. This is usually followed by an amplification step after which one or both ends of the cDNA fragments are sequenced to produce either single or paired-end reads. cDNA synthesis and adapter ligation can be done in a strand-specific manner, in which case the strand of each read is

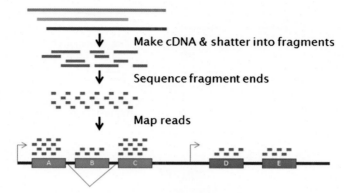

FIGURE 1   A schematic representation of the RNA-Seq protocol, including the first steps of the analysis. RNA transcripts are converted into double-stranded cDNA, which are then fragmented and their ends sequenced. Analysis starts by mapping reads to a reference genome.

known; this is commonly referred to as directional sequencing. In the more common non-directional RNA-Seq protocols, strand specificity is not maintained. The specifics of the sequencing protocols vary from one technology to the other. In particular, the length of produced reads varies depending on the technology, with newer high-throughput technologies typically producing longer reads.

3.2 Exome Sequencing

Exome sequencing is targeted DNA sequencing where the exonic regions are captured and sequenced. Since the size of the human exome is only ~1% of the genome, exome sequencing is a much more cost-effective way to study variations in coding regions, compared to whole-genome sequencing. There are different protocols to capture the exonic regions for sequencing. The two main protocols are array hybridization, and in-solution hybridization. In both protocols, hybridization probes are designed based on sequences of annotated exons available in transcriptome libraries, such as Ensembl [40] or CCDS [41]. In the case of array hybridization (Figure 2), the probes are fixed to a microarray. After extracting and fragmenting the DNA, sequencing adaptors are attached to the DNA fragments. When the DNA fragments are applied to the microarray, only fragments coming from exonic regions will hybridize to the probes on the microarray. The rest of the fragments are washed off. Then the hybridized exonic DNA is stripped off the microarray and sequenced. In-solution hybridization (Figure 3) is a similar approach, except that the probes are attached to magnetic beads and they are hybridized to the DNA in-solution. The probes, which hybridize to the

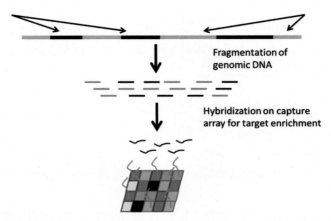

FIGURE 2   A schematic representation of the array hybridization exome extraction protocol. Fragmented DNA is applies to the microarray. Fragments coming from exonic regions (orange) hybridize to the probes on the microarray. The rest of the fragments (black) are washed off.

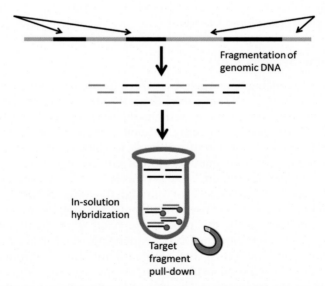

FIGURE 3   A schematic representation of the in-solution hybridization exome extraction protocol. DNA fragments (orange: exonic and black: intronic) are added to a solution with probes designed to hybridize with exonic regions, with magnetic beads attached to each probe. After hybridization, the target fragments (orange) are pulled down.

exonic DNA fragments, are then pulled down, and the DNA is sequenced. Similar to RNA-Seq, the captured exome can be sequenced on any of the next-generation sequencing technologies, such as ION Torrent, Illumina, or 454. Also, the one or both ends of the DNA fragments can be sequenced.

3.3 Bioinformatics Analysis

All analysis pipelines start by mapping the reads to the human reference genome. The choice of mapping software depends on the sequencing protocol (sequencing data type) as well as the sequencing technologies. Exome sequencing can be mapped to the genome using unspliced aligners like Bowtie [42] and Bowtie2 [43]. Bowtie does not allow insertions and deletions in the read alignments, which makes it more suitable for Illumina reads where sequencing errors are expected to be substitution errors. Bowtie2 allows short insertions and deletions in read alignments. Therefore, it can be used for mapping 454 and ION Torrent reads. It is also designed to map longer reads more efficiently than bowtie. ION Torrent reads are more commonly mapped with tmap, a read aligner developed by Life Technologies. On the other hand, RNA Seq must be mapped using spliced aligners, like Tophat [44], or aligners that allow local alignments, like TMAP. Otherwise, reads generated from splice junctions will fail to map.

After aligning reads, the analysis depends on the problem of interest. Next-generation sequencing data are used in different types of cancer genome analysis. Tumor transcriptome profiling is used to identify genes whose expression is associated with a certain type of tumor. For this type of analysis, gene quantification tools [45–47] are used to estimate the expression levels of genes in the tumor RNA-Seq samples. Also, gene differential expression tools [48–50] are used to compare the transcriptome profiles of tumor samples against control samples.

Calling somatic variants is common type of analysis in cancer genomics. To call somatic variants for a human tumor sample, the tumor DNA sequencing sample must be analyzed together with a normal sample from the same patient, in order to differentiate between somatic and germline mutations. The sequencing data can be either whole-genome sequencing samples or exome samples. Exome sequencing is less expensive, and it is sufficient for calling variants within coding regions. Calling variations from whole-genome sequencing makes it possible to identify variants in regulatory regions also. Identifying somatic variants includes calling single point variants as well as structural variations and gene fusions. A number of tools [51–54] have been developed to analyze pairs of tumor-normal DNA samples and call somatic mutations.

Cancer immunology research involves predicting immunogenic epitopes at somatic mutation loci. Predictions are typically done by running epitope prediction tools, such as NetMHC [55] or NetCTL [56] on protein sequences harboring called somatic mutations.

4. ANALYSIS OF OVARIAN CANCER SAMPLES IN THE TCGA DATABASE

We analyzed data for human ovarian cancer from TCGA [57]. We downloaded data for 345 ovarian cancer patients. The racial breakdown of the patients was as follows: 304 white, 18 black or African-American, 10 Asian, 2 American-Indian or Alaska natives, and 11 whose racial information was not available. We downloaded somatic mutations, gene expression profiles, and clinical data for 345 ovarian cancer samples. The somatic mutations were predicted from paired tumor-normal exome sequencing data. Somatic mutation analysis was done at three institutions. Of the 345 samples, 230 were analyzed at The Genome Institute at Washington University, 142 samples were analyzed at The Broad Institute, and 91 were analyzed at The Human Genome Sequencing Center at Baylor College of Medicine. The gene expression profiles were predicted from RNA-Seq data at the Canada's Michael Smith Genome Sciences Centre.

The mutations used in this analysis were predicted from matched normal and tumor exome sequencing. The methods used to call the mutations differ for each of the three research centers that analyzed the samples. Somatic mutations for samples analyzed at The Genome Institute at Washington University were predicted using an automated approach that combines predictions from VarScan2 [51], SomaticSniper [52], and GATK [53]. The Broad Institute used their somatic mutation caller MuTect [54]. Baylor College of Medicine used the *pileup* tool from SamTools [58] suite to predict mutations in the samples analyzed in their Human Genome Sequencing Center, and they considered mutations that were observed in a tumor sample and were absent from the matched normal sample to be somatic. They filtered out mutations that were observed less than five times in the tumor sample or at a locus with coverage less than nine in the normal sample. Details for thresholds for all used algorithms can be found in [57].

We restricted our analysis to predicted SNVs. For each tumor sample, we selected the subset of SNVs in expressed genes. Expressed genes were identified from the gene expression profile of the tumor, which we also downloaded from TCGA. A threshold of 5 fragments per kilobase per million reads was used to declare a gene expressed.

Reference and alternative peptide sequences, of length 21 amino acids, were generated around each expressed somatic mutation. Generated amino acid sequences were then run through the NetMHC 3.0 epitope prediction program [55] and IC50 scores were predicted for 8, 9, and 10 amino acid long peptides. The predicted affinities were for the most common allele in the patient's race [59].

Figure 4 shows the distribution of the number of predicted epitopes versus the number of ovarian cancers. It is clear from these results that for a single HLA allele, 92% EOC patients have between 1 and 64 predicted neo-epitopes (mean = 13, median = 11). Over 70% of ovarian cancers have 5–45 predicted neo-epitopes, just for a single HLA allele. Since each patient has a minimum of 3 and a maximum of 6 distinct alleles, individual human ovarian cancers have tens of predicted neo-epitopes that can be exploited for genomics-guided personalized immunotherapy.

This rather vast trove of potential neo-epitopes is the subject of our ongoing and current efforts. Recently, we have also carried out genomics and bioinformatic studies in mouse models of cancers, which show that [33]:

1. Mouse tumors, like human ovarian cancers shown in Figure 1 have tens to hundreds of neo-epitopes generated by passenger mutations. The ID8 ovarian cancer, for example has about 500 predicted neo-epitopes.
2. A proportion of these neo-epitopes elicit measurable MHC I-restricted CD8+ immune responses; the CD4+ responses are currently being tested.

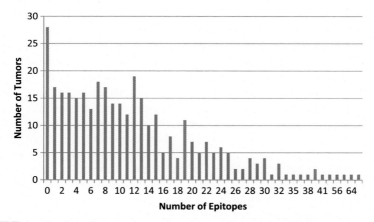

FIGURE 4   Number of tumors (vertical axis) versus the number of HLA-restricted alleles (horizontal axis) in individual human ovarian cancers in the TCGA database (for the single most common allele for the patient's race).

3. ~30% of the neo-epitopes also mediate potent tumor rejection in vivo (Figure 2).
4. We are analyzing the properties of the neo-epitopes that do or do not mediate tumor rejection, in order to better identify the epitopes that may be used for immunotherapy. We have identified one aspect of MHC I-peptide binding (the conformational rigidity of the neo-epitope as it sits in the groove of the MHC I molecule) that correlate with tumor immunity.

Altogether, these data on the neo-epitopes in nearly 400 human ovarian cancers (Figure 4) and the immunological tumor-rejection activity of such neo-epitopes in mice in vivo [33] provide a strong basis for genomics-driven immunotherapy of human EOC. Further, a recent report by Brown et al. [60] has recently concluded that the neo-epitopes predicted by the meta-analysis of over 500 ovarian cancer genomes correlate with increased patient survival.

5. IMPLICATIONS FOR GENOMICS-DRIVEN IMMUNOTHERAPY OF HUMAN OVARIAN CANCER

Based on the identification of a large number of neo-epitopes as discussed above, and based on the hypothesis that the individual tumor-specific antigenic fingerprint of the tumor is the primary driver of antitumor immunity, we plan to initiate the first phase 1 trial in patients with stage III or stage IV EOC. The pipeline for derivation of the vaccine cocktail for each individual patient is shown in Figure 5. Briefly, the patient will be haplotyped, and samples of peripheral blood mononuclear cells (as source of normal cells) and the ovarian cancer tissue shall be obtained. The normal and cancer tissues shall be used for exome sequencing; in addition, the cancer tissues shall also be used for transcriptome sequencing. The normal and the cancer exomes shall be compared to identify SNVs; these will be referenced against the tumor transcriptome in order to identify the variants restricted to the tumor transcripts. This base of cancer-specific variants shall be queried using NetMHC of such tool to create a first list of potential neo-epitopes based on the HLA haplotype of the patient. This list shall be further refined using the differential agretopic index or other criteria that are better predictive of antitumor immunity. The next task would be to select up to 10 neo-epitopes for the purposes of immunization. Our objective would be to select new epitopes that can be presented by a larger number of HLA alleles of the patient; this process shall hopefully reduce the prospect of competition among neo-epitopes for binding to the same HLA allele. Other criteria for selecting the final list

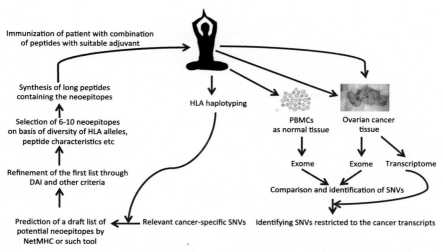

FIGURE 5 The pipeline for the proposed phase 1 clinical trial testing the safety and feasibility of immunization with peptides corresponding to the individual neo-epitopes of each patient's cancer. Immunogenicity of the neo-epitopes shall also be monitored.

of neo-epitopes shall include structural and biophysical properties of peptides that will increase the chances of their presentation by the HLA alleles, as well as characteristics that are compatible with their synthesis. Longer peptides of about 17 amino acids that contain the neo-epitopes in the center shall be chemically synthesized, and the patients shall be immunized with a cocktail of these peptides along with a suitable adjuvant. Patients shall be monitored for safety, any adverse events, and for immune response to the neo-epitopes used for immunization. Clinical endpoints shall not be a part of the formal protocol, but shall nonetheless be monitored. This first clinical study shall form the foundation stone of larger studies that may test for true clinical activity. This approach to individual specific immunotherapy of human cancer was first proposed over 20 years ago [61]; the advances in high-throughput DNA sequencing and bioinformatics have caught up with that idea, and it is now time to test it.

References

[1] Siegel R, Ma J, Zou Z, Jemal A. Cancer statistics, 2014. CA Cancer J Clin 2014;64:9—29.
[2] Fleming GF, Seidman J, Lengyel E. Epithelial ovarian cancer. In: Barakat RR, Markman M, Randall ME, editors. Principles and practice of gynecologic oncology. 6th ed. Philadelphia: Lippincott Williams & Wilkins; 2013. p. 757—847.
[3] American Cancer Society. Cancer facts & figures, 2013. Atlanta (GA): American Cancer Society; 2013.

[4] Armstrong DK, Bundy B, Wenzel L, et al. Intraperitoneal cisplatin and paclitaxel in ovarian cancer. N Engl J Med 2006;354:34–43.

[5] Jelovac D, Armstrong DK. Recent progress in the diagnosis and treatment of ovarian cancer. CA Cancer J Clin 2011;61:183–203.

[6] Holschneider CH, Berek JS. Ovarian cancer: epidemiology, biology, and prognostic factors. Semin Surg Oncol 2000;19:3–10.

[7] Lancaster JM, Powell CB, Kauff ND, et al. Society of Gynecologic Oncologists Education Committee statement on risk assessment for inherited gynecologic cancer predispositions. Gynecol Oncol 2007;107:159–62.

[8] ACOG Practice Bulletin No. 103: hereditary breast and ovarian cancer syndrome. Obstet Gynecol 2009;113:957–66.

[9] Shulman LP. Hereditary breast and ovarian cancer (HBOC): clinical features and counseling for BRCA1 and BRCA2, Lynch syndrome, Cowden syndrome, and Li-Fraumeni syndrome. Obstet Gynecol Clin North Am 2010;37:109–33.

[10] Finch A, Beiner M, Lubinski J, et al. Salpingo-oopherectomy and the risk of ovarian, fallopian tube, and peritoneal cancers in women with a BRCA1 or BRCA2 Mutation. JAMA 2006;296:185–92.

[11] American College of Obstetricians and Gynecologists Committee on Gynecologic Practice. Committee opinion No. 477: the role of the obstetrician-gynecologist in the early detection of epithelial ovarian cancer. Obstet Gynecol 2011;117:742–6.

[12] Goff BA, Mandel LS, Drescher CW, et al. Development of an ovarian cancer symptom index: possibilities for early detection. Cancer 2007;109:221–7.

[13] Hartge P. Designing early detection programs for ovarian cancer. J Natl Cancer Inst 2010;102:3–4.

[14] Gentry-Maharaj A, Menon U. Screening for ovarian cancer in the general population. Best Pract Res Clin Obstet Gynaecol 2012;26:243–56.

[15] Schorge JO, Garrett LA, Goodman A. Cytoreductive surgery for advanced ovarian cancer: quo vadis? Oncology (Williston Park) 2011;25:928–34.

[16] Giede KC, Kieser K, Dodge J, Rosen B. Who should operate on patients with ovarian cancer? an evidence-based review. Gynecol Oncol 2005;99:447–61.

[17] Earle CC, Schrag D, Neville BA, et al. Effect of surgeon specialty on processes of care and outcomes for ovarian cancer patients. J Natl Cancer Inst 2006;98:172–80.

[18] Alett GD, Dowdy SC, Gostout BS, et al. Aggressive surgical effort and improved survival in advanced-stage ovarian cancer. Obstet Gynecol 2006;107:77–85.

[19] Whitney CW, Spirtos N. Gynecologic oncology group surgical procedures manual. Philadelphia: Gynecologic Oncology Group; 2009.

[20] National Comprehensive Cancer Network. NCCN clinical practice guidelines in oncology. Ovarian Cancer 2014.

[21] Luvero D, Milani A, Ledermann J. Treatment options in recurrent ovarian cancer. Ther Adv Med Oncol 2014;6(5):229–39.

[22] Friedlaner ML, Stockler MR, Butow P, et al. Clinical trials of palliative chemotherapy in platinum-resistant or -refractory ovarian cancer: time to think differently? J Clin Oncol 2013;31:2362.

[23] Zhang L, Conejo-Garcia JR, Katsaros D, Gimotty PA, Massobrio M, Regnani G, et al. Intratumoral T cell, recurrence, and survival in epithelial ovarian cancer. N Engl J Med 2003;348:203–13.

[24] Sato E, Olson SH, Ahn J, Bundy B, Nishikawa H, Qian F, et al. Intraepithelial CD8+ tumor-infiltrating lymphocytes and a high CD8+/regulatory T cell ratio are associated with favorable prognosis in ovarian cancer. Proc Natl Acad Sci USA December 20, 2005; 102(51):18538–43.

[25] Curiel TJ, Coukos G, Zou L, Alvarez X, Cheng P, Mottram P, et al. Nat Med 2004;10: 942–9.

[26] Leffers N, et al. Antigen-specific active immunotherapy for ovarian cancer. Cochrane Database Syst Rev 2014;9:Cd007287.

[27] Odunsi K, Matsuzaki J, Karbach J, Neumann A, Mhawech-Fauceglia P, Miller A, et al. Proc Natl Acad Sci USA April 10, 2012;109(15):5797−802.

[28] Integrated genomic analyses of ovarian carcinoma. Nature 2011;474(7353):609−15.

[29] Lennerz V, Fatho M, Gentilini C, Frye RA, Lifke A, Ferel D, et al. The response of autologous T cells to a human melanoma is dominated by mutated neoantigens. Proc Natl Acad Sci USA November 1, 2005;102(44):16013−8.

[30] Segal NH, Parsons DW, Peggs KS, Velculescu V, Kinzler KW, Vogelstein B, et al. Cancer Res February 1, 2008;68(3):889−92.

[31] Castle JC, Kreiter S, Diekmann J, Löwer M, van de Roemer N, de Graaf J, et al. Cancer Res 2012;72(5):1081−91.

[32] Matsushita H, Vesely MD, Koboldt DC, Rickert CG, Uppaluri R, Magrini VJ, et al. Nature February 8, 2012;482(7385):400−4.

[33] Duan F, Duitama J, Al Seesi S, Ayres CM, Corcelli SA, Pawashe AP, Blanchard T, McMahon D, Sidney J, Sette A, Baker BM, Mandoiu II, Srivastava PK. Genomic and bioinformatic profiling of mutational neoepitopes reveals new rules to predict anticancer immunogenicity. J Exp Med 2014;211(11):2231−48.

[34] Yadav M, Jhunjhunwala S, Phung QT, Lupardus P, Tanguay J, Bumbaca S, Franci C, Cheung TK, Fritsche J, Weinschenk T, Modrusan Z, Mellman I, Lill JR, Delamarre L. Predicting immunogenic tumour mutations by combining mass spectrometry and exome sequencing. Nature 2014;515(7528):572−6.

[35] Gubin MM, Zhang X, Schuster H, Caron E, Ward JP, Noguchi T, Ivanova Y, Hundal J, Arthur CD, Krebber W-J, Mulder GE, Toebes M, Vesely MD, Lam SSK, Korman AJ, Allison JP, Freeman GJ, Sharpe AH, Pearce EL, Schumacher TN, Aebersold R, Rammensee H-G, Melief CJM, Mardis ER, Gillanders WE, Artyomov MN, Schreiber RD. Checkpoint blockade cancer immunotherapy targets tumour-specific mutant antigens. Nature 2014;515(7528):577−81.

[36] Pandey V, Nutter RC, Prediger E. Applied Biosystems SOLiD™ System: ligation-based sequencing. In: Next generation genome sequencing (Wiley-VCH Verlag GmbH & Co. KGaA); 2008. p. 29−42.

[37] Thomas RK, Nickerson E, Simons JF, Janne PA, Tengs T, Yuza Y, et al. Sensitive mutation detection in heterogeneous cancer specimens by massively parallel picoliter reactor sequencing. Nat Med 2006;12:852−5.

[38] Bentley DR, Balasubramanian S, Swerdlow HP, Smith GP, Milton J, Brown CG, et al. Accurate whole human genome sequencing using reversible terminator chemistry. Nature 2008;456:53−9.

[39] Rothberg JM, Hinz W, Rearick TM, Schultz J, Mileski W, Davey M, et al. An integrated semiconductor device enabling non-optical genome sequencing. Nature 2011;475: 348−52.

[40] Flicek P, Amode MR, Barrell D, Beal K, Billis K, Brent S, et al. Ensemble 2014. Nucleic Acids Res 2013;42(D1):D749−55.

[41] Farrell CM, O'Leary NA, Harte RA, Loveland JE, Wilming LG, Wallin C, et al. Current status and new features of the Consensus Coding Sequence database. Nucleic Acids Res January 1, 2014;42(1):D865−72.

[42] Langmead B, Trapnell C, Pop M, Salzberg SL. Ultrafast and memory-efficient alignment of short DNA sequences to the human genome. Genome Biol 2009; 10:R25.

[43] Langmead B, Salzberg S. Fast gapped-read alignment with Bowtie 2. Nat Methods 2012;9:357−9.

[44] Trapnell C, Pachter L, Salzberg SL. TopHat: discovering splice junctions with RNA-Seq. Bioinformatics 2009;25(9):1105−11.

[45] Li B, Dewey CN. RSEM: accurate transcript quantification from RNA-Seq data with or without a reference genome. BMC Bioinf 2011;12:323.

[46] Nicolae M, Mangul S, Mandoiu I, Zelikovsky A. Estimation of alternative splicing isoform frequencies from RNA-Seq data. Algorithms Mol Biol 2011;6:9.

[47] Roberts A, Trapnell C, Donaghey J, Rinn JL, Pachter L. Improving RNA-Seq expression estimates by correcting for fragment bias. Genome Biol 2011b;12:R22.

[48] Al Seesi S, Tiagueu YT, Zelikovsky A, Mandoiu II. Bootstrap-based differential gene expression analysis for RNA-Seq data without replicates. BMC Genomics 2014; 15(Suppl. 8):S2.

[49] Robinson MD, McCarthy DJ, Smyth GK. edgeR: a bioconductor package for differential expression analysis of digital gene expression data. Bioinformatics 2010;26(1):139−40.

[50] Feng J, Meyer CA, Wang Q, Liu JS, Liu XS, Zhang Y. Gfold: a generalized fold change for ranking differentially expressed genes from RNA-seq data. Bioinformatics 2012; 28(21):2782−8.

[51] Koboldt DC, Chen K, Wylie T, Larson DE, McLellan MD, Mardis ER, et al. VarScan: variant detection in massively parallel sequencing of individual and pooled samples. Bioinformatics 2009;25(27):2283−5.

[52] Getz G, Hofling H, Mesirov JP, Golub TR, Meyerson M, Tibshirani R, et al. Comment on "The consensus coding sequences of human breast and colorectal cancers". Science 2007;317:1500.

[53] McKenna AH, Hanna M, Banks E, Sivachenko A, Cibulskis K, Kernytsky A, et al. The genome analysis toolkit: a MapReduce framework for analyzing next-generation DNA sequencing data. Genome Res 2010;20:1297−303.

[54] Cibulskis K, Lawrence MS, Carter SL, Sivachenko A, Jaffe D, Sougnez C, et al. Sensitive detection of somatic point mutations in impure and heterogeneous cancer samples. Nat Biotechnol 2013.

[55] Lundegaard C, Lund O, Nielsen M. Accurate approximation method for prediction of class I MHC affinities for peptides of length 8, 10 and 11 using prediction tools trained on 9mers. Bioinformatics 2008;24(11):1397−8.

[56] Larsen MV, Lundegaard C, Lamberth K, Buus S, Lund O, Nielsen M. Large-scale validation of methods for cytotoxic T-lymphocyte epitope prediction. BMC Bioinf October 31, 2007;8:424.

[57] The Cancer Genome Atlas Research Network. Integrated genomic analyses of ovarian carcinoma. Nature 2011;474(7353):609−15.

[58] Li H, Handsaker B, Wysoker A, Fennell T, Ruan J, Homer N, et al. The sequence Alignment/Map format and SAMtools. Bioinformatics 2009;25:2078−9.

[59] Cao K, Hollenbach J, Shi X, Shi W, Chopek M, Fernandez-Vina MA. Analysis of the frequencies of HLA-A, B, and C alleles and haplotypes in the five major ethnic groups of the United States reveals high levels of diversity in these loci and contrasting distribution patterns in these populations. Hum Immunol 2001;62:1009.

[60] Brown SD, Warren RL, Gibb EA, Martin SD, Spinelli JJ, Nelson BH, Holt RA. Neoantigens predicted by tumor genome meta-analysis correlate with increased patient survival. Genome Res 2014;24(5):743−50.

[61] Srivastava PK. Peptide-binding heat shock proteins in the endoplasmic reticulum: role in immune response to cancer and in antigen presentation. Adv Cancer Res 1993;62: 153−77.

Overview of the Intersection of Genomics of Cholesterol Metabolism and Cardiometabolic Disease with Reproductive Health, Especially in Women

Anthony M. DeAngelis, Meaghan Roy-O'Reilly, Annabelle Rodriguez-Oquendo

University of Connecticut Health Center, Farmington, CT, USA

1. INTRODUCTION

About 15% of reproductive aged couples in the United States experience infertility, which is defined as the failure to achieve a successful pregnancy after 12 months or more of appropriate, timed, and unprotected intercourse or therapeutic donor insemination [1–4]. According to the American Society for Reproductive Medicine, about 30% of infertility cases can be attributed to female causes, 30% to male factors, and the remaining cases involve a combination of male and female factors, in addition to unexplained causes [5]. Understanding the various pathologic mechanisms underlying infertility represents an active area of research and serves as an important first step in our ability to treat this health problem.

Genetic mutations leading to functional alterations in proteins involved in cardiometabolic pathways, including steroidogenesis, chronic inflammation, and insulin resistance, have been shown to adversely impact human fertility. In particular, research on the molecular biology, biochemistry, and physiology of human steroidogenesis and its disorders as they relate to fertility have been well characterized and thoroughly reviewed [6–11]. As genetic polymorphisms influencing cholesterol

Translational Cardiometabolic Genomic Medicine
http://dx.doi.org/10.1016/B978-0-12-799961-6.00011-1 **251**

metabolism have been shown to have a profound impact on human fertility, they will serve as the main focus of this review.

Cholesterol is an essential substrate for steroid hormone production and membrane synthesis. Genetic mutations altering the function of proteins involved in cholesterol uptake, mobilization from stored intracellular pools, and synthesis significantly alter normal cellular function and, in turn, impact fertility. In this review, we summarize some of these known genetic mutations. In particular, we highlight work focused on the functional impact of genetic mutations involved in cholesterol uptake, mobilization, and de novo cholesterol synthesis and review how these alterations can affect cholesterol metabolism and human fertility. We also examine how polymorphisms in other cardiometabolic genes, including those involved in steroid signaling, inflammation, and insulin resistance, contribute to the pathogenesis of common fertility problems, including endometriosis, premature ovarian failure (POF), and polycystic ovary syndrome (PCOS).

2. CHOLESTEROL AND INFERTILITY

Dyslipidemia has been implicated in the development of multiple diseases, including cardiovascular disease, diabetes, and polycystic ovarian syndrome [12]. Cholesterol is the principle substrate for steroid production and has been demonstrated to affect the hormonal environment in men and women, suggesting that cholesterol levels may also play an important and complex role in human fertility [6]. Higher HDL levels have been shown to associate with beneficial effects on fertility, including successful oocyte and embryo development [13,14]. In contrast, Gupta et al. [15] demonstrated that diet-induced hypercholesterolemia in rats and rabbits results in reduced spermatogenesis and testicular function.

Recently, Schisterman et al. [12] directly investigated the relationship between serum lipid profiles and fertility in the LIFE cohort (501 couples attempting to achieve pregnancy). The outcome measure of this study was increased time to pregnancy (TTP), a measure of infertility that does not differentiate between male and female causes. Free cholesterol levels in both men and women were significantly higher in couples that did not achieve pregnancy and increased total cholesterol levels in men were also associated with decreased couple fertility. Interestingly, both of these findings were shown to be independent of BMI [12].

While this is the first study correlating lipid levels to TTP, the authors suggest that this research agrees with earlier studies in the field, where male and female patients with metabolic syndrome and women with polycystic ovarian syndrome involving metabolic symptoms also have an increased incidence of infertility [16,17]. This work highlights the importance of serum cholesterol concentration in both male and female fertility and suggests that derangements in cholesterol levels may play a significant role in the development of infertility—one that is independent of increased adiposity. While dietary and lifestyle choices have been cited as the major cause of elevated serum cholesterol, there is an increasing body of evidence demonstrating that genomic variations involving cholesterol uptake, storage, synthesis, and regulation play a significant role in infertility and may represent novel avenues for infertility diagnosis and treatment.

3. CHOLESTEROL METABOLISM

Cholesterol is essential for plasma membrane synthesis and fluidity. In steroidogenic cells of the gonads, adrenal glands, and placenta, cholesterol also acts as a necessary precursor for the biosynthesis of steroid hormones, which in turn, regulate vitally important physiological functions including, but not limited to, the control of reproductive pathways, development of secondary sexual characteristics, and salt balance [14,18]. Activated steroidogenic cells demand high levels of cholesterol substrate. Several mechanisms within the cell work in a coordinated manner to ensure adequate replenishment and availability of the cholesterol pool. These major mechanisms include receptor mediated uptake of lipoprotein derived cholesterol (i.e., LDLR and/or scavenger receptor Class B Type I (SR-BI)), hydrolysis of stored intracellular cholesteryl esters (CEs) by hydrolases such as hormone-sensitive lipase (HSL), and de novo cholesterol synthesis by a multi-step process using acetate as the starting substrate (Figure 1).

Most species, including humans, depend mostly on cellular uptake of lipoprotein cholesterol as the major precursor for steroidogenesis [19,20]. There are various mechanisms that facilitate cholesterol uptake into a cell. The two dominate mechanisms include uptake via the low-density lipoprotein receptor (LDLR) [20–22] and selective CE uptake mediated by SR-BI [23]. LDLR-mediated uptake involves the binding of apolipoprotein B (ApoB)- and apolipoprotein E (ApoE)-enriched cholesterol particles with subsequent endocytosis of the LDLR and uptake of the cholesterol-containing lipoprotein particles into the cell. In contrast, SR-BI mediates high-capacity selective uptake of CEs from the core of HDL, LDL, and VLDL lipoproteins [24]. Selective uptake refers to the process whereby lipoproteins dock on the extracellular loop of SR-BI (see Figure 2), and the CEs within the core of the lipoprotein are transferred into the interior of the cell via a channel that is formed by dimerization of SR-BI receptors. The lipoprotein, now depleted of CEs, dissociates from the receptor and is not internalized. The internalized CE is then hydrolyzed by hydrolases such as

Sources of Cellular Cholesterol

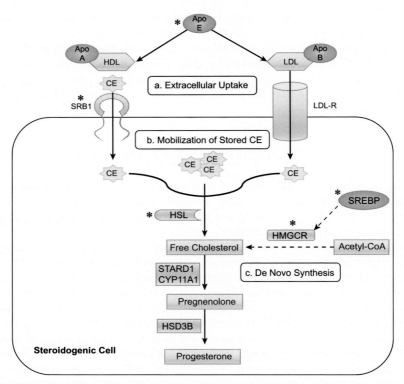

FIGURE 1   **Mechanisms of cellular cholesterol supply.** The three pathways via which the steroidogenic cell maintains an adequate intracellular cholesterol pool for steroidogenesis are outlined here, including uptake (a), mobilization (b), and de novo synthesis (c). Asterisks denote the location of SNPs discussed in this review. HDL = high-density lipoprotein, LDL = low-density lipoprotein, ApoA = apolipoprotein A, ApoE = apolipoprotein E, ApoB = apolipoprotein B, CE = cholesteryl esters, SREBP = sterol regulatory element-binding proteins, HSL = hormone-sensitive lipase, HMGCR = 3-hydroxy-3-methylglutaryl-CoA reductase, STARD1 = steroidogenic acute regulatory protein D1, CYP11A1 = cytochrome P450, family 11, subfamily A, polypeptide 1, HSD3B = 3 beta-hydroxysteroid dehydrogenase.

HSL to generate free or unesterified cholesterol (UC), which is then used for cellular processes such as steroidogenesis or membrane integrity [25].

Following stimulation by tropic hormones, cholesterol may be recruited from the hydrolysis of stored cytoplasmic CEs. This process involves the participation of HSL, an enzyme which is expressed in steroidogenic tissues and functions to convert stored CEs to UC [26]. This mechanism generates UC for the immediate synthesis of steroid hormones following cellular activation [27].

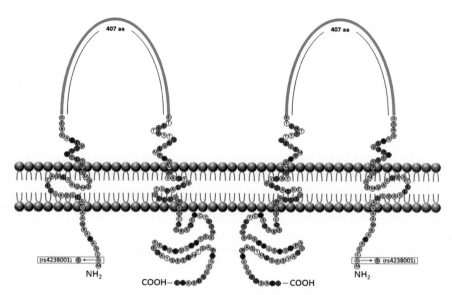

FIGURE 2   **Structure of the scavenger receptor Class B type 1 (SR-BI).** SR-BI is a receptor that mediates high-capacity selective uptake of CEs from the core of HDL, LDL, and VLDL via lipoprotein docking on the extracellular loop. A deficiency in SR-B1 is associated with infertility and the location of the rs4238001 SNP discussed in the paper is highlighted here.

Intracellular cholesterol may also be derived via de novo synthesis in the endoplasmic reticulum (ER) from acetyl CoA, a process regulated by the enzyme 3-hydroxy-3-methylglutaryl coenzyme A reductase (HMG-CoA reductase) [28–30]. The contribution of this pathway in providing cholesterol substrate varies among species, is heavily influenced by the amount of intracellular cholesterol, and is regulated by the activation of the sterol regulatory element-binding proteins (SREBPs). When cells sense abundant intracellular cholesterol stores, the de novo cholesterol synthetic pathway remains inactive or quiescent. During this time, immature or large SREBP remains complexed with two other proteins, SREBP cleavage activating protein and INSIG1, which prevent its translocation to the nucleus. Following a decrease in the intracellular cholesterol pool, as seen in stimulated steroidogenic cells, which in turn would signal a demand for more intracellular cholesterol, there is dissociation of INSIG1 from the complex and subsequent mobilization, cleavage, and translocation of SREBP to the nucleus. In the nucleus, SREBP functions as a transcription factor, resulting in up-regulation of the LDLR and HMG-CoA reductase leading to both increased plasma membrane receptor-derived cholesterol uptake and de novo cholesterol production [31,32].

4. GENETIC ALTERATIONS AFFECTING CHOLESTEROL UPTAKE AND THEIR INFLUENCE ON HUMAN FERTILITY

Lipoprotein particles serve as vehicles for cholesterol transport throughout the body and function to regulate intracellular cholesterol levels within a tight physiological range, as an excess or insufficiency is typically deleterious to normal cellular function [33,34].

Lipoprotein-mediated cholesterol uptake is an important mechanism by which cells obtain cholesterol for steroidogenesis. In fact, extracellular uptake has been shown to provide about 80% or more of cholesterol substrate for steroidogenic cells of the adrenal glands, corpus luteum, and placenta, with the other 20% supplied by de novo cholesterol synthesis from acetate and other precursors and CE hydrolysis [35−37]. In humans, LDL lipoproteins, which are composed of the major apoproteins ApoB100 and ApoE as ligands to the LDLR, bind to the LDLR, and this is followed by endocytosis that leads to internalization of the lipoprotein and the LDLR. In contrast, cholesterol-mediated uptake by SR-BI occurs by selective CE uptake and not by endocytosis. Several known polymorphisms in *APOE* have been discovered and found to significantly impact cholesterol metabolism and human fertility.

4.1 Apolipoprotein E

ApoE is a constituent apoprotein of very low-density lipoproteins (VLDL), intermediate-density lipoproteins, high-density lipoproteins (HDL), and chylomicron particles. It is produced in the liver and in macrophages and plays an important role in lipid transport by mediating the binding of lipoprotein particles to their receptors [38]. ApoE-mediated delivery of cholesterol represents an important mechanism by which steroidogenic cells obtain cholesterol for steroid hormone production.

In rodents, deletion of the *Apoe* gene results in the development of severe hypercholesterolemia and decreased expression of the steroidogenic enzymes cytochrome P450 aromatase (Cyp19a1) and 3β-hydroxysteroid dehydrogenase (3βHSD), essential enzymes involved in the biosynthetic pathway of progesterone and estrogen, leading to a decrease in the level of both hormones [39]. In this study, Zhang et al. [39] showed significant alterations in the ER ultrastructure of *Apoe*-null mice, which are described as swollen and tubby, and we speculate that these ER changes contributed to the reduced expression and function of Cyp19a1 and 3βHSD. Interestingly, no difference was observed in fertility rates between *Apoe*-null and wild-type control mice. However, *Apoe*-null mice did possess greater numbers of ovarian follicles as well as an increased ratio of ovarian follicle atresia and apoptosis compared with their wild-type counterparts [39].

Three important polymorphisms have been discovered in the human *APOE* gene [40,41]. These codominant alleles, designated ε2, ε3, and ε4, give rise to three functionally distinct proteins: ApoE2, ApoE3, and ApoE4 [38], which vary in amino acid residues at positions 112 and 158. ApoE2 is characterized by two cysteine residues—ApoE3 by a cysteine and arginine residue, and ApoE4 by two arginine residues at these sites [42]. While these genetic polymorphisms represent small changes in the DNA sequence, studies have found that they exert a significant impact on the functional properties of their corresponding proteins and result in the alteration of plasma cholesterol levels, apolipoprotein levels [38], and marked differences in reproductive efficiency [43–46].

In one study, Corbo et al. [45] screened 160 European women to investigate the possible impact of *APOE* polymorphisms on reproductive efficiency. Investigators discovered that women harboring the ε3/ε3 genotype exhibited higher reproductive efficiency, whereas carriers of the ε2 allele had lower numbers of children and pregnancies [45]. *APOE* polymorphisms have also been found to impact male fertility. Setarehbadi et al. [47] found significant differences in the distribution of *APOE* allelic combinations in fertile and infertile men, with a higher percentage of fertile males possessing the ε3/ε3 genotype. This study mirrors the results of Gerdes et al. [46] who reported higher birth rates for children from fathers possessing the ε3/ε3 genotype compared with fathers possessing other allelic variations.

Overall, these studies suggested that lower reproductive efficiency is associated with the ε2 allele and higher reproductive efficiency with the ε3 allele.

4.2 LDL Receptor

The LDLR is an important protein that functions to mediate the uptake of LDL cholesterol, the preferred substrate for steroid hormone production [48–52]. Given the important role the LDLR plays in providing adequate substrate for steroidogenesis, one might assume that alterations in its function would lead to significant impairment of fertility. However, LDLR-deficient mice are fertile and produce normal-sized litters, despite exhibiting an altered lipid profile [53]. In contrast, a deficiency of SR-BI is associated with infertility [54,55]. The ability of SR-BI to bind LDL cholesterol may partially explain the normal fertility exhibited by LDLR-null mice. To date, no significant polymorphisms have been discovered in the LDLR which adversely impact human fertility. A recent publication by Rinninger et al. [56] has shown that hepatic and adrenal

SR-BI protein expression was not different between wild-type and LDLR-knockout mice.

4.3 Scavenger Receptor Class B Type I

SR-BI has been identified as the primary lipoprotein receptor for HDL, but it also has the capability to bind LDL, VLDL, and other ligands with high affinity [57–61]. SR-BI is highly expressed in the liver, adrenal gland, and ovarian tissues [55] where it plays an important role in mediating selective CE uptake [25]. This process serves as another lipoprotein receptor-mediated mechanism for providing cholesterol substrate to steroidogenic cells [14,62–68]. In addition, SR-BI has been shown to mediate the bi-directional movement of UC down a concentration gradient between lipoproteins and cells [69–71].

Much of our knowledge of SR-BI function comes from animal studies. In rat ovaries, SR-BI mRNA is present in interstitial cells, thecal cells, corpora lutea, and luteinized granulosa cells [72]. SR-BI protein expression is up-regulated after stimulation with tropic hormones [73,74] and is associated with an increase in CE uptake [75]. SR-BI expression in ovarian tissue plays a particularly important role in female fertility, as substantial amounts of HDL, but not LDL, are present in follicular fluid surrounding oocytes in ovarian follicles [51,76–80]. LDL is absent from the follicular fluid of pre-ovulatory follicles because particles range in size from 18 to 23 nm with a molecular weight of 2.3 million daltons, precluding movement across the basement membrane (size limit about 400,000 Da) [79–82]. In contrast, HDL particles range in size from 8.2 to 13 nm with molecular weights of 200,000–400,000 [81,82]. Due to the absence of other lipoprotein particles in the follicular fluid, SR-BI delivery of HDL may serve as an important mechanism by which cholesterol is delivered to the oocyte and its surrounding steroidogenic support cells. It is not surprising then that genetic alterations in SR-BI have been shown to adversely impact cholesterol metabolism, including that of the ovaries, resulting in alterations in fertility.

Genetically engineered mice lacking SR-BI demonstrate the importance of this protein in reproduction. Female homozygous SR-BI-knockout mice are infertile [23,54,55], most likely secondary to impaired embryonic development of the offspring, as embryos from SR-BI-null females have been shown to develop an abnormal morphology and arrest in culture. Similar abnormalities were found in unfertilized oocytes, but estrus cycles, progesterone levels, and ovulatory patterns in the SR-BI-null mice were found to be normal [55]. However, a more recent study by Jimenez et al. [83] found that SR-BI-null mice have 50% lower plasma progesterone levels compared to wild-type controls, as well as ovaries with small corpora lutea, large follicles with hypertrophied theca cells, and follicular cysts with blood-filled cavities. These later data are more consistent with the phenotypic data observed in humans carrying a single nucleotide polymorphism in the *SCARB1* gene that encodes the SR-BI protein. As a result of altered SR-BI function, null mice have elevated and abnormally large ApoE-enriched circulating HDL particles as well as an elevated ratio of UC to total cholesterol (known as the UC:TC), a phenotype which is most likely secondary to defective removal of the core CE from these lipoprotein particles [84].

Interestingly, fertility in the SR-BI-null mice can be rescued by methods that act to restore normal cholesterol metabolism. Miettinen et al. [54] treated SR-BI-null mice with the HDL cholesterol-lowering drug, Probucol [85,86]. They found that this treatment reduced and remodeled the abnormally large HDL particles and restored fertility. In addition, these investigators also performed a bilateral ovarian transplant in which ovaries were removed from SR-BI-null mice and subsequently transplanted into ovariectomized SR-BI-positive hosts. The transplanted SR-BI-negative ovaries began functioning normally in the SR-BI-positive environment and developed the capacity to produce functional oocytes, resulting in restoration of fertility. In support of these findings, Yesilaltay et al. [84] confirmed the presence of abnormally large HDL particles and an elevated UC:TC ratio in SR-BI-null mice. Investigators demonstrated that hepatic restoration of SR-BI expression alone via adenoviral transduction or stable transgenesis in SR-BI-null mice substantially reduced levels of abnormally large HDL, normalized the UC:TC ratio, and restored fertility [84]. Taken together, these data led the investigators to hypothesize that infertility in SR-BI-null mice is not due to irreversible defects in oocyte or embryonic development, but rather secondary to abnormal lipoprotein metabolism and/or the abnormal transfer of cholesterol to or from the oocyte or supporting cells.

As in rodents, alterations in human SR-BI function have been shown to adversely impact cholesterol metabolism and fertility. In vitro experiments in the HGL-5 immortalized human granulosa cell line in which SR-BI had been knocked down using interfering RNA resulted in significantly reduced progesterone secretion after stimulation with Forskolin (an activator of cyclic AMP) when compared with treated controls [87]. This is in line with recent rodent data by Jimenez et al. [83], which reported a 50% reduction in circulating plasma levels of progesterone in SR-BI-null mice.

Several single nucleotide polymorphisms in the *SCARB1* gene have been identified and found to impact cholesterol metabolism and human fertility (Rev. in [88]) (Figure 1). In a prospective study of adults with hyperalphalipoproteinemia (defined as subjects with fasting HDL-C levels >60 mg/dl), West et al. [89] found that certain polymorphisms in *SCARB1* are associated with increased SR-BI protein degradation resulting in lower receptor protein expression. In addition, these expression levels were inversely proportional to HDL lipoprotein size and positively associated with CE uptake from HDL [89]. In another study, Yates et al. [90] evaluated the association of *SCARB1* polymorphisms and fertility outcomes in women undergoing in vitro fertilization (IVF). In this study, investigators found the missense rs4238001 SNP (encoding an amino acid change from glycine to serine at amino acid 2) was significantly associated with lower follicular fluid progesterone levels and poor fetal viability [90]. Finally, a study by Velasco et al. [91] found that a cohort of women undergoing IVF treatment exhibited low SR-BI RNA expression, a finding which was associated with lower baseline and peak estradiol levels and lower number of retrieved and fertilized oocytes compared with women expressing higher levels of SR-BI [91]. Taken together, research in both animal models and human studies has demonstrated the importance of SR-BI and its role in cholesterol metabolism and fertility.

5. GENETIC ALTERATIONS AFFECTING CHOLESTEROL MOBILIZATION AND THEIR INFLUENCE ON HUMAN FERTILITY

The ability of the cell to store and mobilize intracellular cholesterol is the result of intricate mechanisms which work to regulate the size and availability of the cholesterol pool. Following uptake, UC can be esterified by acyl CoA:cholesterol acyl transferase (ACAT) and stored as CEs in intracellular lipid droplets for later use [92]. Following stimulation with tropic hormones, stored CE can be hydrolyzed by HSL, mobilized to the mitochondria, and subsequently used as a substrate for immediate synthesis of steroid hormones [27] (Rev. in [93]). Single nucleotide polymorphisms in HSL have been discovered and found to impact cholesterol metabolism and fertility.

5.1 Hormone-Sensitive Lipase

HSL is an important protein with multiple physiological functions. It plays an essential role in lipid metabolism and is responsible for mediating the hydrolysis of diacylglycerol and triacylglycerol [26]. In addition, it mediates the hydrolysis of CEs in steroidogenic tissues of the adrenal glands, ovaries, and testis, providing substrate for the immediate synthesis of steroid hormones [94,95].

Disruption of HSL in mice results in an altered metabolic phenotype. HSL-null mice exhibit increased hepatic insulin sensitivity and are resistant to diet-induced obesity [96,97]. With regard to cholesterol metabolism and fertility, the testes of HSL-null mice were found to completely lack neutral cholesterol ester hydrolase activity, resulting in increased accumulation of intracellular CE. This finding was associated with oligospermia with reduced motility and a consequential reduction in fertility [98,99]. Vallet-Erdtmann et al. [100] found that re-introduction of HSL in the testis of HSL-null mice reversed CE accumulation and fully restored fertility.

In humans, several polymorphisms in the *HSL* gene have been identified [101–105]. The C-60 G polymorphism is an *HSL* promoter variant that is associated with a 40% reduction in promoter activity in vitro [104]. A study by Vatannejad et al. [106] observed a significant difference in the *HSL* genotype distribution between fertile and infertile males ($n = 164, 169$). Interestingly, the CC genotype conferred a 2.4-fold greater risk for male infertility compared with male carriers of the GC or GG genotype [106].

Taken together, these data suggest that HSL mobilization of CE plays an essential role in the process of spermatogenesis and that disruption of HSL function adversely impacts fertility.

6. GENETIC ALTERATIONS AFFECTING DE NOVO CHOLESTEROL SYNTHESIS AND THEIR INFLUENCE ON HUMAN FERTILITY AND EMBRYONIC VIABILITY

De novo synthesis represents another major source of cellular cholesterol, which is required for structural integrity, steroidogenesis, and ooctye maturation. We will refer readers to a recent comprehensive review of malformations caused by enzymatic defects in cholesterol synthesis [107], and we would like to highlight some of these major enzymatic defects that result in embryonic lethality and/or developmental abnormalities.

As shown in Figure 3, there are many enzymatic steps in the conversion of acetate to cholesterol, and multiple enzymatic defects have been reported, particularly involving steps downstream of squalene production [107].

6.1 HMGCoA Reductase

HMGCoA reductase (HMGCR) is the rate-limiting enzyme in the early steps of cholesterol synthesis and acts to catalyze the formation of mevalonate from 3-hydroxy-3-methylglutaryl coenzyme A. In 2003, Ohashi et al. [108] showed that mouse embryos genetically engineered to have complete deficiency of *Hmgcr* did not develop past the blastocyst stage; thus, no viable *Hmgcr*-null mice were produced from heterozygotic

De Novo Cholesterol Synthesis

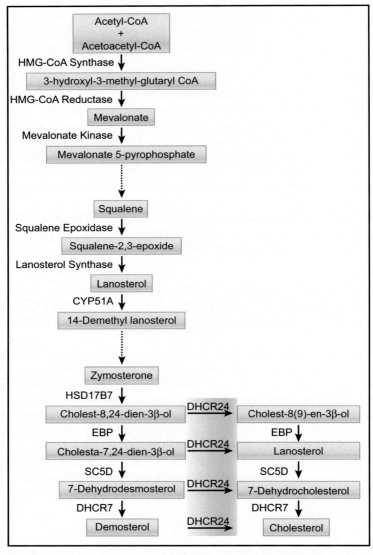

FIGURE 3 **Enzymatic steps in the conversion of acetate to cholesterol.** Mutations in these enzymes have been shown to result in impairment of embryonic viability.

crossings. Investigators determined that preimplantation embryos (E3.5) appeared normal and the defect in viability likely occurred at the level of implantation. An attempt to infuse replacement mevalonate via a subcutaneous osmotic pump did not overcome the defect in *Hmgcr*-null embryos, as the embryos were able to implant but did not progress to viable offspring. Mice heterozygous for this mutation appeared normal in development, gross anatomy, fertility, and cholesterol synthesis capacity despite decreased HMGCR activity. Overall cholesterol biosynthesis remained unchanged in the liver of these mice and lipoprotein profiles were similar to wild-type mice profiles, suggesting that one copy of the *Hmgcr* gene produces sufficient HMGCR for hepatic cholesterol synthesis.

HMGCR is regulated both transcriptionally and post-transcriptionally. It is well established that when cells sense a reduction in intracellular cholesterol mass, activation of the ER-residing SREBPs (1a, 1c, and 2) occurs which leads to its transformation from a larger ER protein to a smaller nuclear transcription factor [109]. Cholesterol genes known to be activated by SREBP2 include *HMGCR*, *LDLR*, and *INSIG*, among others [110]. Complete knockout of SREBP2 in mice is 100% embryonic lethal [109].

Surprisingly, little is known regarding the functional effect of variants in the SREBP2 gene in relation to reproduction or other medical diseases. Liu et al. [111] reported a significant increase in LDL cholesterol levels in Chinese Han female carriers of the CC genotype for the rs4822063 SREBP2 variant. Incrementally, more is known regarding the functional effects of variations in the HMGCR gene in human reproduction. In 2007, Steffen et al. [112] examined genetic etiologies in the mother and infant associated with preterm delivery (PTD). These investigators were particularly interested in examining the effects of genetic variations in cholesterol genes in the mother and infant on the risk of PTD. Cholesterol genes were genotyped in infants born before 37 weeks of gestation ($n = 414$) and at least one parent (253 mothers and 351 fathers). Selection of candidate genes was based on a previous publication from Knoblauch et al. [113] which identified haplotypes responsible for the majority of the genetic variance in LDL, HDL, and LDL/HDL in a German population. The results in the infant dataset showed a significant association of variations in the *APOAI* gene (apoA-I is the major apolipoprotein in HDL fractions) and *HMGCR* (rs2303152). In the maternal dataset, the *HMGCR* variant, rs2303152, was not analyzed due to low allele frequency; however, the results showed a significant association of PTD with *ABCA1*, *LCAT*, *LIPC*, and *APOE*, all genes associated with HDL metabolism. The authors concluded that there might a benefit in assessing the effect of nutritional risk on PTD by genetic testing for common variations in cholesterol genes.

6.2 Lanosterol

Lanosterol represents the first step in sterol formation and can be converted by lanosterol 14 α-demethylase (CYP51, a member of the cytochrome P450 family), to follicular fluid meiosis activating factor

(FF-MAF), a sterol intermediate that has been extensively studied and shown to activate meiosis in gametes [114]. Recently, Lewinska et al. [115] examined polymorphisms within the *CYP51A1* gene and found significant associations with birth weight and maternal lipid levels. Genotyping for CYP51A1 variants was performed in samples isolated from four different populations of mothers experiencing preterm birth or neonates affected by growth retardation or preterm birth. In a cohort of women and neonates of Western European descent, the investigators identified 22 sequence variants and reported 10 as novel and rare. Of the polymorphisms identified, the minor allele frequencies (MAF) for the common variants, rs6465348 and rs12673910, were significantly higher in neonates and mothers affected by preterm birth as compared with the general population. Based on its high MAF and its location in the CYP51A1 3'UTR region, the investigators chose rs6465348 for further study and found a significant association between low body weight for gestational age and lower LDL cholesterol and total cholesterol in women. These genomic approaches underscore the effort to identify genetic causes for infertility problems related to alterations in lipid metabolism.

In addition to its well-known role in generating meiosis activating factor, other studies have suggested an intriguing role for lanosterol and dihydrolanosterol (a reduction product of lanosterol) in the post-translational regulation of HMGCoA reductase [110,116]. As mentioned previously, Insig is an ER-resident protein that complexes with SREBP. Song et al. [110] showed that lanosterol enhanced Insig-mediated ubiquitination and degradation of HMGCoA reductase. Lange et al. [116] subsequently showed that 24, 25 dihydrolanosterol could potently inhibit HMGCoA reductase activity in fibroblasts. Together, these studies strongly suggest that lanosterol exerts a negative feedback inhibition on the rate-limiting enzyme of cholesterol synthesis.

6.3 Desmosterol

In the Bloch pathway, the enzyme 3β-hydroxysteroid-Δ24 reductase (DHCR24) catalyzes the conversion of desmosterol in the final steps of cholesterol formation [117]. A recent review by Zerenturk et al. [117] highlights the renewed interest in studying the biological effects of desmosterol. DHCR24 was first identified due to high homology with the DIMINUTO/DWARF1 gene in *Arabidopsis thaliana*; it was also separately identified later as Seladin-1 (selective Alzheimer's disease indicator 1) due to its association with Alzheimer's disease. The *DCHR24* gene is highly expressed in brain, liver, and steroidogenic tissues; missense mutations in this gene lead to desmosterolosis, a rare autosomal-recessive disorder characterized by congenital anomalies and elevated desmosterol levels [118].

The first report of a child with congenital malformations and confirmed elevated levels of desmosterol were reported in 1998 by FitzPatrick et al. [118]. The child was born at 34 weeks gestation with extensive developmental abnormalities (including macrocephaly, cleft palate, and short limbs) and died within 1 h of birth due to respiratory distress. Desmosterol levels were measured using gas chromatography-mass spectrometry (GC–MS) from extracts isolated from liver, kidney, and brain tissues. The highest concentrations of desmosterol were found in the brain tissue of the deceased child, while plasma levels were considered normal in the mother but markedly elevated in the father. The authors also noted that triparanol, a highly teratogenic inhibitor of DHCR24, results in similar malformations, underlining the importance of this enzyme in human development [118].

More recently, Schaaf et al. [119] reported a third case of desmosterolosis and also examined the functional effects of polymorphisms in the DCHR24 gene in this case report. These investigators determined that the affected infant was a compound heterozygote for two novel missense mutations, c. 281G>A (p.R94H, which was inherited from the mother) and c.1438G>A (p.E480K, which was inherited from the father). In yeast expression assays, the two mutations significantly lowered DHCR24 enzymatic activity as measured by the conversion of desmosterol to cholesterol compared with wild-type DHCR24. The mutations in this infant were not fatal and a trial of therapy with riboflavin, nicotinamide, and thiamine was initiated, but the investigators reported concerns regarding noncompliance with use of the prescribed treatment. A subsequent analysis of sterol levels at 14 months showed no significant changes in desmosterol or cholesterol levels [109].

Admittedly, the clinical condition of desmosterolosis is rare; however, there have been a number of in vitro studies examining other aspects of desmosterol effects on inflammation and cell proliferation. For instance, McGrath et al. [120] reported an interesting relationship between SR-BI and DHCR24. Using human coronary artery endothelial cells (HCAEC), these investigators incubated the cells with recombinant or native HDLs and found that preincubation of the cells with the HDL particles suppressed NF-κB-mediated DNA transcription. One of the genes transcriptionally regulated by preincubation with HDL was DHCR24, which had previously been shown to be a hydrogen peroxide scavenger [121] and has antioxidant properties [122]. Furthermore, after SR-BI expression was silenced in the HCAEC, basal and stimulated mRNA levels of DHCR24 were significantly reduced. These results showed that the effect of DHCR24 in modulating inflammatory responses in endothelial cells was dependent on SR-BI. It is unknown if there is a direct interaction between SR-BI and DHCR24, and this should be explored further.

6.4 7-Dehydrocholesterol

Of the enzymatic defects identified in the de novo cholesterol synthesis pathway, the mutation resulting in Smith—Lemli—Opitz syndrome (SLOS) is the best characterized. SLOS is an autosomal-recessive disease caused by a defect in the 3β-hydroxysterol Δ7 cholesterol reductase (DHCR7) gene, leading to reduced cholesterol end-product and accumulation of 7-dehydrocholesterol and 27-hydroxy-7-dehydrocholesterol [123]. This deficiency results in morphological abnormalities, mental retardation, and reduced lifespan (including embryonic lethality) [123]. Knowledge of cholesterol metabolism has proven crucial to both the early diagnosis and treatment of SLOS.

A number of diagnostic approaches have been used to identify affected fetuses, including analysis of synthetic cholesterol intermediates. In 2005, Chevy et al. [124] argued against the utility of molecular diagnostic testing and instead advocated for routine GC—MS profiling of amniotic fluid (AF) in the diagnoses of developmental abnormalities attributed to defects in sterol intermediate synthesis. Clinical indications for AF sampling included fetal growth retardation, external male genitalia defect, and limb defects. Using GC—MS, these investigators were able to detect significant differences in AF levels of cholesterol, 7-dehydrocholesterol, and 8-dehydrocholesterol between normal and SLO-affected infants. Of these sterol intermediates, these investigators concluded that 8-dehydrocholesterol levels were the best marker for SLO given its long-term stability. The results from this larger clinical population confirmed previous results by Abuelo et al. [125] and Tint et al. [126]. In addition, a urine analysis can also be performed as early as 12 weeks to diagnose SLOS [127].

Since the defect in DHCR7 leads to low plasma cholesterol levels and elevated 7- and 8-dehydrocholesterol levels, the current treatment standards include dietary cholesterol supplementation with statin therapies [128]. There are also studies underway that are examining the benefits of antioxidants in the treatment of SLOS (Dr Ellen Elias, Children's Hospital, Aurora CO).

7. THE GENOMICS OF HUMAN FERTILITY DISORDERS

There are several common disorders that negatively impact human fertility, including PCOS, POF, and endometriosis. All three of these diseases are known to have a strong heritable component, yet their complex and multifactorial nature has made it difficult to develop accurate methods for diagnosis and therapy. The use of new genetic screening tools, particularly genome-wide association study (GWAS), is showing incredible promise for the identification of candidate risk genes with genome-wide significance. In this section, we examine the current state of knowledge regarding genes that may predispose toward the development of common diseases of infertility. In particular, we focus on genes involved in steroid production, response, and regulation as they represent particularly strong candidates in the search for novel therapeutic targets.

8. POLYCYSTIC OVARY SYNDROME

PCOS, a complex endocrine disorder affecting 7% of reproductive-aged women, is a well-known cause of reduced fertility [129]. This disorder is defined by the presence of two or more of the following characteristics: hyperandrogenism, polycystic ovaries, and ovulatory derangements leading to reduced fertility. Importantly, PCOS has also been associated with increased risk of obesity, insulin resistance, early-onset atherosclerosis, and type 2 diabetes mellitus. As a result of these associations, current clinical guidelines recommend early screening for cardiometabolic disease in PCOS patients [130]. Although environmental conditions are known to play a role in PCOS development, this disorder has been demonstrated to have a strong genetic component. A study conducted in monozygotic and dizygotic twins of Dutch origin suggests that 72% of PCOS pathogenesis may be due to genetic influences [131]. Multiple cardiometabolic candidate genes have been implicated in the risk for developing PCOS, including genes encoding inflammatory cytokines and proteins involved in steroidogenesis and insulin signaling.

Chronic inflammation is thought to be a key component in the pathogenesis of obesity-associated PCOS. Adipose tissue shows increased expression of inflammatory cytokines, which are known to play a role in insulin resistance [132]. Current inflammatory genes of interest include TNFA (encoding tumor necrosis factor-α) and the IL-6 and IL-6 receptor gene [132]. Several steroidogenic genes have also been examined for association, including a repeat polymorphism in the CYP11A1 gene promoter region. While an association was found between carriers of this SNP and PCOS with elevated testosterone levels in a small Chinese population ($n = 12$), the result failed to replicate in a large-scale study ($n = 898$) conducted in the UK and Finland [133,134].

Genes involved in insulin signaling are also thought to be likely candidate genes for PCOS. Insulin stimulates ovarian growth, prevents apoptosis of follicles, and induces the production and secretion of androgens from the ovaries. While a linkage study of 17 families showed that a variable number tandem repeat polymorphism in the insulin gene was associated with PCOS, larger studies conducted in several populations failed to replicate this association [135,136]. Of the PCOS candidate genes that have been studied, the insulin receptor (INSR) gene has been the most promising. When Goodarzi et al. [137] examined a Caucasian population for 295 SNPs in genes associated with insulin signaling, the rs2252673 SNP in the INSR was found to be associated with PCOS in both the discovery and replication cohort, a finding also replicated in a separate study conducted in a Korean population [138]. While further investigation into the INSR as a candidate gene for PCOS is necessary, the conflicting results of these association studies indicate that the complicated nature of the genetics underlying PCOS risk, as well as the contributory nature of the environment, may be best examined via a more comprehensive study method like GWAS.

In the first GWAS of PCOS, Chen et al. [139] examined a cohort of 744 PCOS cases and found 28 SNPs within three loci (2p16.3, 2p21, and 9q33) with strong associations to PCOS; these loci were then replicated in two additional independent cohorts. The SNP with the highest significant was rs13405728, located at 2p16.3, which is located near a linkage disequilibrium block containing the GTF2A1L and LHCGR genes [139]. GTF2A1L is specific to germ cells and encodes TIIA-alpha and beta-like factor, believed to be important in testicular function. LHCGR encodes a receptor that recognizes both human chorionic gonadotropin, a hormone necessary for pregnancy maintenance, and luteinizing hormone (LH). This receptor is expressed by granulosa cells in late pre-ovulatory follicles, allowing a surge in LH to stimulate ovulation; mutations inactivating the LHCGR gene have been demonstrated to result in oligomenorrhea, enlarged ovaries, and infertility [140]. Interestingly, the researchers found that this SNP was not associated with other common PCOS phenotypes such as insulin resistance or increased BMI [139]. Notably, the rs13405728 SNP was found to be located 211 kb upstream of the FSHR gene, which has been previously shown to associate with the ovarian response to follicle-stimulating hormone (FSH), suggesting it may represent a strong candidate gene for PCOS. Of the 65 SNPs examined within this FSHR region, 13 were just below the threshold of significance [139].

An additional locus of interest was 2p21, which contained 21 SNPs of significance including two intronic SNPs (rs13429458 and rs12478601) found within the THADA gene, which has been previously shown to associate with type 2 diabetes mellitus risk in a European population [141]. The third locus of interest was located on chromosome 9q33.3, two independently associated SNPs (rs10818854 and rs2479106) within an LD block in the DENND1A gene, which is expressed ubiquitously and can bind and negatively regulate endoplasmic reticulum aminopeptidase 1

(ERAP1) [142]. Studies by Eriksen et al. [143] and Lerchbaum et al. [144] analyzing women of European descent found that the G allele at rs2479106 in the *DENND1A* gene was associated with increased risk of developing PCOS, but did not find significance for the *THADA* or *LHCGR* loci. Interestingly, two other studies conducted in European women found an association between PCOS and the rs10818854 SNP in *DENND1A*, but not within rs2479106 SNP [145,146]. Goodarzi et al. also reported a significant association with the *THADA* SNP rs12468394, but the European studies did not demonstrate significance for SNPs within the *LHCGR* locus, suggested to be due in part to the low allele frequency in European populations [145,146]. The identification of PCOS susceptibility genes is made particularly difficult by the heterogeneity of PCOS phenotypes and the complex nature of the disorder. It has been suggested that investigation into associations within networks of genes, such as groups of those responsible for sex steroid signaling or inflammation, may be required in order to truly understand the complex nature of PCOS susceptibility [130].

9. PREMATURE OVARIAN FAILURE

POF affects 1–3% of reproductive aged women, resulting in amenorrhea and ovarian failure before the age of 40 years [147]. The disorder is characterized by menstrual problems and elevation of FSH levels into the menopausal range, with a concurrent loss in estrogen [148]. Although some sources have been identified, including chromosomal abnormalities, autoimmunity, iatrogenic, metabolic, and environmental causes, 90% of cases are of unknown etiology [149].

Although the majority of POF cases are sporadic, it has been suggested that 4–31% are familial in nature, most commonly due to chromosomal abnormalities [150]. Gene mutations have also been shown to play a role in the development of POF, including genes involved in ovarian steroidogenesis and folliculogenesis. The Gly146Ala polymorphism in the *SF1* nuclear receptor gene, which plays a role in reproductive development and ovarian steroidogenesis, has been shown to have an association with POF, believed to be due to a parallel reduction in plasma estradiol levels [151,152].

Another interesting candidate is the methylenetetrahydrofolate reductase (*MTHFR*) gene, an important enzyme that functions in DNA synthesis and repair and has been shown to be crucial for female fertility and pregnancy viability in human and animal studies [153]. Two polymorphisms of the *MTHFR* gene, C677T and A1298C, have been shown to associate with POF risk in different populations [153,154]. Both of these polymorphisms have been demonstrated to correlate with reduced in vitro enzyme activity [155]. Defects in MTHFR result in the build-up of an intermediate amino acid called homocysteine, which has been well characterized as a non-traditional risk factor for cardiovascular disease [156]. Polymorphisms in this gene have been shown to result in higher homocysteine levels and increase the risk of cardiometabolic disease in a variety of populations, including diabetes (A1298C) and cardiovascular disease (C677T) [157,158]. These polymorphisms have been postulated to decrease fertility via a variety of mechanisms, including excess homocysteine in the blood (increasing the risk of atherosclerosis and thromboembolism during pregnancy), reduction in MTHFR end-products (impeding DNA synthesis during pregnancy and leading to birth defects and early pregnancy loss), and direct impairment of folliculogenesis [154,159,160].

Genetic variations in the estrogen receptor α 1 (*ESR1*) have also been implicated in the development of POF. Two commonly studied polymorphisms of the *ESR1* are the Pvull (rs2234693) and Xba1 (rs9340799), located at restriction sites within the intron. These variants have been shown to play a role in idiopathic female infertility, IVF outcomes, and the onset of menarche and menopause [161]. Three studies conducted in Korean, Canadian, and Chinese populations determined that these polymorphisms were also found to associate with risk of POF [161–163]. Although the exact mechanism by which these polymorphisms may accelerate menopause remains unknown, it has been speculated that they may alter transcription factor binding, mRNA splicing, or be involved in linkage disequilibrium with other ESR1 mutations [164-166].

POF is a complex disease that can be devastating to fertility and can result in other systemic consequences secondary to the loss of sex steroids, including cardiovascular disease [167]. Further investigation into candidate genes is still required, but it is clear that understanding the genetic risk factors that predispose to POF will be crucial to allow the development of early identification and novel therapeutics.

10. ENDOMETRIOSIS

Endometriosis is a gynecological disease characterized by the implantation of ectopic endometrial tissue outside the uterus, resulting in symptoms including infertility, dysmenorrhea, dyspareunia, and pelvic pain [168]. This condition affects 10% of reproductive aged women and represents $22 billion in annual United States health-care costs [169]. Human endometrial tissue is a hormone-responsive tissue that can also produce its own steroids, including estrogen, which has been demonstrated to play a role in the pathogenesis of endometriosis [170]. Excess estrogens and aberrant progesterone signaling in the endometrium have been implicated in both endometriosis and female infertility, suggesting the two conditions may share common pathogenic mechanisms [171].

Extensive studies of patients with endometriosis have confirmed that it has a strong genetic component, with a heritability of about 51% [172]. Endometriosis is characterized as a chronic inflammatory disease that is dependent on estrogen, which is known to be pro-inflammatory. Thus, genes involved in inflammation and steroid hormone signaling have been nominated as possible candidate risk genes. Genetic association studies in endometriosis patients have examined polymorphisms in many genes within these pathways, including those encoding steroid metabolic proteins, steroid hormone receptors, and cytokines. To date, these studies have produced disparate results, due in part to disease complexity, limited sample size, and failure to replicate findings in different ethnic populations [168].

GWAS has recently been applied to the study of endometriosis in hopes of providing a more complete risk profile for this multifactorial disease. The first endometriosis GWAS conducted in a Japanese population by Uno et al. [173] showed a strong association between endometriosis and rs10965235, an intronic SNP in the CDKN2BAS gene on chromosome 9p21. CDKN2BAS encodes the cyclin-dependent kinase inhibitor 2B antisense RNA, which controls a cell-cycle kinase inhibitor that in turn regulates endometrial proliferation and is believed to play a role in the development of endometriosis [174]. Remarkably, variants in 9p21 loci have also been linked to an increased risk for coronary artery disease (CAD), an association that is further exacerbated by poor glycemic control in type 2 diabetic patients [175,176]. A subsequent GWAS in a European population found an association between endometriosis and rs12700667, an intergenic SNP located on chromosome 7p15.2 [177]. This polymorphism lies upstream of HOXA10 and HOXA11, two endometriosis candidate genes that encode homeobox transcription factors involved in uterine development [178]. HOXA10 and HOXA11 are expressed cyclically in response to sex steroid signaling, with increased expression during implantation that facilitates fertility [179,180].

Nyholt et al. [181] conducted a meta-analysis of these two studies and confirmed seven SNPs associated with endometriosis, three of which (rs7521902, rs10859871, and rs1270067) reached genome-wide significance. They found that the rs7521902 SNP located on chromosome 1p36.12 near the WNT4 gene was associated with endometriosis in the European cohort and suggestively associated in the Japanese cohort [181]. The WNT4 gene, encoding the wingless-type MMTV integration site family, plays an important role in female reproductive development and steroidogenesis and is required for normal female fertility [182,183]. An association between endometriosis and the rs13394619 SNP in GREB1, a gene located on 2p25.1 involved in the estrogen regulation pathway, was also confirmed in the Nyholt study [181].

A more recent two-stage GWAS by Albertsen et al. [184] identified associations between endometriosis and novel SNPs within four genomic regions that displayed genome-wide significance, including rs2235529 encoding LINC00339-WNT4 on 1p36.12. A 150 kb associated region containing WNT4 was also found to include the HSPC157 and CDC24 genes [184]. CDC24 is a small Rho-GTPase that is regulated by estrogen and plays an important role in a variety of cell functions, including cell cycle, endocytosis, and migration [185]. Analysis of patient samples revealed that both CDC24 and HSP157 (a non-protein coding RNA) have differential expression in endometriosis [185,186]. Although much progress has been made in recent years, further work is needed to elucidate the associations between these candidate genes and the pathogenesis of endometriosis.

11. CONCLUSIONS

Based on current evidence, it is clear that certain genetic polymorphisms leading to functional alterations of proteins involved in cholesterol metabolism, steroidogenesis, and steroid regulation and response can significantly impact human fertility. With the advent of personalized medicine, the identification and functional importance of polymorphisms and other genomic alterations related to reproductive function in men and women remains an important area of research. It is expected and anticipated that in the coming years, cholesterol and steroid-related diagnostic studies and therapeutic interventions will be refined based on the principles of personalized medicine.

References

[1] Hull MG, Glazener CM, Kelly NJ, Conway DI, Foster PA, Hinton RA, et al. Population study of causes, treatment, and outcome of infertility. Br Med J Clin Res Ed 1985;291: 1693–7.
[2] Menken J, Larsen U. Estimating the incidence and prevalence and analyzing the correlates of infertility and sterility. Ann N Y Acad Sci 1994;709:249–65.
[3] Practice Committee of American Society for Reproductive M. Definitions of infertility and recurrent pregnancy loss: a committee opinion. Fertil Steril 2013;99:63.
[4] Thoma ME, McLain AC, Louis JF, King RB, Trumble AC, Sundaram R, et al. Prevalence of infertility in the United States as estimated by the current duration approach and a traditional constructed approach. Fertil Steril 2013;99:1324–31. e1321.
[5] ASRM. In: Causes of infertility, vol. 2014; 2012.
[6] Miller WL, Auchus RJ. The molecular biology, biochemistry, and physiology of human steroidogenesis and its disorders. Endocr Rev 2011;32:81–151.
[7] Rosenwaks Z, Adashi EY. Introduction: fertility in the face of genetically determined steroidogenic dysfunction. Fertil Steril 2014;101:299–300.

[8] Marsh CA, Auchus RJ. Fertility in patients with genetic deficiencies of cytochrome P450c17 (CYP17A1): combined 17-hydroxylase/17,20-lyase deficiency and isolated 17,20-lyase deficiency. Fertil Steril 2014;101:317—22.

[9] Kang HJ, Imperato-McGinley J, Zhu YS, Rosenwaks Z. The effect of 5alpha-reductase-2 deficiency on human fertility. Fertil Steril 2014;101:310—6.

[10] Bulun SE. Aromatase and estrogen receptor alpha deficiency. Fertil Steril 2014;101: 323—9.

[11] Reichman DE, White PC, New MI, Rosenwaks Z. Fertility in patients with congenital adrenal hyperplasia. Fertil Steril 2014;101:301—9.

[12] Schisterman EF, Mumford SL, Browne RW, Boyd Barr D, Chen Z, Buck Louis GM. Lipid concentrations and couple fecundity: the LIFE study. J Clin Endocrinol Metab 2014;99:2786—94. doi:jc20133936.

[13] Fujimoto VY, Kane JP, Ishida BY, Bloom MS, Browne RW. High-density lipoprotein metabolism and the human embryo. Hum Reprod Update 2010;16:20—38.

[14] Gwynne JT, Strauss 3rd JF. The role of lipoproteins in steroidogenesis and cholesterol metabolism in steroidogenic glands. Endocr Rev 1982;3:299—329.

[15] Gupta RS, Dixit VP. Effect of dietary cholesterol on spermatogenesis. Z Ernahrungswiss 1988;27:236—43.

[16] Kasturi SS, Tannir J, Brannigan RE. The metabolic syndrome and male infertility. J Androl 2008;29:251—9.

[17] Essah PA, Wickham EP, Nestler JE. The metabolic syndrome in polycystic ovary syndrome. Clin Obstet Gynecol 2007;50:205—25.

[18] Stocco DM, Clark BJ. Regulation of the acute production of steroids in steroidogenic cells. Endocr Rev 1996;17:221—44.

[19] Brown MS, Goldstein JL. A receptor-mediated pathway for cholesterol homeostasis. Science 1986;232:34—47.

[20] Brown MS, Kovanen PT, Goldstein JL. Receptor-mediated uptake of lipoprotein-cholesterol and its utilization for steroid synthesis in the adrenal cortex. Recent Prog Horm Res 1979;35:215—57.

[21] Goldstein JL, Brown MS. The low-density lipoprotein pathway and its relation to atherosclerosis. Annu Rev Biochem 1977;46:897—930.

[22] Brown MS, Goldstein JL. Receptor-mediated control of cholesterol metabolism. Science 1976;191:150—4.

[23] Rigotti A, Trigatti BL, Penman M, Rayburn H, Herz J, Krieger M. A targeted mutation in the murine gene encoding the high density lipoprotein (HDL) receptor scavenger receptor class B type I reveals its key role in HDL metabolism. Proc Natl Acad Sci USA 1997;94:12610—5.

[24] Thuahnai ST, Lund-Katz S, Anantharamaiah GM, Williams DL, Phillips MC. A quantitative analysis of apolipoprotein binding to SR-BI: multiple binding sites for lipid-free and lipid-associated apolipoproteins. J Lipid Res 2003;44:1132—42.

[25] Acton S, Rigotti A, Landschulz KT, Xu S, Hobbs HH, Krieger M. Identification of scavenger receptor SR-BI as a high density lipoprotein receptor. Science 1996;271:518—20.

[26] Kraemer FB, Shen WJ. Hormone-sensitive lipase: control of intracellular tri-(di-)acylglycerol and cholesteryl ester hydrolysis. J Lipid Res 2002;43:1585—94.

[27] Bisgaier CL, Chanderbhan R, Hinds RW, Vahouny GV. Adrenal cholesterol esters as substrate source for steroidogenesis. J Steroid Biochem 1985;23:967—74.

[28] Andersen JM, Dietschy JM. Relative importance of high and low density lipoproteins in the regulation of cholesterol synthesis in the adrenal gland, ovary, and testis of the rat. J Biol Chem 1978;253:9024—32.

[29] Spady DK, Dietschy JM. Rates of cholesterol synthesis and low-density lipoprotein uptake in the adrenal glands of the rat, hamster and rabbit in vivo. Biochim Biophys Acta 1985;836:167—75.

[30] Spady DK, Dietschy JM. Sterol synthesis in vivo in 18 tissues of the squirrel monkey, guinea pig, rabbit, hamster, and rat. J Lipid Res 1983;24:303–15.

[31] Espenshade PJ, Hughes AL. Regulation of sterol synthesis in eukaryotes. Annu Rev Genet 2007;41:401–27.

[32] Brown MS, Goldstein JL. The SREBP pathway: regulation of cholesterol metabolism by proteolysis of a membrane-bound transcription factor. Cell 1997;89:331–40.

[33] Stein JH, Rosenson RS. Lipoprotein Lp(a) excess and coronary heart disease. Arch Intern Med 1997;157:1170–6.

[34] Lloyd JK. Lipoprotein deficiency disorders. Clin Endocrinol Metabolism 1973;2:127–47.

[35] Borkowski AJ, Levin S, Delcroix C, Mahler A, Verhas V. Blood cholesterol and hydrocortisone production in man: quantitative aspects of the utilization of circulating cholesterol by the adrenals at rest and under adrenocorticotropin stimulation. J Clin Invest 1967;46:797–811.

[36] Bolte E, Coudert S, Lefebvre Y. Steroid production from plasma cholesterol. II. In vivo conversion of plasma cholesterol to ovarian progesterone and adrenal C19 and C21 steroids in humans. J Clin Endocrinol Metab 1974;38:394–400.

[37] Hellig H, Gattereau D, Lefebvre Y, Bolte E. Steroid production from plasma cholesterol. I. Conversion of plasma cholesterol to placental progesterone in humans. J Clin Endocrinol Metab 1970;30:624–31.

[38] Mahley RW, Rall Jr SC. Apolipoprotein E: far more than a lipid transport protein. Annu Rev Genomics Hum Genet 2000;1:507–37.

[39] Zhang T, Dai P, Cheng D, Zhang L, Chen Z, Meng X, et al. Obesity occurring in apolipoprotein E-knockout mice has mild effects on fertility. Reproduction 2014;147:141–51.

[40] Utermann G, Hees M, Steinmetz A. Polymorphism of apolipoprotein E and occurrence of dysbetalipoproteinaemia in man. Nature 1977;269:604–7.

[41] Wenham PR, Price WH, Blandell G. Apolipoprotein E genotyping by one-stage PCR. Lancet 1991;337:1158–9.

[42] Hatters DM, Peters-Libeu CA, Weisgraber KH. Apolipoprotein E structure: insights into function. Trends Biochem Sci 2006;31:445–54.

[43] Corbo RM, Scacchi R, Cresta M. Differential reproductive efficiency associated with common apolipoprotein e alleles in postreproductive-aged subjects. Fertil Steril 2004;81:104–7.

[44] Corbo RM, Ulizzi L, Piombo L, Scacchi R. Study on a possible effect of four longevity candidate genes (ACE, PON1, PPAR-gamma, and APOE) on human fertility. Biogerontology 2008;9:317–23.

[45] Corbo RM, Ulizzi L, Scacchi R, Martinez-Labarga C, De Stefano GF. Apolipoprotein E polymorphism and fertility: a study in pre-industrial populations. Mol Hum Reprod 2004;10:617–20.

[46] Gerdes LU, Gerdes C, Hansen PS, Klausen IC, Faergeman O. Are men carrying the apolipoprotein epsilon 4- or epsilon 2 allele less fertile than epsilon 3 epsilon 3 genotypes? Hum Genet 1996;98:239–42.

[47] Setarehbadi R, Vatannejad A, Vaisi-Raygani A, Amiri I, Esfahani M, Fattahi A, et al. Apolipoprotein E genotypes of fertile and infertile men. Syst Biol Reprod Med 2012;58:263–7.

[48] Brannian JD, Stouffer RL. Native and modified (acetylated) low density lipoprotein-supported steroidogenesis by macaque granulosa cells collected before and after the ovulatory stimulus: correlation with fluorescent lipoprotein uptake. Endocrinology 1993;132:591–7.

[49] Brannian JD, Shiigi SM, Stouffer RL. Gonadotropin surge increases fluorescent-tagged low-density lipoprotein uptake by macaque granulosa cells from preovulatory follicles. Biol Reprod 1992;47:355–60.

[50] Parinaud J, Perret B, Ribbes H, Chap H, Pontonnier G, Douste-Blazy L. High density lipoprotein and low density lipoprotein utilization by human granulosa cells for progesterone synthesis in serum-free culture: respective contributions of free and esterified cholesterol. J Clin Endocrinol Metab 1987;64:409−17.

[51] Volpe A, Coukos G, Uccelli E, Droghini F, Adamo R, Artini PG. Follicular fluid lipoproteins in preovulatory period and their relationship with follicular maturation and progesterone production by human granulosa-luteal cells in vivo and in vitro. J Endocrinol Invest 1991;14:737−42.

[52] Carr BR, MacDonald PC, Simpson ER. The role of lipoproteins in the regulation of progesterone secretion by the human corpus luteum. Fertil Steril 1982;38:303−11.

[53] Ishibashi S, Brown MS, Goldstein JL, Gerard RD, Hammer RE, Herz J. Hypercholesterolemia in low density lipoprotein receptor knockout mice and its reversal by adenovirus-mediated gene delivery. J Clin Invest 1993;92:883−93.

[54] Miettinen HE, Rayburn H, Krieger M. Abnormal lipoprotein metabolism and reversible female infertility in HDL receptor (SR-BI)-deficient mice. J Clin Invest 2001;108:1717−22.

[55] Trigatti B, Rayburn H, Vinals M, Braun A, Miettinen H, Penman M, et al. Influence of the high density lipoprotein receptor SR-BI on reproductive and cardiovascular pathophysiology. Proc Natl Acad Sci USA 1999;96:9322−7.

[56] Rinninger F, Heine M, Singaraja R, Hayden M, Brundert M, Ramakrishnan R, et al. High density lipoprotein metabolism in low density lipoprotein receptor-deficient mice. J Lipid Res 2014;55:1914−24.

[57] Murao K, Terpstra V, Green SR, Kondratenko N, Steinberg D, Quehenberger O. Characterization of CLA-1, a human homologue of rodent scavenger receptor BI, as a receptor for high density lipoprotein and apoptotic thymocytes. J Biol Chem 1997; 272:17551−7.

[58] Rigotti A, Acton SL, Krieger M. The class B scavenger receptors SR-BI and CD36 are receptors for anionic phospholipids. J Biol Chem 1995;270:16221−4.

[59] Calvo D, Gomez-Coronado D, Lasuncion MA, Vega MA. CLA-1 is an 85-kD plasma membrane glycoprotein that acts as a high-affinity receptor for both native (HDL, LDL, and VLDL) and modified (OxLDL and AcLDL) lipoproteins. Arterioscler Thromb Vasc Biol 1997;17:2341−9.

[60] Swarnakar S, Temel RE, Connelly MA, Azhar S, Williams DL. Scavenger receptor class B, type I, mediates selective uptake of low density lipoprotein cholesteryl ester. J Biol Chem 1999;274:29733−9.

[61] Acton SL, Scherer PE, Lodish HF, Krieger M. Expression cloning of SR-BI, a CD36-related class B scavenger receptor. J Biol Chem 1994;269:21003−9.

[62] Gwynne JT, Hess B. The role of high density lipoproteins in rat adrenal cholesterol metabolism and steroidogenesis. J Biol Chem 1980;255:10875−83.

[63] Glass C, Pittman RC, Weinstein DB, Steinberg D. Dissociation of tissue uptake of cholesterol ester from that of apoprotein A-I of rat plasma high density lipoprotein: selective delivery of cholesterol ester to liver, adrenal, and gonad. Proc Natl Acad Sci USA 1983;80:5435−9.

[64] Pittman RC, Knecht TP, Rosenbaum MS, Taylor Jr CA. A nonendocytotic mechanism for the selective uptake of high density lipoprotein-associated cholesterol esters. J Biol Chem 1987;262:2443−50.

[65] Azhar S, Tsai L, Reaven E. Uptake and utilization of lipoprotein cholesteryl esters by rat granulosa cells. Biochim Biophys Acta 1990;1047:148−60.

[66] Reaven E, Chen YD, Spicher M, Azhar S. Morphological evidence that high density lipoproteins are not internalized by steroid-producing cells during in situ organ perfusion. J Clin Invest 1984;74:1384−97.

[67] Reaven E, Tsai L, Azhar S. Cholesterol uptake by the 'selective' pathway of ovarian granulosa cells: early intracellular events. J Lipid Res 1995;36:1602−17.

[68] Reaven E, Tsai L, Azhar S. Intracellular events in the "selective" transport of lipoprotein-derived cholesteryl esters. J Biol Chem 1996;271:16208—17.

[69] Jian B, de la Llera-Moya M, Ji Y, Wang N, Phillips MC, Swaney JB, et al. Scavenger receptor class B type I as a mediator of cellular cholesterol efflux to lipoproteins and phospholipid acceptors. J Biol Chem 1998;273:5599—606.

[70] de La Llera-Moya M, Connelly MA, Drazul D, Klein SM, Favari E, Yancey PG, et al. Scavenger receptor class B type I affects cholesterol homeostasis by magnifying cholesterol flux between cells and HDL. J Lipid Res 2001;42:1969—78.

[71] Azhar S, Reaven E. Scavenger receptor class BI and selective cholesteryl ester uptake: partners in the regulation of steroidogenesis. Mol Cell Endocrinol 2002;195:1—26.

[72] Li X, Peegel H, Menon KM. In situ hybridization of high density lipoprotein (scavenger, type 1) receptor messenger ribonucleic acid (mRNA) during folliculogenesis and luteinization: evidence for mRNA expression and induction by human chorionic gonadotropin specifically in cell types that use cholesterol for steroidogenesis. Endocrinology 1998;139:3043—9.

[73] Rigotti A, Edelman ER, Seifert P, Iqbal SN, DeMattos RB, Temel RE, et al. Regulation by adrenocorticotropic hormone of the in vivo expression of scavenger receptor class B type I (SR-BI), a high density lipoprotein receptor, in steroidogenic cells of the murine adrenal gland. J Biol Chem 1996;271:33545—9.

[74] Landschulz KT, Pathak RK, Rigotti A, Krieger M, Hobbs HH. Regulation of scavenger receptor, class B, type I, a high density lipoprotein receptor, in liver and steroidogenic tissues of the rat. J Clin Invest 1996;98:984—95.

[75] Azhar S, Nomoto A, Leers-Sucheta S, Reaven E. Simultaneous induction of an HDL receptor protein (SR-BI) and the selective uptake of HDL-cholesteryl esters in a physiologically relevant steroidogenic cell model. J Lipid Res 1998;39:1616—28.

[76] Jaspard B, Fournier N, Vieitez G, Atger V, Barbaras R, Vieu C, et al. Structural and functional comparison of HDL from homologous human plasma and follicular fluid. A model for extravascular fluid. Arterioscler Thromb Vasc Biol 1997;17:1605—13.

[77] Le Goff D. Follicular fluid lipoproteins in the mare: evaluation of HDL transfer from plasma to follicular fluid. Biochim Biophys Acta 1994;1210:226—32.

[78] Perret BP, Parinaud J, Ribbes H, Moatti JP, Pontonnier G, Chap H, et al. Lipoprotein and phospholipid distribution in human follicular fluids. Fertil Steril 1985;43:405—9.

[79] Shalgi R, Kraicer P, Rimon A, Pinto M, Soferman N. Proteins of human follicular fluid: the blood-follicle barrier. Fertil Steril 1973;24:429—34.

[80] Simpson ER, Rochelle DB, Carr BR, MacDonald PC. Plasma lipoproteins in follicular fluid of human ovaries. J Clin Endocrinol Metab 1980;51:1469—71.

[81] Bambauer R, Bambauer C, Lehmann B, Latza R, Schiel R. LDL-apheresis: technical and clinical aspects. Sci World J 2012;2012:314283.

[82] Mora S, Szklo M, Otvos JD, Greenland P, Psaty BM, Goff Jr DC, et al. LDL particle subclasses, LDL particle size, and carotid atherosclerosis in the Multi-Ethnic Study of Atherosclerosis (MESA). Atherosclerosis 2007;192:211—7.

[83] Jimenez LM, Binelli M, Bertolin K, Pelletier RM, Murphy BD. Scavenger receptor-B1 and luteal function in mice. J Lipid Res 2010;51:2362—71.

[84] Yesilaltay A, Morales MG, Amigo L, Zanlungo S, Rigotti A, Karackattu SL, et al. Effects of hepatic expression of the high-density lipoprotein receptor SR-BI on lipoprotein metabolism and female fertility. Endocrinology 2006;147:1577—88.

[85] Hirano K, Ikegami C, Tsujii K, Zhang Z, Matsuura F, Nakagawa-Toyama Y, et al. Probucol enhances the expression of human hepatic scavenger receptor class B type I, possibly through a species-specific mechanism. Arterioscler Thromb Vasc Biol 2005;25:2422—7.

[86] Miida T, Seino U, Miyazaki O, Hanyu O, Hirayama S, Saito T, et al. Probucol markedly reduces HDL phospholipids and elevated prebeta1-HDL without delayed conversion into alpha-migrating HDL: putative role of angiopoietin-like protein 3 in probucol-induced HDL remodeling. Atherosclerosis 2008;200:329—35.

[87] Kolmakova A, Wang J, Brogan R, Chaffin C, Rodriguez A. Deficiency of scavenger receptor class B type I negatively affects progesterone secretion in human granulosa cells. Endocrinology 2010;151:5519–27.
[88] Christianson MS, Yates M. Scavenger receptor class B type 1 gene polymorphisms and female fertility. Curr Opin Endocrinol Diabetes Obes 2012;19:115–20.
[89] West M, Greason E, Kolmakova A, Jahangiri A, Asztalos B, Pollin TI, et al. Scavenger receptor class B type I protein as an independent predictor of high-density lipoprotein cholesterol levels in subjects with hyperalphalipoproteinemia. J Clin Endocrinol Metab 2009;94:1451–7.
[90] Yates M, Kolmakova A, Zhao Y, Rodriguez A. Clinical impact of scavenger receptor class B type I gene polymorphisms on human female fertility. Hum Reprod 2011;26:1910–6.
[91] Velasco M, Alexander C, King J, Zhao Y, Garcia J, Rodriguez A. Association of lower plasma estradiol levels and low expression of scavenger receptor class B, type I in infertile women. Fertil Steril 2006;85:1391–7.
[92] Brown MS, Goldstein JL, Krieger M, Ho YK, Anderson RG. Reversible accumulation of cholesteryl esters in macrophages incubated with acetylated lipoproteins. J Cell Biol 1979;82:597–613.
[93] Kraemer FB, Khor VK, Shen WJ, Azhar S. Cholesterol ester droplets and steroidogenesis. Mol Cell Endocrinol 2013;371:15–9.
[94] Yeaman SJ. Hormone-sensitive lipase—a multipurpose enzyme in lipid metabolism. Biochim Biophys Acta 1990;1052:128–32.
[95] Kraemer FB, Tavangar K, Hoffman AR. Developmental regulation of hormone-sensitive lipase mRNA in the rat: changes in steroidogenic tissues. J Lipid Res 1991;32:1303–10.
[96] Park SY, Kim HJ, Wang S, Higashimori T, Dong J, Kim YJ, et al. Hormone-sensitive lipase knockout mice have increased hepatic insulin sensitivity and are protected from short-term diet-induced insulin resistance in skeletal muscle and heart. Am J Physiol Endocrinol Metab 2005;289:E30–9.
[97] Kraemer FB, Shen WJ. Hormone-sensitive lipase knockouts. Nutr Metab (Lond) 2006;3:12.
[98] Osuga J, Ishibashi S, Oka T, Yagyu H, Tozawa R, Fujimoto A, et al. Targeted disruption of hormone-sensitive lipase results in male sterility and adipocyte hypertrophy, but not in obesity. Proc Natl Acad Sci USA 2000;97:787–92.
[99] Hermo L, Chung S, Gregory M, Smith CE, Wang SP, El-Alfy M, et al. Alterations in the testis of hormone sensitive lipase-deficient mice is associated with decreased sperm counts, sperm motility, and fertility. Mol Reprod Dev 2008;75:565–77.
[100] Vallet-Erdtmann V, Tavernier G, Contreras JA, Mairal A, Rieu C, Touzalin AM, et al. The testicular form of hormone-sensitive lipase HSLtes confers rescue of male infertility in HSL-deficient mice. J Biol Chem 2004;279:42875–80.
[101] Garenc C, Perusse L, Chagnon YC, Rankinen T, Gagnon J, Borecki IB, et al. The hormone-sensitive lipase gene and body composition: the HERITAGE Family Study. Int J Obes Relat Metab Disord 2002;26:220–7.
[102] Shimada F, Makino H, Hashimoto N, Iwaoka H, Taira M, Nozaki O, et al. Detection of an amino acid polymorphism in hormone-sensitive lipase in Japanese subjects. Metabolism 1996;45:862–4.
[103] Magre J, Laurell H, Fizames C, Antoine PJ, Dib C, Vigouroux C, et al. Human hormone-sensitive lipase: genetic mapping, identification of a new dinucleotide repeat, and association with obesity and NIDDM. Diabetes 1998;47:284–6.
[104] Talmud PJ, Palmen J, Luan J, Flavell D, Byrne CD, Waterworth DM, et al. Variation in the promoter of the human hormone sensitive lipase gene shows gender specific effects on insulin and lipid levels: results from the Ely study. Biochim Biophys Acta 2001;1537:239–44.

[105] Talmud PJ, Palmen J, Walker M. Identification of genetic variation in the human hormone-sensitive lipase gene and 5' sequences: homology of 5' sequences with mouse promoter and identification of potential regulatory elements. Biochem Biophys Res Commun 1998;252:661−8.

[106] Vatannejad A, Khodadadi I, Amiri I, Vaisi-Raygani A, Ghorbani M, Tavilani H. Genetic variation of hormone sensitive lipase and male infertility. Syst Biol Reprod Med 2011;57:288−91.

[107] Porter FD, Herman GE. Malformation syndromes caused by disorders of cholesterol synthesis. J Lipid Res 2011;52:6−34.

[108] Ohashi M, Mizushima N, Kabeya Y, Yoshimori T. Localization of mammalian NAD(P)H steroid dehydrogenase-like protein on lipid droplets. J Biol Chem 2003;278:36819−29.

[109] Horton JD, Goldstein JL, Brown MS. SREBPs: activators of the complete program of cholesterol and fatty acid synthesis in the liver. J Clin Invest 2002;109:1125−31.

[110] Song BL, Javitt NB, DeBose-Boyd RA. Insig-mediated degradation of HMG CoA reductase stimulated by lanosterol, an intermediate in the synthesis of cholesterol. Cell Metab 2005;1:179−89.

[111] Liu X, Li Y, Lu X, Wang L, Zhao Q, Yang W, et al. Interactions among genetic variants from SREBP2 activating-related pathway on risk of coronary heart disease in Chinese Han population. Atherosclerosis 2010;208:421−6.

[112] Steffen KM, Cooper ME, Shi M, Caprau D, Simhan HN, Dagle JM, et al. Maternal and fetal variation in genes of cholesterol metabolism is associated with preterm delivery. J Perinatol 2007;27:672−80.

[113] Knoblauch H, Bauerfeind A, Toliat MR, Becker C, Luganskaja T, Gunther UP, et al. Haplotypes and SNPs in 13 lipid-relevant genes explain most of the genetic variance in high-density lipoprotein and low-density lipoprotein cholesterol. Hum Mol Genet 2004;13:993−1004.

[114] Rozman D, Seliskar M, Cotman M, Fink M. Pre-cholesterol precursors in gametogenesis. Mol Cell Endocrinol 2005;234:47−56.

[115] Lewinska M, Zelenko U, Merzel F, Golic Grdadolnik S, Murray JC, Rozman D. Polymorphisms of CYP51A1 from cholesterol synthesis: associations with birth weight and maternal lipid levels and impact on CYP51 protein structure. PLoS One 2013;8:e82554.

[116] Lange Y, Ory DS, Ye J, Lanier MH, Hsu FF, Steck TL. Effectors of rapid homeostatic responses of endoplasmic reticulum cholesterol and 3-hydroxy-3-methylglutaryl-CoA reductase. J Biol Chem 2008;283:1445−55.

[117] Zerenturk EJ, Sharpe LJ, Ikonen E, Brown AJ. Desmosterol and DHCR24: unexpected new directions for a terminal step in cholesterol synthesis. Prog Lipid Res 2013;52:666−80.

[118] FitzPatrick DR, Keeling JW, Evans MJ, Kan AE, Bell JE, Porteous ME, et al. Clinical phenotype of desmosterolosis. Am J Med Genet 1998;75:145−52.

[119] Schaaf CP, Koster J, Katsonis P, Kratz L, Shchelochkov OA, Scaglia F, et al. Desmosterolosis-phenotypic and molecular characterization of a third case and review of the literature. Am J Med Genet A 2011;155A:1597−604.

[120] McGrath KC, Li XH, Puranik R, Liong EC, Tan JT, Dy VM, et al. Role of 3beta-hydroxysteroid-delta 24 reductase in mediating antiinflammatory effects of high-density lipoproteins in endothelial cells. Arterioscler Thromb Vasc Biol 2009;29:877−82.

[121] Lu X, Kambe F, Cao X, Kozaki Y, Kaji T, Ishii T, et al. 3beta-Hydroxysteroid-delta24 reductase is a hydrogen peroxide scavenger, protecting cells from oxidative stress-induced apoptosis. Endocrinology 2008;149:3267−73.

[122] Greeve I, Hermans-Borgmeyer I, Brellinger C, Kasper D, Gomez-Isla T, Behl C, et al. The human DIMINUTO/DWARF1 homolog seladin-1 confers resistance to Alzheimer's disease-associated neurodegeneration and oxidative stress. J Neurosci 2000;20:7345−52.

[123] Tint GS, Irons M, Elias ER, Batta AK, Frieden R, Chen TS, et al. Defective cholesterol biosynthesis associated with the Smith-Lemli-Opitz syndrome. N Engl J Med 1994; 330:107—13.

[124] Chevy F, Humbert L, Wolf C. Sterol profiling of amniotic fluid: a routine method for the detection of distal cholesterol synthesis deficit. Prenat Diagn 2005;25:1000—6.

[125] Abuelo DN, Tint GS, Kelley R, Batta AK, Shefer S, Salen G. Prenatal detection of the cholesterol biosynthetic defect in the Smith-Lemli-Opitz syndrome by the analysis of amniotic fluid sterols. Am J Med Genet 1995;56:281—5.

[126] Tint GS, Abuelo D, Till M, Cordier MP, Batta AK, Shefer S, et al. Fetal Smith-Lemli-Opitz syndrome can be detected accurately and reliably by measuring amniotic fluid dehydrocholesterols. Prenat Diagn 1998;18:651—8.

[127] Shackleton CH, Roitman E, Kratz L, Kelley R. Dehydro-oestriol and dehydropregnanetriol are candidate analytes for prenatal diagnosis of Smith-Lemli-Opitz syndrome. Prenat Diagn 2001;21:207—12.

[128] Svoboda MD, Christie JM, Eroglu Y, Freeman KA, Steiner RD. Treatment of Smith-Lemli-Opitz syndrome and other sterol disorders. Am J Med Genet C Semin Med Genet 2012;160C:285—94.

[129] Knochenhauer ES, Key TJ, Kahsar-Miller M, Waggoner W, Boots LR, Azziz R. Prevalence of the polycystic ovary syndrome in unselected black and white women of the southeastern United States: a prospective study. J Clin Endocrinol Metab 1998;83: 3078—82.

[130] Kosova G, Urbanek M. Genetics of the polycystic ovary syndrome. Mol Cell Endocrinol 2013;373:29—38.

[131] Vink JM, Sadrzadeh S, Lambalk CB, Boomsma DI. Heritability of polycystic ovary syndrome in a Dutch twin-family study. J Clin Endocrinol Metab 2006;91:2100—4.

[132] Deligeoroglou E, Kouskouti C, Christopoulos P. The role of genes in the polycystic ovary syndrome: predisposition and mechanisms. Gynecol Endocrinol 2009;25:603—9.

[133] Liu Y, Jiang H, He LY, Huang WJ, He XY, Xing FQ. Abnormal expression of uncoupling protein-2 correlates with CYP11A1 expression in polycystic ovary syndrome. Reprod Fertil Dev 2011;23:520—6.

[134] Gaasenbeek M, Powell BL, Sovio U, Haddad L, Gharani N, Bennett A, et al. Large-scale analysis of the relationship between CYP11A promoter variation, polycystic ovarian syndrome, and serum testosterone. J Clin Endocrinol Metab 2004;89:2408—13.

[135] Waterworth DM, Bennett ST, Gharani N, McCarthy MI, Hague S, Batty S, et al. Linkage and association of insulin gene VNTR regulatory polymorphism with polycystic ovary syndrome. Lancet 1997;349:986—90.

[136] Powell BL, Haddad L, Bennett A, Gharani N, Sovio U, Groves CJ, et al. Analysis of multiple data sets reveals no association between the insulin gene variable number tandem repeat element and polycystic ovary syndrome or related traits. J Clin Endocrinol Metab 2005;90:2988—93.

[137] Goodarzi MO, Louwers YV, Taylor KD, Jones MR, Cui J, Kwon S, et al. Replication of association of a novel insulin receptor gene polymorphism with polycystic ovary syndrome. Fertil Steril 2011;95:1736—41. e1731-1711.

[138] Lee EJ, Oh B, Lee JY, Kimm K, Lee SH, Baek KH. A novel single nucleotide polymorphism of INSR gene for polycystic ovary syndrome. Fertil Steril 2008;89:1213—20.

[139] Chen ZJ, Zhao H, He L, Shi Y, Qin Y, Li Z, et al. Genome-wide association study identifies susceptibility loci for polycystic ovary syndrome on chromosome 2p16.3, 2p21 and 9q33.3. Nat Genet 2011;43:55—9.

[140] Latronico AC, Chai Y, Arnhold IJ, Liu X, Mendonca BB, Segaloff DL. A homozygous microdeletion in helix 7 of the luteinizing hormone receptor associated with familial testicular and ovarian resistance is due to both decreased cell surface expression and impaired effector activation by the cell surface receptor. Mol Endocrinol 1998;12:442—50.

[141] Zeggini E, Scott LJ, Saxena R, Voight BF, Marchini JL, Hu T, et al. Meta-analysis of genome-wide association data and large-scale replication identifies additional susceptibility loci for type 2 diabetes. Nat Genet 2008;40:638–45.

[142] Yoshimura S, Gerondopoulos A, Linford A, Rigden DJ, Barr FA. Family-wide characterization of the DENN domain Rab GDP-GTP exchange factors. J Cell Biol 2010;191:367–81.

[143] Eriksen MB, Brusgaard K, Andersen M, Tan Q, Altinok ML, Gaster M, et al. Association of polycystic ovary syndrome susceptibility single nucleotide polymorphism rs2479106 and PCOS in Caucasian patients with PCOS or hirsutism as referral diagnosis. Eur J Obstet Gynecol Reprod Biol 2012;163:39–42.

[144] Lerchbaum E, Trummer O, Giuliani A, Gruber HJ, Pieber TR, Obermayer-Pietsch B. Susceptibility loci for polycystic ovary syndrome on chromosome 2p16.3, 2p21, and 9q33.3 in a cohort of Caucasian women. Horm Metab Res 2011;43:743–7.

[145] Goodarzi MO, Jones MR, Li X, Chua AK, Garcia OA, Chen YD, et al. Replication of association of DENND1A and THADA variants with polycystic ovary syndrome in European cohorts. J Med Genet 2012;49:90–5.

[146] Welt CK, Styrkarsdottir U, Ehrmann DA, Thorleifsson G, Arason G, Gudmundsson JA, et al. Variants in DENND1A are associated with polycystic ovary syndrome in women of European ancestry. J Clin Endocrinol Metab 2012;97:E1342–7.

[147] Goswami D, Conway GS. Premature ovarian failure. Hum Reprod Update 2005;11:391–410.

[148] Goswami D, Conway GS. Premature ovarian failure. Horm Res 2007;68:196–202.

[149] Vujovic S. Aetiology of premature ovarian failure. Menopause Int 2009;15:72–5.

[150] Pouresmaeili F, Fazeli Z. Premature ovarian failure: a critical condition in the reproductive potential with various genetic causes. Int J Fertil Steril 2014;8:1–12.

[151] Jeyasuria P, Ikeda Y, Jamin SP, Zhao L, De Rooij DG, Themmen AP, et al. Cell-specific knockout of steroidogenic factor 1 reveals its essential roles in gonadal function. Mol Endocrinol 2004;18:1610–9.

[152] Lakhal B, Ben-Hadj-Khalifa S, Bouali N, Braham R, Hatem E, Saad A. Mutational screening of SF1 and WNT4 in Tunisian women with premature ovarian failure. Gene 2012;509:298–301.

[153] Rah H, Jeon YJ, Choi Y, Shim SH, Yoon TK, Choi DH, et al. Association of methylenetetrahydrofolate reductase (MTHFR 677C>T) and thymidylate synthase (TSER and TS 1494del6) polymorphisms with premature ovarian failure in Korean women. Menopause 2012;19:1260–6.

[154] Rosen MP, Shen S, McCulloch CE, Rinaudo PF, Cedars MI, Dobson AT. Methylenetetrahydrofolate reductase (MTHFR) is associated with ovarian follicular activity. Fertil Steril 2007;88:632–8.

[155] Weisberg I, Tran P, Christensen B, Sibani S, Rozen R. A second genetic polymorphism in methylenetetrahydrofolate reductase (MTHFR) associated with decreased enzyme activity. Mol Genet Metab 1998;64:169–72.

[156] Clarke R, Daly L, Robinson K, Naughten E, Cahalane S, Fowler B, et al. Hyperhomocysteinemia: an independent risk factor for vascular disease. N Engl J Med 1991;324:1149–55.

[157] Frosst P, Blom HJ, Milos R, Goyette P, Sheppard CA, Matthews RG, et al. A candidate genetic risk factor for vascular disease: a common mutation in methylenetetrahydrofolate reductase. Nat Genet 1995;10:111–3.

[158] Yan Y, Liang H, Yang S, Wang J, Xie L, Qin X, et al. Methylenetetrahydrofolate reductase A1298C polymorphism and diabetes risk: evidence from a meta-analysis. Ren Fail 2014;36:1013–7.

[159] Herrmann W. The importance of hyperhomocysteinemia as a risk factor for diseases: an overview. Clin Chem Lab Med 2001;39:666–74.

[160] Gueant JL, Gueant-Rodriguez RM, Anello G, Bosco P, Brunaud L, Romano C, et al. Genetic determinants of folate and vitamin B12 metabolism: a common pathway in neural tube defect and Down syndrome? Clin Chem Lab Med 2003;41:1473–7.

[161] Liu L, Tan R, Cui Y, Liu J, Wu J. Estrogen receptor alpha gene (ESR1) polymorphisms associated with idiopathic premature ovarian failure in Chinese women. Gynecol Endocrinol 2013;29:182–5.

[162] Bretherick KL, Hanna CW, Currie LM, Fluker MR, Hammond GL, Robinson WP. Estrogen receptor alpha gene polymorphisms are associated with idiopathic premature ovarian failure. Fertil Steril 2008;89:318–24.

[163] Yang JJ, Cho LY, Lim YJ, Ko KP, Lee KS, Kim H, et al. Estrogen receptor-1 genetic polymorphisms for the risk of premature ovarian failure and early menopause. J Womens Health (Larchmt) 2010;19:297–304.

[164] Hill SM, Fuqua SA, Chamness GC, Greene GL, McGuire WL. Estrogen receptor expression in human breast cancer associated with an estrogen receptor gene restriction fragment length polymorphism. Cancer Res 1989;49:145–8.

[165] Herrington DM, Howard TD, Brosnihan KB, McDonnell DP, Li X, Hawkins GA, et al. Common estrogen receptor polymorphism augments effects of hormone replacement therapy on E-selectin but not C-reactive protein. Circulation 2002;105:1879–82.

[166] Yaich L, Dupont WD, Cavener DR, Parl FF. Analysis of the PvuII restriction fragment-length polymorphism and exon structure of the estrogen receptor gene in breast cancer and peripheral blood. Cancer Res 1992;52:77–83.

[167] Kalantaridou SN, Nelson LM. Premature ovarian failure is not premature menopause. Ann N Y Acad Sci 2000;900:393–402.

[168] Dun EC, Taylor RN, Wieser F. Advances in the genetics of endometriosis. Genome Med 2010;2:75.

[169] Simoens S, Hummelshoj L, D'Hooghe T. Endometriosis: cost estimates and methodological perspective. Hum Reprod Update 2007;13:395–404.

[170] Bulun SE, Cheng YH, Pavone ME, Xue Q, Attar E, Trukhacheva E, et al. Estrogen receptor-beta, estrogen receptor-alpha, and progesterone resistance in endometriosis. Semin Reprod Med 2010;28:36–43.

[171] Burney RO, Talbi S, Hamilton AE, Vo KC, Nyegaard M, Nezhat CR, et al. Gene expression analysis of endometrium reveals progesterone resistance and candidate susceptibility genes in women with endometriosis. Endocrinology 2007;148:3814–26.

[172] Treloar SA, O'Connor DT, O'Connor VM, Martin NG. Genetic influences on endometriosis in an Australian twin sample. sueT@qimr.edu.au. Fertil Steril 1999;71:701–10.

[173] Uno S, Zembutsu H, Hirasawa A, Takahashi A, Kubo M, Akahane T, et al. A genome-wide association study identifies genetic variants in the CDKN2BAS locus associated with endometriosis in Japanese. Nat Genet 2010;42:707–10.

[174] Goumenou AG, Arvanitis DA, Matalliotakis IM, Koumantakis EE, Spandidos DA. Loss of heterozygosity in adenomyosis on hMSH2, hMLH1, p16Ink4 and GALT loci. Int J Mol Med 2000;6:667–71.

[175] Horne BD, Carlquist JF, Muhlestein JB, Bair TL, Anderson JL. Association of variation in the chromosome 9p21 locus with myocardial infarction versus chronic coronary artery disease. Circ Cardiovasc Genet 2008;1:85–92.

[176] Doria A, Wojcik J, Xu R, Gervino EV, Hauser TH, Johnstone MT, et al. Interaction between poor glycemic control and 9p21 locus on risk of coronary artery disease in type 2 diabetes. Jama 2008;300:2389–97.

[177] Painter JN, Anderson CA, Nyholt DR, Macgregor S, Lin J, Lee SH, et al. Genome-wide association study identifies a locus at 7p15.2 associated with endometriosis. Nat Genet 2011;43:51–4.

[178] Taylor HS, Bagot C, Kardana A, Olive D, Arici A. HOX gene expression is altered in the endometrium of women with endometriosis. Hum Reprod 1999;14:1328–31.

[179] Andersson KL, Bussani C, Fambrini M, Polverino V, Taddei GL, Gemzell-Danielsson K, et al. DNA methylation of HOXA10 in eutopic and ectopic endometrium. Hum Reprod 2014;29:1906–11.
[180] Taylor HS, Arici A, Olive D, Igarashi P. HOXA10 is expressed in response to sex steroids at the time of implantation in the human endometrium. J Clin Invest 1998;101:1379–84.
[181] Nyholt DR, Low SK, Anderson CA, Painter JN, Uno S, Morris AP, et al. Genome-wide association meta-analysis identifies new endometriosis risk loci. Nat Genet 2012;44:1355–9.
[182] Vainio S, Heikkila M, Kispert A, Chin N, McMahon AP. Female development in mammals is regulated by Wnt-4 signalling. Nature 1999;397:405–9.
[183] Boyer A, Lapointe E, Zheng X, Cowan RG, Li H, Quirk SM, et al. WNT4 is required for normal ovarian follicle development and female fertility. FASEB J 2010;24:3010–25.
[184] Albertsen HM, Chettier R, Farrington P, Ward K. Genome-wide association study link novel loci to endometriosis. PLoS One 2013;8:e58257.
[185] Goteri G, Ciavattini A, Lucarini G, Montik N, Filosa A, Stramazzotti D, et al. Expression of motility-related molecule Cdc42 in endometrial tissue in women with adenomyosis and ovarian endometriomata. Fertil Steril 2006;86:559–65.
[186] Hu WP, Tay SK, Zhao Y. Endometriosis-specific genes identified by real-time reverse transcription-polymerase chain reaction expression profiling of endometriosis versus autologous uterine endometrium. J Clin Endocrinol Metab 2006;91:228–38.

12

Overview of Intersection of Genomics of Cardiometabolic Disease and Other Disease States, Such as Eye Health (Macular Degeneration)

Gareth J. McKay

Centre for Public Health, Queen's University Belfast, Belfast, Northern Ireland, UK

1. INTRODUCTION

Cardiovascular disease (CVD) is the leading single cause of global mortality, accounting for 15.8 million deaths per year [1]. At the age of 40 years, the lifetime risk for developing CVD in Western Society is 50% in men and 33% in women [2], with atherosclerotic vascular disease also accounting for more deaths and disability than all types of cancer [3]. Atherosclerosis results from chronic inflammation of the arterial blood vessels and is characterized by a build-up of lipid-derived plaques or lesions within the blood vessels. Unstable atherosclerotic plaques recruit circulating immune cells propagating a cycle of inflammation with modification of the endothelial cell phenotype to become pro-atherogenic [4]. Such lesions may form a large low-density-lipoprotein pool, which is prone to rupture leading to occlusion of the coronary or cerebral circulation, resulting in a myocardial infarction or stroke, respectively.

Age-related macular degeneration (AMD; MIM# 603075) is the leading cause of visual impairment among older people of European descent, accounting for greater than half of all new cases of registered blindness [6]. The socioeconomic consequences associated with AMD provide increasing challenges to our aging society with almost one-third of those aged 75 years and older showing early signs of disease [137]. By definition, AMD specifically affects the macular region of the retina, an area which is responsible for detailed central vision. It is a disease continuum with degenerative changes found at the level of the outer retina, retinal pigment epithelium (RPE), Bruch's membrane, and the choriocapillaris that commences as degenerative changes at the level of the RPE (early AMD) progressing to the visually disabling late phenotypes of geographic atrophy (GA) and/or neovascular AMD (nvAMD). The degenerative features are visible clinically as drusen, which are yellowish deposits, focal pigmentary irregularities (hypopigmentation and hyperpigmentation) in the macular fundus, or both together [5].

AMD is a common multifactorial disorder of complex etiology with multiple genetic, environmental, and lifestyle risk factors. Although the specifics of the etiology remain poorly defined, there is evidence to suggest common mechanistic processes and risk factors shared with CVD [7−9,19]. Risk for both AMD and CVD is modified by smoking, hypertension, inflammatory markers (e.g., C-reactive protein) [10,11], and common genetic variants including but not limited to apolipoprotein E gene and complement factor H (CFH) [12−15]. However, consistent evidence provided through cross-sectional studies identifying common association between AMD and CVD, or its multiple risk factors (e.g., plasma triglycerides) remains elusive [9,16−20]. The presence of atherosclerotic lesions, determined by ultrasound, was examined in relation to risk of AMD in a large population-based study conducted in the Netherlands [21]. Results obtained from this cross-sectional study showed a 4.5-fold increased risk of late AMD (defined as GA or CNV) associated with plaques in the carotid bifurcation and a twofold increased risk associated with plaques in the common carotid artery. Lower-extremity arterial disease (as measured by the ratio of the systolic blood pressure level of the ankle to the arm) was also associated with a 2.5-times increased risk of AMD. Other studies of the association between AMD and CVD risk factors have shown associations between a higher pulse pressure, higher systolic blood pressure, and increased carotid wall thickness and incident AMD [16,22].

Although environmental and lifestyle risk factors which include smoking, blood pressure, age, diet, gender, body mass index, and physical activity are known to exert a major influence on the overall risk associated with both CVD and AMD (Figure 1), the underlying genetic architecture that modulates this overall risk remains to be fully resolved, despite recent scientific advances. In recent years, our understanding of the genetic architecture that underpins AMD has been greatly improved through genetic discovery with the implication of multiple biological processes including systemic dysregulation of chronic inflammation, lipid metabolism, and extracellular matrix deposition pathways involved

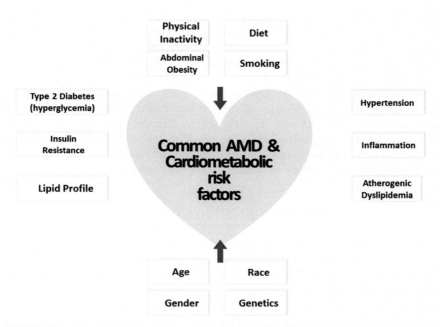

FIGURE 1 Shared factors common to both age-related macular degeneration (AMD) and cardiometabolic risk.

in the pathogenesis of the condition [23]. The recent identification of genetic risk factors associated with AMD has created new opportunities to better understand the disease pathophysiology. Of note, these findings may enable the discovery of molecular changes in the tissues of interest even before the onset of disease by detectable means.

2. CHARACTERISTIC AMD LESIONS

Abnormal extracellular deposition of proteins may contribute to AMD pathogenesis and progression, in a manner similar to that observed for both Alzheimer's disease and atherosclerosis. Of note, the respective pathophysiological deposits for each of these conditions contain many shared constituents such as ApoE, complement, and Aβ peptides [24]. For instance, in AMD, Aβ peptide deposition is observed in association with the characteristic pathological hallmarks of AMD, drusen, where it has been shown to accumulate and co-localize with activated components of the complement pathway [25–27].

Early-stage AMD is characterized by these hallmark drusen lesions which are also reported to contain cholesterol and lipid-rich basal linear deposits that accumulate with age between the RPE and the choroid

[28,29]. Bruch's membrane forms the inner margin of the choroid and effectively acts in a manner similar to that of a vessel wall [29]. During the process of atherosclerosis, lipoproteins traverse the vascular endothelium, bind with proteoglycans, and accumulate within the arterial wall, which culminates in the deleterious processes that include inflammation and neovascularization [30]. In comparative terms, these atherosclerotic plaques mimic the accumulation of drusen and basal laminar deposits in a manner similar to that seen in early AMD. Drusen deposition can occur through normal physiological aging via slow degenerative mechanisms possibly through localized choroidal vascular ischemia and may be influenced by risk factors that also commonly induce atherosclerosis [7].

3. THE GENETICS OF AMD

The investigation of ApoE was one of the first significant reported genetic associations for AMD [12,31] and indicated a potential role for immunoregulation and cell signaling in the disease etiology [32]. More recently, an improved understanding of the genetic basis of AMD has enabled the identification of risk and protective variants within several genes associated with the complement pathway and chronic inflammation including factor H (*CFH*) [15,33−36], component 2 (*CC2*)/factor B (*CFB*) region [37,38], component 3 (*C3*) [39,40] and complement factor I (*CFI*) [41]. Figure 2 is a composite figure illustrating the Manhattan plots

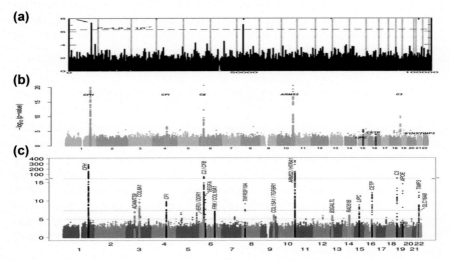

FIGURE 2 Amalgamation of the genome-wide association study outputs for AMD: (a) Ref. [15]; 96 cases and 50 controls. (b) Ref. [48]; 2157 cases and 1150 controls. (c) Ref. [49]; 17,100 cases and >60,000 controls.

from each of the three AMD genome-wide association studies (GWAS) published to date and illustrates the ability to identify additional disease-associated variants with improved statistical power through increasing sample size (Figure 2).

Beyond the complement pathway, chromosome 10q26 and specifically the Age-Related Maculopathy Susceptibility 2 (*ARMS2*) locus has been implicated as a second major genetic contributor to the AMD disease process, although the precise mechanistic effects remain to be resolved [42–45]. The high linkage disequilibrium that exists across the 10q26 region between *ARMS2* and the serine protease *HTRA1* gene confound identification of the genetic source or causal variants within this locus, although previous reports have been supportive of mitochondrial involvement through interaction with translocase of outer mitochondrial membrane proteins and co-localization to the mitochondrial-rich ellipsoid region of the retinal photoreceptors [46,47].

Mitochondrial dysfunction and oxidative stress have also been implicated as major processes that contribute to AMD disease etiology with increased mitochondrial damage reported within both the neural retina and the RPE with aging. Genetic variation within the mitochondrial genome has been reported in association with increased AMD risk [50,51] providing further support for the involvement of oxidative stress in AMD pathophysiology. A metalloproteinase, *TIMP3*, which is involved in the degradation of the extracellular matrix, and hepatic lipase C (*LIPC*) and cholesteryl ester transfer protein (*CETP*), key genes involved in triglyceride and high-density lipoprotein (HDL) metabolism, have been implicated in AMD pathogenesis through GWAS [48,52]. Most recently, the largest GWAS undertaken to date in AMD, has examined more than two and a half million single nucleotide polymorphisms (SNPs) in 17,000 advanced AMD cases and >60,000 controls of European and Asian ancestry [49]. The study investigators identified 19 genomic loci associated with AMD at a level which exceeded genome-wide significance ($P < 5 \times 10^{-8}$) including genes enriched in regulation of complement activity, lipid metabolism, extracellular matrix remodeling, and angiogenesis (Figure 3). Seven of these loci were novel reaching $P < 5 \times 10^{-8}$ for the first time, and were located near the genes COL8A1/FILIP1L, IER3/DDR1, SLC16A8, TGFBR1, RAD51B, ADAMTS9/MIR548A2, and B3GALTL.

4. AGE, OXIDATIVE STRESS, AND REVERSE CHOLESTEROL TRANSPORT

In a recent systematic review, both early and late AMD were associated with a modest increased incidence of CVD when data from both

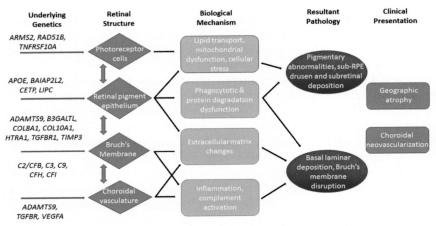

FIGURE 3   Reproduced from ref. [138]. A multihit "threshold" model of AMD. Advanced age and environmental factors have an impact on all components (in green) associated with the photoreceptor support system. The assignment of genes to distinct groups is based on published literature and not on functional studies. Each gene group can be associated with a specific function/pathway that may influence one or more components (e.g., alterations in extracellular matrix, ECM) associated genes can affect both the retinal pigment epithelium (RPE) and Bruch's membrane. Some of the genetic variants (e.g., those in the complement pathway) are expected to have a stronger impact on specific clinical findings. The cellular changes (in blue) are shown next to (or near) components (in green) that are likely impacted. Some of the changes in one or more components can have a domino effect, leading to pathology phenotypes (red) in AMD. The appearance of late-stage disease (geographic atrophy, GA or choroidal neovascularization CNV) would depend on the acquisition of threshold levels that can be reached by the cumulative effect of multiple hits, including genetic susceptibility, aging-associated changes, and environmental factors.

prospective and retrospective cohort studies were summarized [53]. In data assessed from prospective cohort studies only a 66% increase in risk of CVD was observed in those with late AMD. Subgroup analyses revealed the association observed between late AMD and CVD risk was only present in populations with a mean age <75 years. Wu and colleagues [53] have suggested that this observation may result from increased prevalence of other CVD risk factors with aging and/or in the absolute risk of CVD. On the other hand, those with a greater underlying genetic risk for late AMD may present earlier in life, and may coincidentally have increased risk of CVD due to genetic or other factors. In addition, the potential for selective mortality bias cannot be excluded. These findings were consistent with previous reports that late AMD was significantly associated with cardiovascular mortality in those aged <75 years, but not in those who were older [54].

Cholesterol levels are a well-established biomarker for CVD [55], and its high abundance in drusen deposits have been noted [29]. Comparisons between cholesterol deposition within the arterial intima and Bruch's membrane have supported an association between cholesterol and AMD. Despite this, the findings have not always been consistent [56]. One possible explanation that has been proposed for the inconsistencies observed is that cholesterol found in both Bruch's membrane and drusen deposits may have been derived from lipoprotein of intraocular origin and, as a consequence, association with plasma lipoproteins might not be detected [57]. The genes that contribute to the HDL-cholesterol (HDL-C) pathway that have been associated with AMD have been reported to have discordant effects between the HDL-increasing alleles and their protective or risk effects for AMD [56]. For example, the HDL-raising allele of the *LIPC* gene was associated with reduced AMD risk, while the HDL-increasing alleles of *CETP* and *ABCA1* increased AMD risk [48,52]. In addition, the association reported between *LIPC* and AMD was independent of HDL, suggesting its effect on AMD pathogenesis may not directly relate to HDL concentration [58].

5. APOLIPOPROTEIN E

The human apolipoprotein E (*APOE*, MIM 107741) gene, located on chromosome 19q13.2, is central to the metabolism of low-density lipoprotein cholesterol (LDL-C) and triglycerides and has been shown to modulate risk in several common complex and age-related disorders [59,60,63]. These include CVD [61], atherosclerosis [62], AMD [60,63], Alzheimer's disease (AD) [64], and other dementias [65]. Studies have reported a rise in mortality associated with the $\varepsilon 4$ allele, and this has been attributed to a corresponding increase in risk associated with CHD, atherosclerosis, AD, and other dementias. The small ApoE lipid transport protein is multifunctional and binds primarily to the LDL receptor facilitating the maintenance and repair of neuronal cell membranes in the central and peripheral nervous system.

The protective effect exerted by the $\varepsilon 4$ isoform of ApoE was one of the first significant genetic associations reported for AMD [12,31] suggestive of a role for immunoregulation and cell signaling within AMD pathogenesis [32]. Variation at two SNPs within the coding sequence of the *APOE* gene, rs429358 and rs7412, are commonly referred to as $\varepsilon 2$, $\varepsilon 3$, and $\varepsilon 4$ and are differentiated on the basis of cysteine (Cys) and arginine (Arg) residue interchanges at positions 112 and 158 in the amino acid sequence. These amino acid substitutions have strong physiological consequences with regard to protein function. Various biological mechanisms have been proposed in relation to the functionality of ApoE, with the different isoforms generated reported to interact differently with the lipoprotein receptors leading to altered cholesterol levels.

These isoforms are believed to exert opposing effects in relation to the metabolism of CVD-related blood products such as LDL-C and triglycerides [62,66,67]. As such, ε2 has a much reduced binding affinity leading to lower total cholesterol levels compared to ε3 and ε4, which have a much greater binding affinity leading to subsequent elevated total cholesterol levels [68].

A delicate balance exists between cholesterol homeostasis and the development, maintenance, and repair of neuronal cells within the central nervous system (CNS), given the cholesterol requirement of neuronal cells [69]. Apolipoprotein E is central to serum cholesterol homeostasis through its ability to bind cholesterol and other lipids and facilitate their subsequent transportation into cells [70,71]. ApoE is the primary protein component of CNS lipoproteins and is produced by glial cells within the CNS [72]. The neural retina has the body's second highest level of APOE production after the liver, and it has been hypothesized that APOE is central to the maintenance of normal retinal function [73]. Given the protective effect observed in association with ε4 in AMD, several possible mechanisms have been suggested, including variable receptor binding affinities or changing isoform dimerization potential associated with lipid cholesterol transport [12]. In addition, the positively charged ε4 isoform has been suggested to improve permeability of the ocular Bruch's membrane, for lipid transport and reducing debris accumulation associated with drusen deposition [31,74]. Evidence observed through the examination of older eyes supports a reduced lipoprotein transportation mechanism across Bruch's membrane associated with the aging process, which leads to increased deposition of drusen with subsequent RPE insult [75]. Additional evidence has suggested that ε4 may function as a potential lipoprotein transporter of the macular pigments lutein (L) and zeaxanthin (Z) [76], with strong evidence indicating that a reduced dietary intake of these carotenoids is associated with an increased risk of AMD [77]. Consequently, variation in genes that modulate retinal cholesterol levels or attenuate lipoprotein transportation are likely to influence AMD risk [78].

Cross-sectional analyses have estimated the frequency of the ε4 allelic distribution to halve between the ages of 60 and 85 years [79]. This decrease in the ε4 frequency observed is likely to result as a consequence of the increased risk associated with the ε4 allele in relation to CHD, atherosclerosis, AD, and other dementias [59,60,63]. However, the exact mechanistic processes regarding its association with increased mortality and indeed the association of APOE with a variety of disease and pathological processes remain to be determined. More importantly, why the ε4 allele is associated with decreasing risk in AMD, yet increasing risk associated with many other common chronic conditions, remains to be resolved.

Issues detailing the role and implications of ApoE in AMD pathogenesis and indeed other common chronic conditions remain largely unresolved. Improved understanding of the mechanistic effects of ApoE help improve the accuracy of disease-risk prediction models, and may eventually offer therapeutic potential for the identification of individuals with increased risk for chronic disease. While the complexity surrounding the role of ApoE and cholesterol modulation within the retina remains unclear, the potential benefit from statin therapy for the treatment of CVD warrants further investigation in AMD, given the overlap in risk factors associated with both conditions [7]. While ε4 significantly increases risk associated with CVD, atherosclerosis, AD, and other dementias [59], it clearly plays a protective role in the context of AMD [60,63]. The mechanisms by which ApoE and cholesterol levels modulate AMD risk, especially with regard to the opposing effects observed in other complex diseases such as AD and CVD, remain poorly defined and in need of further investigation.

6. LIPC AND CETP GENES

Recent GWAS have identified novel variants within the *LIPC* and *CETP* genes to be significantly associated with AMD and variation in cholesterol levels, and concluded that some genetic variants may influence cholesterol levels both in the blood and the macula, but possibly not always in the same direction [48,52]. LIPC and CETP are expressed in the subretinal space and may contribute to rapid cholesterol transfer from the RPE to the neural retina [80,81]. Possible gene—environment interactions have been proposed for *LIPC*, which is a member of the HDL pathway, a process involved in the transportation of L and Z in the body. It is possible that changes in HDL metabolism may be influenced by *LIPC* genotype which influences the transportation of these carotenoids to the macula [56].

7. NUTRITIONAL INFLUENCES

Given that AMD is a common late-onset condition in older people, a potential therapeutic strategy proposed to ameliorate associated risk includes modification of nutrient intake. Several studies have suggested increased macular carotenoid intake of lutein (L) and zeaxanthin (Z) through foods rich in these nutrients (e.g., spinach and egg yolk) may reduce the risk of the development and/or progression of late AMD [77,82]. Macular pigment (MP) is known to accumulate in the macula region of the central retina and is composed of L, Z, and meso-zeaxanthin (meso-Z), which give the macula a characteristic yellow color. L and Z are both dietary derived (mostly fruit and vegetables) as they are not synthesized de novo in humans. However, meso-Z is believed to be predominantly non-dietary and is formed after conversion from L in the retina [83], although the precise mechanisms of this process are still unclear [84]. Gastrointestinal absorption of dietary L and Z occurs together with fats in the gut and with transportation to the liver where the formation of carotenoid—lipoprotein complexes occurs that enable their passage through the vascular system [85]. MP provide powerful antioxidant protection given their ability to filter actinic short wavelength, blue light, thereby minimizing the (photo-)oxidative damage caused to retinal cells in the macular region of the retina, regarded as one of the most metabolically active tissues in the body [86]. Given the protective properties exerted by MP, they have been hypothesized to limit the development and/or progression of AMD [87]. Although MP is of dietary origin, its transportation and concentration within the macular region of the retina is determined, in part, by the underlying the genetic architecture [88,89,93].

Genetic variation in the lipoprotein scavenger receptor class B type I gene *SCARB1* has been previously reported to be associated with AMD, further supporting a role for both cholesterol and MP metabolism in the pathophysiological disease process of AMD [90,93]. SCARB1 is a multiligand cell surface receptor that mediates selective cholesterol uptake and efflux [91,92]. Reverse cholesterol transport (RCT) is a major biological process essential for the removal of excess cholesterol from the body and the HDL-C pathway genes, *LIPC* and *CETP*, have been implicated in this process. Variants within these genes have also been reported previously to modify AMD-associated risk [48,52]. The identification of these genetic risk variants through AMD GWAS further reinforce the involvement of RCT mechanistic processes in AMD disease pathogenesis, particularly given the previously reported association of genetic variants within these genes with HDL-C levels in blood [94,95]. In addition, further studies have identified common variants in the *SCARB1* gene to be independently associated with the development of coronary heart disease [96] and variation of lipid profiles [97,98], with some evidence to suggest this may occur in a sex-specific manner [99—101]. Furthermore, the protein encoded by *SCARB1*, SR-BI, has been detected within the RPE in the eye [102,103], and within intestinal cells and has been implicated in the mediation of cholesterol efflux and xanthophyll uptake [104]. When changes at a genetic level in *SCARB1* lead to proteomic variability that attenuate carotenoid uptake into both the body and the eye, SR-BI has been further implicated in the transportation of carotenoids to the retina [103] and intestinal uptake [104]. Association between rs11057841 in *SCARB1* and serum L and Z levels were reported in three independent cohorts with carotenoid transportation [93].

Previous studies have also reported an association between rs10846744 and common carotid intimal−medial artery thickness in the Multi-Ethnic Study of Atherosclerosis. Common carotid intimal−medial artery thickness is a surrogate marker for sub-clinical atherosclerosis and subsequent increasing risk of CVD [101]. In populations of European descent, the rs10846744 genetic variant has been shown to share a high degree of linkage disequilibrium with the variant reported by McKay and colleagues (rs11957841; $r^2 = 0.93$) in association with elevated serum L concentration [93]. Manichaikul and colleagues have suggested that the genetic variation observed within *SCARB1* at this variant may lead to a regulatory cis- or trans-effect that may ultimately influence endothelial function or modulation of the inflammatory pathway. As such, given the common pathways shared by both AMD and CVD, these associations implicate *SCARB1* variants in both disease processes.

8. INFLAMMATION AND THE COMPLEMENT PATHWAY

Previous studies have provided evidence in support of an epidemiological association between AMD and CVD [7,9,19] with an indication that the complement pathway may contribute to both [105,106]. Hard evidence implicating the complement pathway in AMD has been derived from both immunohistochemistry and genetic studies. In contrast, evidence for a role of complement in CVD comes mainly from non-genetic observational studies. Circulating complement component C5a has been reported to be associated with incident cardiovascular events (including myocardial infarction, MI, stroke, and coronary and carotid artery disease) in individuals with lower limb atherosclerosis [107]. In addition, markers indicative of complement activation have been reported to increase in individuals with acute coronary syndrome [108], and C3 levels have been reported in association with traditional risk factors for MI in population-based studies [109]. Furthermore, proteomic analyses have provided supporting evidence that circulating complement components complex with HDL-C [110], which is itself a well-recognized CVD-risk factor and several complement proteins have been identified as present within atherosclerotic lesions in humans [111]. However, although the associations reported support complement involvement in CVD, they fail to provide a conclusive causal link.

Some experts suggest obesity, and excess abdominal fat in particular, is a primary underlying cause of insulin resistance. Basic understanding on the role for fat tissue has centered on its role as a repository to facilitate energy storage. However, more recently, studies have shown that abdominal fat facilitates the production of hormones and other substances that can lead to serious health problems including insulin resistance, hypertension, cholesterol imbalances, and CVD. Abdominal fat is involved in the development of chronic, long-lasting, inflammation within the body that often results in long-term damage, without any signs or symptoms. Research studies have identified complex interactions within fat tissue that attracts immune cells to the target area activating low-level chronic inflammation. This inflammation can contribute to the subsequent development of insulin resistance, type 2 diabetes, and longer-term CVD. Numerous studies have shown that weight loss can reduce insulin resistance and prevent or delay the onset of type 2 diabetes.

Chronic inflammation has been shown to underpin the foundations of the disease mechanisms central to AMD. In 2005, four independent studies identified an association between a specific genetic variant in CFH and AMD [15,33−35]. The SNP variant identified at codon position 402 (Y402H) was associated with a non-synonymous change in the amino acid sequence from tyrosine to histidine in the factor H protein and was located within an important functional region of the key regulator of the alternative complement cascade. Since 2005, many studies have replicated these findings in multiple populations, underpinning the importance of complement activation within the pathogenesis of AMD. Indeed, many geneticists refer to the seminal paper by Klein and colleagues in 2005 as an example of the efficacy of the GWAS approach in identifying an extremely common variant (as high as 40% in many populations) with a large effect size (producing an odds ratio in excess of 2.8) in a small study cohort (96 cases and 50 controls; Figure 1(a)). Few, if any, GWAS have identified common variants with such large effect size in other common disease traits. Further investigation of the CFH gene has identified additional multiple variants within the gene that exert effects in opposing directions with concomitant increase and decrease risk in disease risk accordingly. Identification of associated variants within CFH has focused interest in other members of the complement pathway and their role in inflammation leading to the identification of additional AMD associated variants located within other complement pathway regulators, such as complement component B (a complement activator) and which likely play a significant role in the etiology of the condition [37,39,40].

The complement system is an evolutionarily conserved defense mechanism designed to address the detrimental effects of pathogenic infections that have evolved over time. It consists of multiple circulatory proteins that become localized to target tissues. Activation of the complement system can occur through multiple pathways which always results in the formation of a membrane-spanning pore called the membrane attack complex (MAC), regardless of the initial pathway activated and whose function is to lyse the target pathogens. Critically, when regulation of the process is left unchecked, the processes designed to lyse microbes can similarly damage eukaryotic cells, and this is why almost half of the complement pathway components have some form of regulatory function [112].

Activation of the complement pathway results in a cascade of sequential events that leads to proteolytic cleavage of a series of proteins that are commonly referred to as C1 to C9. The complement pathway components are commonly found throughout the body and possess multiple functional roles. Their main role includes the initiation and promotion of the immune response to foreign particles or microorganisms within the body and clearance of apoptotic or necrotic debris. Investigation of complement pathway dysregulation in AMD has not been limited solely to genetic association studies. Complement components including C5 have been identified in ocular drusen and within the retina–choroid complex of eyes with AMD by immunohistochemical staining of AMD drusen deposits [113], together with complement regulators such as vitronectin [74] and potential activators of the complement cascade such as beta-amyloid [27]. Messenger ribonucleic acid derived from complement pathway components such as C5 have been found in RPE cells, offering support that C5 is produced locally within the retina and may act as a local cellular source for complement components present in drusen [114]. Animal models provide further support of the complement cascade as a major contributory factor in laser-induced choroidal neovascularization, possibly through the induction of increased levels of angiogenic growth factors [115].

The evidence implicating a dysfunctional complement system in AMD have provided considerable momentum in the development of novel therapeutic interventions that may attenuate complement activity in the eye, with many of these at an advanced stage in various clinical trials across the world. As the complement cascade is composed of a series of steps from initial activation right through to production of end terminal components, transition from one stage to the next, in theory may provide a potential therapeutic target for inhibition. A fine balance between permitting and dampening complement activity exists. Given the constitutive nature of the alternative pathway, strict regulation of the cascade is essential. The amino acid at codon 402 in CFH, the position of the SNP most strongly associated with AMD, is located in a heparin-binding region of CFH, and as such there is evidence to suggest that the resultant amino acid change may impair the binding efficiency or some other aspect of the molecule functionality and hence reduce its concomitant inhibitory abilities [116].

The Y402H variant in *CFH* was identified initially by GWAS in association with AMD as opposed to CVD, and although the association observed with AMD has been robustly replicated in multiple study populations, this has not been the case with CVD despite initial reports of a positive association with CVD, albeit with a smaller effect size than that identified for AMD in a population-based study [117]. A nested case–control study among two large prospective cohorts identified Y402H to be inversely associated with CVD among women, but not men [118]. Fewer studies have evaluated the effect of *CFH* SNPs on established CVD risk factors or markers of inflammation that have been linked to CVD, which could offer possible mechanistic effects for the reported association observed between CFH and CVD. A well-powered subsequent systematic review/meta-analysis and association study did not provide supporting evidence that SNPs in the CFH gene are involved in the pathogenesis of CVD, despite strong evidence of a role for this gene in AMD, and evidence of shared risk factors between both conditions [119,120].

As the consensus of data published indicates no association between the presence of AMD or associated genetic variants in *CFH* and a history of CVD, the possibility remains that both conditions have different underlying pathogenic mechanisms. As such, novel therapeutic strategies preventing the development of AMD may considerably differ from those developed for the treatment of cardiovascular disorders. If there is truly a negative correlation between both disorders, the possibility arises that the treatment of AMD through complement inhibition therapies could cause side effects that may worsen CVD outcomes and vice versa. This scenario has manifested to an extent in the treatment of nvAMD using vascular endothelial growth factor (VEGF) inhibition therapy that appears to increase CVD risk [121].

9. ANGIOGENESIS

Several studies have implicated a number of cellular growth factors including VEGF, in a variety of ocular neovascular processes, which has led to the development of a number of anti-angiogenic drugs for the treatment of AMD. Results from the ANCHOR trial of Lucentis™ (ranibizumab) has demonstrated preservation of vision in almost 95% of patients after year 1 and, more importantly, has shown an improvement in vision in a significant proportion of individuals with nvAMD [122]. Indeed, regular intravitreal injections of VEGF inhibition therapy have become a standard treatment of care for nvAMD for many. However, concern exists with regard to all antiangiogenic drugs in relation to their potential for systemic inhibition that may increase risk to cardiovascular or cerebrovascular complications or inhibit wound healing [123]. Moreover, response to therapy is not uniform and the reasons for this are poorly understood.

Previous studies have provided conflicting conclusions on the genetic influence of VEGF and nvAMD with initial studies suggesting that polymorphic variation within the *VEGF* gene and the promoter region in particular to be significantly associated with nvAMD implicating it as a primary genetic causal factor [38,124—126]. However, other studies have failed to replicate this finding [127—129]. Elucidation of any underlying genetic association of VEGF with nvAMD would help determine whether the therapeutic effect conferred by anti-angiogenic drugs address the genetic basis of the disease or treats the symptoms associated with the fundamental disease etiology observed.

Increased expression of VEGF in the RPE and outer nuclear layer in maculae taken post-mortem from individuals with AMD has been reported [130] and there is perhaps little doubt that VEGF has a significant role to play in the pathogenic process of AMD. Secretion of the VEGF-A isoform by the RPE helps to maintain the choriocapillaris and is enhanced by hypoxia. Changes in Bruch's membrane due to aging can affect the maintenance of the choriocapillaris leading to its atrophy and as such, decreased diffusion of oxygen toward the retina [131]. Outer retinal hypoxia may be a major influence on CNV formation by stimulating increased expression of VEGF by the RPE. Senescence of the RPE in combination with hypoxia may also reduce the expression of angiogenesis inhibitors, further shifting toward a pro-angiogenic state in the aging eye [132]. With the recent promise shown by anti-angiogenic therapies and their increased use, it is important to determine whether the effectiveness of these treatments is targeted at the symptoms of nvAMD or whether they might address the primary underlying cause. There is strong evidence to support the fact that angiogenesis is initiated by a variety of physiological triggers and in turn is regulated by a number of pro- and anti-angiogenic growth factors released in response to hypoxia, hypoglycaemia, mechanical stress, release of inflammatory proteins, and genetic variation [131,132].

More recently, SNPs in other genes involved in angiogenic processes have been identified by GWAS both the *TGFBR1* and *ADAMTS9* genes in association with AMD [49]. While the effect size reported was relatively small in comparison with other AMD genetic risk factors (an increase in risk of about 10%), both exceeded genome-wide significance. *ADAMTS9* has been implicated in coronary artery calcification [133] and *TGFBR1* has been identified as a novel prognostic biomarker after acute MI [134].

10. CONCLUSION

Multiple studies have considered the possible link between AMD and CHD generating strong evidence in support of common, broad underlying mechanisms between both conditions. Both diseases are strongly age related, with striking overlaps in several features including the presence of extracellular deposits and chronic inflammatory processes at the sites of pathogenesis, in particular, lipid deposition through drusen formation in AMD and the atherosclerotic process in CVD [135]. Inflammatory processes also impact on the pathogenesis of both conditions with implications for both the retinal and choroidal circulation and underwritten by the complement pathway and inflammation. CVD has long been postulated as a risk factor for the occurrence of AMD but results from epidemiological case—control studies and genetic analyses have not always been consistent. However, the data generated through systematic review suggest the relationship between AMD and CVD is probably more complex than shared risk factors or even single shared biological pathways, and more work is clearly needed.

Genes associated with genetic risk have been implicated in both disease processes, but the size and direction of effect, has often differed between AMD and CVD, suggesting that while common processes are involved, the mechanistic detail differs. Atherosclerosis may contribute to the development of AMD and the ensuing pathological processes, in part through its effect on the choroidal circulation and lipid deposition at Bruch's membrane. Subsequent lipid oxidation and macrophage binding lead to the activation of a downstream inflammatory cascade that results in the production of atherosclerotic lesions [136]. Collectively, the risk factors contributing to AMD represent a complex disorder through interactions among inflammation, atherosclerosis, and oxidative stress, pathogenic processes that are also related to CVD.

References

[1] Mensah GA. Public health: cardiovascular disease insights—something new out of Africa. Nat Rev Cardiol 2013;10:433—4.
[2] Lloyd-Jones, et al. Lifetime risk of developing coronary heart disease. Lancet 1999;353: 89—92.
[3] Budoff, Gul. Expert review on coronary calcium. Vasc Health Risk Manage 2008;4: 315—24.
[4] Businaro, et al. Cellular and molecular players in the atherosclerotic plaque progression. Ann NY Acad Sci 2012;1262:134—41.
[5] Bird AC, Bressler NM, Bressler SB, et al., International ARM Epidemiological Study Group. An international classification and grading system for age-related maculopathy and age-related macular degeneration. Surv Ophthalmol 1995;39: 367—74.
[6] Bunce C, Xing W, Wormald R. Causes of blind and partial sight certifications in England and Wales: April 2007—March 2008. Eye (Lond) 2010;24:1692—9.
[7] Snow KK, Seddon JM. Do age-related macular degeneration and cardiovascular disease share common antecedents? Ophthalmic Epidemiol 1999;6:125—43.
[8] Wong TY, Mitchell P. The eye in hypertension. Lancet 2007;369:425—35.
[9] Klein R, Deng Y, Klein BE, Hyman L, Seddon J, et al. Cardiovascular disease, its risk factors and treatment, and age-related macular degeneration: Women's Health Initiative Sight Exam ancillary study. Am J Ophthalmol 2007;143:473—83.

[10] Seddon JM, Gensler G, Milton RC, et al. Association between C-reactive protein and age-related macular degeneration. JAMA 2004;291:704—10.

[11] Cao JJ, Arnold AM, Manolio TA, et al. Association of carotid artery intima-media thickness, plaques, and C-reactive protein with future cardiovascular disease and all-cause mortality: the Cardiovascular Health Study. Circulation 2007;116: 32—8.

[12] Klaver CC, Kliffen M, van Duijn CM, et al. Genetic association of apolipoprotein E with age-related macular degeneration. Am J Hum Genet 1998;63:200—6.

[13] McCarron MO, Delong D, Alberts MJ. APOE genotype as a risk factor for ischemic cerebrovascular disease: a meta-analysis. Neurology 1999;53:1308—11.

[14] Song Y, Stampfer MJ, Liu S. Meta-analysis: apolipoprotein E genotypes and risk for coronary heart disease. Ann Intern Med 2004;141:137—47.

[15] Klein RJ, Zeiss C, Chew EY, et al. Complement factor H polymorphism in age-related macular degeneration. Science 2005;308:385—9.

[16] Klein R, Klein BE, Marino EK, et al. Early age-related maculopathy in the Cardiovascular Health Study. Ophthalmology 2003;110:25—33.

[17] Klein R, Klein BE, Jensen SC, et al. Age-related maculopathy in a multiracial United States population: the National Health and Nutrition Examination Survey III. Ophthalmology 1999;106:1056—65.

[18] Hyman L, Schachat AP, He Q, Leske MC, Age-Related Macular Degeneration Risk Factors Study Group. Hypertension, cardiovascular disease, and age-related macular degeneration. Arch Ophthalmol 2000;118:351—8.

[19] Klein R, Klein BE, Knudtson MD, et al. Subclinical atherosclerotic cardiovascular disease and early age-related macular degeneration in a multiracial cohort: the Multiethnic Study of Atherosclerosis. Arch Ophthalmol 2007;125:534—43.

[20] Smith W, Mitchell P, Leeder SR, Wang JJ. Plasma fibrinogen levels, other cardiovascular risk factors, and age-related maculopathy: the Blue Mountains Eye Study. Arch Ophthalmol 1998;116:583—7.

[21] Vingerling JR, Dielemans I, Bots ML, Hofman A, Grobbee DE, de Jong PT. Age-related macular degeneration is associated with atherosclerosis. The Rotterdam Study. Am J Epidemiol August 15, 1995;142(4):404—9.

[22] van Leeuwen R, Ikram MK, Vingerling JR, Witteman JC, Hofman A, de Jong PT. Blood pressure, atherosclerosis, and the incidence of age-related maculopathy: the Rotterdam Study. Invest Ophthalmol Vis Sci September 2003;44(9):3771—7.

[23] Baird PN, Chakrabarti S. How genetic studies have advanced our understanding of age-related macular degeneration and their impact on patient care: a review. Clin Exp Ophthalmol January—February 2014;42(1):53—64.

[24] Ding JD, Johnson LV, Herrmann R, Farsiu S, Smith SG, Groelle M, et al. Anti-amyloid therapy protects against retinal pigmented epithelium damage and vision loss in a model of age-related macular degeneration. Proc Natl Acad Sci USA July 12, 2011; 108(28):E279—87.

[25] Anderson DH, et al. Characterization of beta amyloid assemblies in drusen: the deposits associated with aging and age-related macular degeneration. Exp Eye Res 2004;78:243—56.

[26] Dentchev T, Milam AH, Lee VM, Trojanowski JQ, Dunaief JL. Amyloid-beta is found in drusen from some age-related macular degeneration retinas, but not in drusen from normal retinas. Mol Vis 2003;9:184—90.

[27] Johnson LV, et al. The Alzheimer's A beta-peptide is deposited at sites of complement activation in pathologic deposits associated with aging and age-related macular degeneration. Proc Natl Acad Sci USA 2002;99:11830—5.

[28] Mullins RF, Russell SR, Anderson DH, Hageman GS. Drusen associated with aging and age-related macular degeneration contain proteins common to extracellular

deposits associated with atherosclerosis, elastosis, amyloidosis, and dense deposit disease. FASEB J 2000;14:835—46.

[29] Curcio CA, Johnson M, Rudolf M, Huang JD. The oil spill in ageing Bruch membrane. Br J Ophthalmol 2011;95:1638—45.

[30] Tabas I, Williams KJ, Borén J. Subendothelial lipoprotein retention as the initiating process in atherosclerosis: update and therapeutic implications. Circulation 2007;116: 1832—44.

[31] Souied EH, Benlian P, Amouyel P, Feingold J, Lagarde JP, Munnich A, et al. The epsilon4 allele of the apolipoprotein E gene as a potential protective factor for exudative age-related macular degeneration. Am J Ophthalmol March 1998;125(3):353—9.

[32] Zarbin MA. Current concepts in the pathogenesis of age-related macular degeneration. Arch Ophthalmol 2004;122:598—614.

[33] Haines JL, Hauser MA, Schmidt S, Scott WK, Olson LM, Gallins P, et al. Complement factor H variant increases the risk of agerelated macular degeneration. Science 2005; 308:419—21.

[34] Edwards AO, Ritter III R, Abel KJ, Manning A, Panhuysen C, Farrer LA. Complement factor H polymorphism and age related macular degeneration. Science 2005;308: 421—4.

[35] Hageman GS, Anderson DH, Johnson LV, Hancox LS, Taiber AJ, Hardisty LI, et al. A common haplotype in the complement regulatory gene factor H (HF1/CFH) predisposes individuals to age-related macular degeneration. Proc Natl Acad Sci USA 2005; 102:7227—32.

[36] Hughes AE, Orr N, Esfandiary H, az-Torres M, Goodship T, Chakravarthy U. A common CFH haplotype, with deletion of CFHR1 and CFHR3, is associated with lower risk of agerelated macular degeneration. Nat Genet 2006;38:1173—7.

[37] Gold B, Merriam JE, Zernant J, Hancox LS, Taiber AJ, Gehrs K, et al., AMD Genetics Clinical Study Group, Hageman GS, Dean M, Allikmets R. Variation in factor B (BF) and complement component 2 (C2) genes is associated with age-related macular degeneration. Nat Genet 2006;38:458—62.

[38] McKay GJ, Silvestri G, Patterson CC, Hogg RE, Chakravarthy U, Hughes AE. Further assessment of the complement component 2 and factor B region associated with age-related macular degeneration. Invest Ophthalmol Vis Sci 2009;50:533—9.

[39] Yates JR, Sepp T, Matharu BK, Khan JC, Thurlby DA, Shahid H, et al., Genetic Factors in AMD Study Group. Complement C3 variant and the risk of age-related macular degeneration. N Engl J Med 2007;357:553—61.

[40] Maller JB, Fagerness JA, Reynolds RC, Neale BM, Daly MJ, Seddon JM. Variation in complement factor 3 is associated with risk of age-related macular degeneration. Nat Genet 2007;39:1200—1.

[41] Fagerness JA, Maller JB, Neale BM, Reynolds RC, Daly MJ, Seddon JM. Variation near complement factor I is associated with risk of advanced AMD. Eur J Hum Genet January 2009;17(1):100—4.

[42] Jakobsdottir J, Conley YP, Weeks DE, Mah TS, Ferrell RE, Gorin MB. Susceptibility genes for age-related maculopathy on chromosome 10q26. Am J Hum Genet 2005; 77:389—407.

[43] Rivera A, Fisher SA, Fritsche LG, Keilhauer CN, Lichtner P, Meitinger T, et al. Hypothetical LOC387715 is a second major susceptibility gene for age-related macular degeneration, contributing independently of complement factor H to disease risk. Hum Mol Genet 2005;14:3227—36.

[44] Dewan A, Liu M, Hartman S, Zhang SS, Liu DT, Zhao C, et al. HTRA1 promoter polymorphism in wet age-related macular degeneration. Science 2006;314:989—92.

[45] Yang Z, Camp NJ, Sun H, Tong Z, Gibbs D, Cameron DJ, et al. A variant of the HTRA1 gene increases susceptibility to age-related macular degeneration. Science 2006;314:992—3.

[46] Kanda A, Chen W, Othman M, Branham KE, Brooks M, Khanna R, et al. A variant of mitochondrial protein LOC387715/ARMS2, not HTRA1, is strongly associated with age-related macular degeneration. Proc Natl Acad Sci USA 2007;104:16227−32.

[47] Fritsche LG, Loenhardt T, Janssen A, Fisher SA, Rivera A, Keilhauer CN, et al. Age-related macular degeneration is associated with an unstable ARMS2 (LOC387715) mRNA. Nat Genet 2008;40:892−6.

[48] Chen W, Stambolian D, Edwards AO, et al. Genetic variants near *TIMP3* and high-density lipoprotein-associated loci influence susceptibility to age-related macular degeneration. Proc Natl Acad Sci USA 2010;107:7401−6.

[49] Fritsche LG, Chen W, Schu M, et al. Seven new loci associated with age-related macular degeneration. Nat Genet April 2013;45(4):433−9.

[50] Canter JA, Olson LM, Spencer K, Schnetz-Boutaud N, Anderson B, Hauser MA, et al. Mitochondrial DNA polymorphism A4917G is independently associated with age-related macular degeneration. PLoS One 2008;3:e2091.

[51] SanGiovanni JP, Arking DE, Iyengar SK, Elashoff M, Clemons TE, Reed GF, et al. Mitochondrial DNA variants of respiratory complex I that uniquely characterize haplogroup T2 are associated with increased risk of age-related macular degeneration. PLoS One 2009;4:e5508.

[52] Neale BM, Fagerness J, Reynolds R, et al. Genome-wide association study of advanced age-related macular degeneration identifies a role of the hepatic lipase gene (LIPC). Proc Natl Acad Sci USA 2010;107:7395−400.

[53] Wu J, Uchino M, Sastry SM, Schaumberg DA. Age-related macular degeneration and the incidence of cardiovascular disease: a systematic review and meta-analysis. PLoS One March 28, 2014;9(3):e89600.

[54] Tan JS, Wang JJ, Liew G, Rochtchina E, Mitchell P. Age-related macular degeneration and mortality from cardiovascular disease or stroke. Br J Ophthalmol April 2008;92(4): 509−12.

[55] Gordon T, Kannel WB, Castelli WP, Dawber TR. Lipoproteins, cardiovascular disease, and death. The Framingham study. Arch Intern Med 1981;141:1128−31.

[56] Sobrin L, Seddon JM. Nature and nurture- genes and environment- predict onset and progression of macular degeneration. Prog Retin Eye Res May 2014;40C:1−15.

[57] Curcio CA, Presley JB, Malek G, Medeiros NE, Avery DV, Kruth HS. Esterified and unesterified cholesterol in drusen and basal deposits of eyes with age-related maculopathy. Exp Eye Res 2005;81:731e741.

[58] Reynolds R, Rosner B, Seddon JM. Serum lipid biomarkers and hepatic lipase gene associations with age-related macular degeneration. Ophthalmology October 2010; 117(10):1989−95.

[59] Ang LS, Cruz RP, Hendel A, et al. Apolipoprotein E, an important player in longevity and age-related diseases. Exp Gerontol 2008;43(7):615−22.

[60] McKay GJ, Silvestri G, Chakravarthy U, et al. Variations in apolipoprotein E frequency with age in a pooled analysis of a large group of older people. Am J Epidemiol June 15, 2011;173(12):1357−64.

[61] Ward H, Mitrou PN, Bowman R, et al. APOE genotype, lipids, and coronary heart disease risk: a prospective population study. Arch Intern Med 2009;169(15):1424−9.

[62] Davignon J, Gregg RE, Sing CF. Apolipoprotein E polymorphism and atherosclerosis. Arteriosclerosis 1988;8(1):1−21.

[63] McKay GJ, Patterson CC, Chakravarthy U, et al. Evidence of association of APOE with age-related macular degeneration: a pooled analysis of 15 studies. Hum Mutat December 2011;32(12):1407−16.

[64] Saunders AM, Schmader K, Breitner JC, et al. Apolipoprotein E epsilon 4 allele distributions in late-onset Alzheimer's disease and in other amyloid-forming diseases. Lancet 1993;342(8873):710−1.

[65] Azad NA, Al Bugami M, Loy-English I. Gender differences in dementia risk factors. Gend Med 2007;4(2):120−9.
[66] Schaefer EJ, Lamon-Fava S, Johnson S, et al. Effects of gender and menopausal status on the association of apolipoprotein E phenotype with plasma lipoprotein levels. Results from the Framingham Offspring Study. Arterioscler Thromb 1994;14(7):1105−13.
[67] Sing CF, Davignon J. Role of the apolipoprotein E polymorphism in determining normal plasma lipid and lipoprotein variation. Am J Hum Genet 1985;37(2):268−85.
[68] Siest G, Pillot T, Régis-Bailly A, Leininger-Muller B, Steinmetz J, Galteau MM, et al. Apolipoprotein E: an important gene and protein to follow in laboratory medicine. Clin Chem 1995;41:1068−86.
[69] Ong JM, Zorapapel NC, Rich KA, Wagstaff RE, Lambert RW, Rosenberg SE, et al. Effects of cholesterol and apolipoprotein E on retinal abnormalities in ApoE-deficient mice. Invest Ophthalmol Vis Sci July 2001;42(8):1891−900.
[70] Weisgraber KH, Mahley RW. Human apolipoprotein E: the Alzheimer's disease connection. FASEB J 1996;10:1485−94.
[71] Mahley RW. Apolipoprotein E: cholesterol transport protein with expanding role in cell biology (review). Science 1988;240:622−30.
[72] Elshourbagy NA, Liao WS, Mahley RW, Taylor JM. Apolipoprotein E mRNA is abundant in the brain and adrenals, as well as in the liver, and is present in other peripheral tissues of rats and marmosets. Proc Natl Acad Sci USA 1985;82:203−7.
[73] Anderson DH, Ozaki S, Nealon M, Neitz J, Mullins RF, Hageman GS, et al. Local cellular sources of apolipoprotein E in the human retina and retinal pigmented epithelium: implications for the process of drusen formation. Am J Ophthalmol 2001;131:767−81.
[74] Crabb JW, Miyagi M, Gu X, Shadrach K, West KA, Sakaguchi H, et al. Drusen proteome analysis: an approach to the etiology of age-related macular degeneration. Proc Natl Acad Sci USA 2002;99:14682−7.
[75] Curcio CA, Johnson M, Huang JD, Rudolf M. Aging, age-related macular degeneration, and the response-to-retention of apolipoprotein B-containing lipoproteins. Prog Retin Eye Res 2009;28:393−422.
[76] Loane E, McKay GJ, Nolan JM, Beatty S. Apolipoprotein E genotype is associated with macular pigment optical density. Invest Ophthalmol Vis Sci 2010;51:2636−43.
[77] Seddon JM, Ajani UA, Sperduto RD, et al. Dietary carotenoids, vitamins A, C, and E, and advanced age-related macular degeneration. Eye Disease Case-Control Study Group. JAMA 1994;272:1413−20. Erratum in: JAMA 1995;273:622.
[78] Connor WE, Duell PB, Kean R, Wang Y. The primerole of HDL to transport lutein into the retina: evidence from HDL-deficient WHAM chicks having a mutant ABCA1 transporter. Invest Ophthalmol Vis Sci 2007;48:4226−31.
[79] Payami H, Zhu M, Montimurro J, et al. One step closer to fixing association studies: evidence for age- and genderspecific allele frequency variations and deviations from Hardy-Weinberg expectations in controls. Hum Genet 2005;118(3−4):322−30.
[80] Tserentsoodol N, Gordiyenko NV, Pascual I, Lee JW, Fliesler SJ, Rodriguez IR. Intraretinal lipid transport is dependent on high density lipoprotein-like particles and class B scavenger receptors. Mol Vis 2006;12:1319−33.
[81] Tserentsoodol N, Sztein J, Campos M, Gordiyenko NV, Fariss RN, et al. Uptake of cholesterol by the retina occurs primarily via a low density lipoprotein receptor-mediated process. Mol Vis 2006;12:1306−18.
[82] Loane E, Kelliher C, Beatty S, Nolan JM. The rationale and evidence base for a protective role of macular pigment in age-related maculopathy. Br J Ophthalmol 2008;92:1163−8.
[83] Bone RA, Landrum JT, Hime GW, et al. Stereochemistry of the human macular carotenoids. Invest Ophthalmol Vis Sci 1993;34:2033−40.

[84] Thurnham DI, Trémel A, Howard AN. A supplementation study in human subjects with a combination of meso-zeaxanthin, (3R,3'R)-zeaxanthin and (3R,3'R,6'R)-lutein. Br J Nutr 2008;100:1307–14.
[85] Parker RS. Absorption, metabolism, and transport of carotenoids. FASEB J 1996;10: 542–51.
[86] Snodderly DM. Evidence for protection against age-related macular degeneration by carotenoids and antioxidant vitamins. Am J Clin Nutr 1995;62:S1448–61.
[87] Beatty S, Murray IJ, Henson DB, et al. Macular pigment and risk for age-related macular degeneration in subjects from a Northern European population. Invest Ophthalmol Vis Sci 2001;42:439–46.
[88] Liew SHM, Gilbert C, Spector TD, et al. Heritability of macular pigment: a twin study. Invest Ophthalmol Vis Sci 2005;46:4430–6.
[89] Hammond CJ, Liew SM, Van Kuijk FJ, et al. The heritability of macular response to supplemental lutein and zeaxanthin: a classical twin study. Invest Ophthalmol Vis Sci 2012;53:4963–8.
[90] Zerbib J, Seddon JM, Richard F, et al. rs5888 variant of SCARB1 gene is a possible susceptibility factor for age-related macular degeneration. PLoS One 2009;4:e7341.
[91] Ji Y, Jian B, Wang N, et al. Scavenger receptor BI promotes high density lipoprotein-mediated cellular cholesterol efflux. J Biol Chem 1997;272:20982–5.
[92] Acton S, Rigotti A, Landschulz KT, et al. Identification of scavenger receptor SR-BI as a high density lipoprotein receptor. Science 1996;271:518–20.
[93] McKay GJ, Loane E, Nolan JM, Patterson CC, Meyers KJ, Mares JA, et al. Investigation of genetic variation in scavenger receptor class B, member 1 (SCARB1) and association with serum carotenoids. Ophthalmology August, 2013;120(8):1632–40.
[94] Willer CJ, Sanna S, Jackson AU, et al. Newly identified loci that influence lipid concentrations and risk of coronary artery disease. Nat Genet 2008;40:161–9.
[95] Kathiresan S, Willer CJ, Peloso GM, et al. Common variants at 30 loci contribute to polygenic dyslipidemia. Nat Genet 2009;41:56–65.
[96] Rodríguez-Esparragón F, Rodríguez-Pérez JC, Hernández-Trujillo Y, et al. Allelic variants of the human scavenger receptor class B type 1 and paraoxonase 1 on coronary heart disease: genotype-phenotype correlations. Arterioscler Thromb Vasc Biol 2005; 25:854–60.
[97] Acton S, Osgood D, Donoghue M, et al. Association of polymorphisms at the SR-BI gene locus with plasma lipid levels and body mass index in a white population. Arterioscler Thromb Vasc Biol 1999;19:1734–43.
[98] Morabia A, Ross BM, Costanza MC, et al. Population-based study of SR-BI genetic variation and lipid profile. Atherosclerosis 2004;175:159–68.
[99] McCarthy JJ, Somji A, Weiss LA, et al. Polymorphisms of the scavenger receptor class B member 1 are associated with insulin resistance with evidence of gene by sex interaction. J Clin Endocrinol Metab 2009;94:1789–96.
[100] Chiba-Falek O, Nichols M, Suchindran S, et al. Impact of gene variants on sex-specific regulation of human Scavenger receptor class B type 1 (SR-BI) expression in liver and association with lipid levels in a population-based study. BMC Med Genet 2010;11:9.
[101] Manichaikul A, Naj AC, Herrington D, et al. Association of SCARB1 variants with subclinical atherosclerosis and incident cardiovascular disease: the multi-ethnic study of atherosclerosis. Arterioscler Thromb Vasc Biol 2012;32:1991–9.
[102] Duncan KG, Bailey KR, Kane JP, Schwartz DM. Human retinal pigment epithelial cells express scavenger receptors BI and BII. Biochem Biophys Res Commun 2002;292: 1017–22.
[103] During A, Doraiswamy S, Harrison EH. Xanthophylls are preferentially taken up compared with beta-carotene by retinal cells via a SRBI-dependent mechanism. J Lipid Res 2008;49:1715–24.

[104] Reboul E, Abou L, Mikail C, et al. Lutein transport by Caco-2 TC-7 cells occurs partly by a facilitated process involving the scavenger receptor class B type I (SR-BI). Biochem J 2005;387:455—61.

[105] Donoso LA, Kim D, Frost A, Callahan A, Hageman G. The role of inflammation in the pathogenesis of age-related macular degeneration. Surv Ophthalmol 2006;51:137—52.

[106] Haskard DO, Boyle JJ, Mason JC. The role of complement in atherosclerosis. Curr Opin Lipidol 2008;19:478—82.

[107] Speidl WS, Exner M, Amighi J, et al. Complement component C5a predicts future cardiovascular events in patients with advanced atherosclerosis. Eur Heart J 2005;26: 2294—9.

[108] Iltumur K, Karabulut A, Toprak G, Toprak N. Complement activation in acute coronary syndromes. APMIS 2005;113:167—74.

[109] Muscari A, Massarelli G, Bastagli L, et al. Relationship between serum C3 levels and traditional risk factors for myocardial infarction. Acta Cardiol 1998;53:345—54.

[110] Vaisar T, Pennathur S, Green PS, et al. Shotgun proteomics implicates protease inhibition and complement activation in the antiinflammatory properties of HDL. J Clin Invest 2007;117:746—56.

[111] Lagrand WK, Niessen HW, Wolbink GJ, et al. C-reactive protein colocalizes with complement in human hearts during acute myocardial infarction. Circulation 1997;95: 97—103.

[112] Neher MD, Weckbach S, Flierl MA, Huber-Lang MS, Stahel PF. Molecular mechanisms of inflammation and tissue injury after major trauma—is complement the "bad guy"? J Biomed Sci 2011;18:90.

[113] Hageman GS, Luthert PJ, Chong NHV, Johnson LV, Anderson DH, Mullins RF. An integrated hypothesis that considers drusen as biomarkers of immune-mediated processes at the RPE-Bruch's membrane interface in aging and age-related macular degeneration. Prog Retin Eye Res 2001;20(6):705—32.

[114] Johnson LV, Ozaki S, Staples MK, Erickson PA, Anderson DH. A potential role for immune complex pathogenesis in drusen formation. Exp Eye Res 2000;70(4):441—9.

[115] Bora PS, Sohn JH, Cruz JM, Jha P, Nishihori H, Wang Y, et al. Role of complement and complement membrane attack complex in laser-induced choroidal neovascularization. J Immunol 2005;174(1):491—7.

[116] Rodriguez de Cordoba S, Esparza-Gordillo J, Goicoechea de Jorge E, Lopez-Trascasa M, Sanchez-Corral P. The human complement factor H: functional roles, genetic variations and disease associations. Mol Immunol 2004;41(4):355—67.

[117] Kardys I, Klaver CC, Despriet DD, et al. A common polymorphism in the complement factor H gene is associated with increased risk of myocardial infarction: the Rotterdam study. J Am Coll Cardiol 2006;47:1568—75.

[118] Pai JK, Manson JE, Rexrode KM, Albert CM, Hunter DJ, Rimm EB. Complement factor H (Y402H) polymorphism and risk of coronary heart disease in US men and women. Eur Heart J 2007;28:1297e1303.

[119] Sofat R, Casas JP, Kumari M, Talmud PJ, Ireland H, Kivimaki M, et al. Genetic variation in complement factor H and risk of coronary heart disease: eight new studies and a meta-analysis of around 48,000 individuals. Atherosclerosis November 2010;213(1):184—90.

[120] Keilhauer CN, Fritsche LG, Guthoff R, Haubitz I, Weber BH. Age-related macular degeneration and coronary heart disease: evaluation of genetic and environmental associations. Eur J Med Genet February 2013;56(2):72—9.

[121] Curtis LH, Hammill BG, Schulman KA, Cousins SW. Risks of mortality, myocardial infarction, bleeding, and stroke associated with therapies for age related macular degeneration. Arch Ophthalmol 2010;128:1273e1279.

[122] Rosenfeld PJ, Rich RM, Lalwani GA. Ranibizumab: Phase III clinical trial results. Ophthalmol Clin North Am 2006;19:361—72.

[123] Gehrs KM, Anderson DH, Johnson LV, Hageman GS. Age-related macular degeneration-emerging pathogenetic and therapeutic concepts. Ann Med 2006;38: 450−71.

[124] Haines JL, Schnetz-Boutaud N, Schmidt S, et al. Functional candidate genes in age-related macular degeneration: significant association with VEGF, VLDLR, and LRP6. Invest Ophthalmol Vis Sci 2006;47:329−35.

[125] Churchill AJ, Carter JG, Lovell HC, et al. VEGF polymorphisms are associated with neovascular age-related macular degeneration. Hum Mol Genet 2006;15:2955−61.

[126] McKay GJ, Silvestri G, Orr N, Chakravarthy U, Hughes AE. VEGF and age-related macular degeneration. Ophthalmology June 2009;116(6):1227.e1−3.

[127] Richardson AJ, Islam FM, Guymer RH, Cain M, Baird PN. A tag-single nucleotide polymorphisms approach to the vascular endothelial growth factor-A gene in age-related macular degeneration. Mol Vis 2007;13:2148−52.

[128] Smith SR, Tong ZZ, Constantine R, et al. Invest Ophthalmol Vis Sci 2007;48. ARVO e-abstract 2097.

[129] de Jong PT, Boekhoorn SS, Vingerling JR, Uitterlinden AG, Hofman A. Invest Ophthalmol Vis Sci 2007;48. e-abstract 2119.

[130] Kliffen M, Sharma HS, Mooy CM, et al. Increased expression of angiogenic growth factors in age-related maculopathy. Br J Ophthalmol 1997;81:154−62.

[131] Schlingemann RO. Role of growth factors and the wound healing response in age-related macular degeneration. Graefes Arch Clin Exp Ophthalmol 2004;242:91−101.

[132] Zondor SD, Medina PJ. Bevacizumab: an angiogenesis inhibitor with efficacy in colorectal and other malignancies. Ann Pharmacother 2004;38:1258−64.

[133] Polfus LM, Smith JA, Shimmin LC, Bielak LF, Morrison AC, Kardia SL, et al. Genome-wide association study of gene by smoking interactions in coronary artery calcification. PLoS One October 3, 2013;8(10):e74642.

[134] Devaux Y, Bousquenaud M, Rodius S, Marie PY, Maskali F, Zhang L, et al. Transforming growth factor β receptor 1 is a new candidate prognostic biomarker after acute myocardial infarction. BMC Med Genomics December 5, 2011;4:83.

[135] Friedman E. The role of the atherosclerotic process in the pathogenesis of age-related macular degeneration. Am J Ophthalmol 2000;130:658−63.

[136] Shaw PX, Zhang L, Zhang M, Du H, Zhao L, et al. Complement factor H genotypes impact risk of age-related macular degeneration by interaction with oxidized phospholipids. Proc Natl Acad Sci USA 2012;109:13757−62.

[137] Klein R, Klein BE. The prevalence of age-related eye diseases and visual impairment in aging: current estimates. Invest Ophthalmol Vis Sci. 2013;54(14):ORSF5−13. http://dx.doi.org/10.1167/iovs.13-12789. Review. PubMed PMID: 24335069; PubMed Central PMCID: PMC4139275.

[138] Fritsche LG, Fariss RN, Stambolian D, Abecasis GR, Curcio CA, Swaroop A. Age-related macular degeneration: genetics and biology coming together. Annu Rev Genomics Hum Genet. 2014;15:151−71. http://dx.doi.org/10.1146/annurev-genom-090413-025610. Epub 2014 Apr 16. Review. PubMed PMID: 24773320; PubMed Central PMCID: PMC4217162.

13

Translating Genomic Research to the Marketplace

Steven Myint

Center for Enterprise and Development, Duke-NUS Medical School,
Singapore; Nanyang Business School, Nanyang Technological University,
Singapore; Inex Private Ltd, Singapore; Plexpress Oy, Finland

1. INTRODUCTION

Whether explicit or not, the main aim of most medical research is to contribute to societal benefit. The primary customer is the patient and this benefit can be directly beneficial through publicly funded research or indirectly through the intermediary of the commercial marketplace. In this chapter, we outline the principles of commercializing genomics research.

The commercial potential of the human genome has always been expected to be realized and indeed the initial race was between public and private sector organizations. The cost of the 1988–2003 human genome project was not trivial, so the economic and social return needed to be substantial. Indeed, it has been calculated that there has been $141 returned for every dollar of the $5.4 billion spent on this project [1]. This will only grow as more diagnostic tests and new therapeutics are advanced.

2. THE CARDIOVASCULAR GENOMICS MARKETPLACE

The cardiovascular genomics marketplace consists of both therapeutics and diagnostics (and theranostics). As the therapeutics arena which has been derived from the human genome is in its infancy (the first gene therapy, Glybera, was approved only in 2013) and none exists for CVD, this chapter focuses on the diagnostics space, particularly the development of biomarkers.

The overall global market for diagnostics was valued at $45.6 billion in 2012 and is expected to grow at about 7% annually over the following five years to reach a market size of $64.6 billion in 2017. The United States and Europe account for about 60% of that market, with Asia—Pacific forecast as the highest growth region. The US diagnostics market was about $15.5 billion in 2012, with a forecasted growth rate of about 6% over the next following years.

The molecular diagnostics market is a relatively small portion of the entire diagnostics market (10—11%) but in the fastest growing segment. In 2012, it was worth $5 billion of which 11% was based on specific gene diagnostics. It is expected to grow at 12% annually so that by 2017 it will reach $8 billion.

The cardiac biomarker market is increasing rapidly as cardiac disease continues to be an ever-growing problem worldwide as obesity and aging take their toll. A cardiac biomarker is defined as a measurable entity that is a useful indicator in the diagnosis and risk stratification of cardiovascular diseases (CVDs) such as acute coronary syndrome, coronary syndromes, myocardial infarction (MI), and heart failure. Some of the established cardiac biomarkers include creatine kinase MB, troponins (cTnI and cTnT), basic natreutic peptide (BNP), NT-proBNP, GFAP, H-FABP, LDH isoenzymes, D-dimer, IMA, myeloperoxidase, myoglobin, and sCD40L. In addition, biomarkers useful as companion diagnostics for therapy are growing at an even faster pace. Currently Roche, Ortho, Abbot, and Alere are the dominant players. It is currently valued at over $4 billion and is expected to grow to over $7 billion by 2018. Immunoassays currently dominate this space, but gene-based diagnostics are an increasingly large fraction of this market.

The vast majority of genomic-based tests actually used in clinical practice in CVD are **laboratory-derived tests** (LDTs), the so-called home brews that have been developed by scientists and physicians. Regulated **In-vitro diagnostics** (IVD) are a minority but a common diagnostics (Figure 1). Reimbursement is a particular issue for cardiovascular genomic testing. For example, warfarin is used to reduce risk of death, stroke, or further heart attack after an initial MI. There is much individual variation in patient response to warfarin dosing, however, and two gene variations (VKORC1 and CYP2C9) have been shown to be responsible for 30—50% of individual variation. The Centers for Medicare and Medicaid Services (CMS) responsible for reimbursement in the United States declined to reimburse the cost because the "available evidence does not demonstrate that pharmacogenetic testing to predict warfarin responsiveness improves health outcome in Medicare beneficiaries." This is not an attractive situation for commercialization. Another common area where both LDTs and IVDs exist is the application of pharmacogenetic

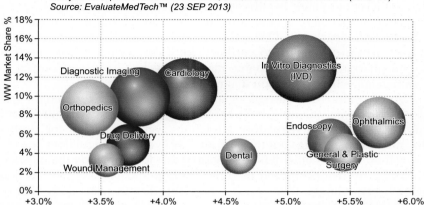

FIGURE 1   Market share of IVD.

biomarkers in statin use. Tests based on variants of the solute carrier organic anion transporter gene, SLCO1B1, which confers risk of myopathy with certain statins, have been developed [2]. This same gene variation seems to affect adherence to statin therapy and could be useful. Current evidence, based on cost-effectiveness does not, however, support universal testing but companies such as 23andMe and deCODEME have still seen commercial potential with direct to consumer approaches.

Biomarkers as objective and measurable indicators of biological processes are an important consequence of the human genome project. Biomarker identification is expanding at an unprecedented rate as they offer the promise of changing the current health care model to one of "personalized" or "precision" medicine. They can be used for disease diagnostics or as "companion" diagnostics to predict success or failure, continuing effectiveness, or adverse reaction to a specific drug or combination of drugs. **Companion diagnostics**, where the biomarker is used to predict or monitor effectiveness or safety of a drug, is the most commercially attractive proposition for industry currently, but requires a pharmaceutical partner.

Commercialization of gene-based tests depends heavily on health systems adopting them in most major economies. There are many barriers to adoption. The first is that most health care systems tend to be slow adoption of new technologies and often require champions in clinical practice who convince others of their value. Physicians themselves are often slow adopters and training in genomics medicine is not universal. Second, there are often social, ethical, and political issues over the interpretation of genomic tests. Perhaps the most important, however, is that there is not a single health care system in the world that is not tightly budgeted. Thus, there is often a requirement for a lot of data to support introduction of such tests. One of the advantages of companion diagnostics is for start-up companies to partner with a diagnostics developer, that is, the pharmaceutical company, that usually has greater resources to commercialize the companion diagnostic.

Future commercial success of a new biomarker or companion test also depends on adapting both current regulatory and reimbursement approaches. Current regulatory guidelines were developed for simple diagnostic test or drugs but they are already evolving. Regulators need to know more than whether it is effective and safe for biomarker diagnostics. They require demonstration of clinical utility and validity (see section below). Increasingly, they are also requiring health technology assessments (HTAs) to show cost-effectiveness. In the UK, the National Institute of Health and Clinical Excellence makes such assessments and makes them available to health care providers. One recent proposal from the US Food and Drug Administration (FDA) that offers the opportunity to streamline processes is the requirement to get approval for LDTs in the same way as IVDs are regulated. Detractors suggest it will stifle innovation but from a commercial viewpoint it should lead to better products and expand the market. There are thousands of biomarkers which have been developed as LDTs where the clinical benefit and validity are left to the discretion of individual physicians and laboratories. In Europe, the European Medicines Agency (EMA) published guidelines on genomic biomarkers in 2009. International harmonization of the data for genomic biomarker regulatory submission have also been considered under an International Conference on Harmonisation (ICH) project called "Genomic Biomarkers Related to Drug Response: Context, Structure and format of Qualification Submissions."

Even if there is unmet clinical and market need, a new test will not make it if it is beyond the affordability of health care systems. Reimbursement of new technologies is increasingly dependent on HTA. This is being adopted in many, but not all countries, and requires evidence to support both clinical utility and health care value. In the United States, this is helped by reimbursement with codes 83,900 (amplification of patient nucleic acid) and 83,914 (mutation identification by enzymatic ligation or primer extension) being the most relevant. Reimbursement of laboratory tests in the United States has, however, been based on a cost recovery basis rather than value-based assessment. New test methodologies will no doubt lead to more sophisticated HTA. There have been several initiatives such as that of the Secretary's Advisory Committee on Genetics, Health, and Society (SACGHS) and the Personalized Medicine coalition (2010) that have attempted to address these issues and are likely to lead to the required changes.

3. INITIAL CONSIDERATIONS IN COMMERCIALIZING RESEARCH

The first question the researchers need to identify is whether their research has any market potential. The drive from many academic institutions is to encourage researchers to publish and then perhaps commercialize the intellectual property (IP). Traditionally science has been about having an inquisitive mind and developing ideas for the public good. Much money has been wasted by technology transfer offices around the world that really are not commercially viable and it would have been better to make the technology publicly available so that it is useful to a wider group more quickly. The costs of technology transfer are not inconsiderable, and the costs of product development and then sales and marketing are even greater. There are many considerations for deciding whether there will be a good investment on time and money to take a research to the market.

First, and the most important consideration is whether there is an unmet need. If the technology is no better than what is already available then its commercial value would be rare. Most **technology transfer organizations** (TTOs) will use market reports to identify what is commercially available. This leaves three very large gaps in the analysis: those technologies that are publicly available and non-profit making; better technologies that are under development; and disruptive technologies that will make what is being developed obsolete rapidly.

The second consideration is whether, even if what is developed is an advance scientifically, it is likely to be used. Users (in most cases, physicians) and patient preferences need to be considered. In general, physicians are often consulted (and, indeed, are often the initiators) but patients rarely so. This is particularly important when cultural differences need to be taken into consideration. For example, an automatic body washing device that encloses an elderly patient in a cocoon-like structure which sells in the United States proved unacceptable in Asia as the latter likened it to entering a coffin!

Researchers need also to consider whether their idea is protectable. In an ideal world this would be assured under patent or copyright protection but even this is not foolproof. It is well known that in some global territories patents are not always upheld. It is important to have a holistic patent strategy that not only allows "freedom to operate" but is also defensible commercially.

A consideration needs to be made as to whether a technology should be best licensed (or sold) to a commercial partner or spun out as a separate company. There are several factors that need to be taken into account to help make this decision. Expertise to do the spinout is probably the most important—many companies fail because a scientific founder without business experience appoints himself/herself as the CEO and does not realize quickly enough that he/she does not have the most important skillsets to run a company. If the technology is merely an add-on to something that is currently commercialized then it may be easier to license than attempt to go into direct competition. Sometimes the financials would suggest that the cost of development in a spinout outweighs the likely profits and for the individual investigators, a critical decision is whether they have the desire and resource available to get involved in a spinout.

4. COMMERCIALIZATION MODEL

As a general rule, the financial returns are in the order of selling the technology, **licensing** the technology to a third party, or spinout of a company. The relative risk of failure is in the opposite order! Selling or licensing would be preferable if there are major companies that dominate a market such that it would be difficult for a small player to enter it. Licensing has an advantage in that the technology may continue to be developed "in-house." There are two basic forms: "exclusive" and "non-exclusive." Exclusive licenses, which give the rights to the exclusion of all other parties may be narrowed to a geography or specific application. Non-exclusive licenses can be narrowed similarly. Most companies, naturally, prefer exclusive licenses as it reduces competitive risk.

Starting a **company** offers the most potential reward but requires dedication. Most scientists do not have the business acumen to make a success (technical superiority is only a minor factor in likelihood of success) so partnering with an experienced business person is highly recommended. The key skills that a scientist often lacks are financial acumen (particularly ability to raise money), business development, marketing, and business planning.

There are many sources of capital for a start-up:

- Government technology transfer: In advanced economies, often the easiest and least restrictive but generally small funding for a life science start up. In the United States, small business innovation research grants are attractive and have a good track record of return on investment. Other countries are following suit; in the UK a similarly named small business research initiative provides 100% funding.

- Venture capital: Traditional mode, but less than 10% of what they investigate gets an investment. Typically a few million dollars can be invested but a return on investment is sought in 3—5 years.
- Business angels: Generally, high net worth individuals who have a personal interest in a specific area. They often offer the business experience in addition to funding.
- Crowdfunding: A growing source of funding in life sciences. The sophistication tends to range from web-based advertising ("Kickstarter") to sophisticated equity investment instruments.
- Friends and family: A reality of life is that these are the folks who most believe in you.

Once funded, it is important for a start-up to then do all the things that a commercial licensee would also do. Successful commercialization depends on consideration on four important areas:

- IP management
- Product development
- Regulatory approval
- Marketing/sales development

These are equally important and should all be considered in parallel.

5. INTELLECTUAL PROPERTY

IP is an important part of commercializing biomedical research. It not only provides some commercial protection, in the early research it provides much of the commercial value. IP is often defined as "legally exclusive rights to creations of the mind." There are four main types of IP: patents, copyrights, trademarks, and trade secrets. In addition, in some parts of the world layout designs, registered designs, geographical indications, and plant varieties are considered as IP. In the United States, patents cover inventions and discoveries that satisfy the following criteria: (1) statutory, (2) novel, (3) useful, and (4) non-obvious. Most other patent systems, including the European Patent Office, have similar requirements.

A copyright is a legal right created by the law of a country, which grants the creator of an original work exclusive rights to its use and distribution, usually for a limited time, with the intention of enabling the creator to receive compensation for their intellectual effort. There are international agreements in place that often allow the copyright to protect IP in more than one country. Typically, the duration of copyright is the whole life of the creator plus 50—100 years from the creator's death, or a finite period for anonymous or corporate creations.

A **trademark** is a sign or symbol that distinguishes goods or services from those of others. There are different systems used globally but the

"International Classification of Good and Services (Nice agreement)" is commonly used. This sets out 34 different classes of goods and 11 classes of service. A trademark lasts indefinitely with a 10-year renewal period.

A **trade secret** is defined variously but has the essential elements of information that is not generally known to the public; confers some economic benefit, based specifically on it being a secret, to its holder; and is the subject of reasonable efforts to maintain its secrecy. There is no defined protection period and legal recourse tends to be based on "breach of confidence" from those with whom the secret has been shared.

A **patent** is a legal right granted to an inventor (or more specifically the agreed owner of an invention) to prevent others from making, using, importing, or selling that invention without the owner's permission. An invention can be a novel product or a process. It can be a completely new method, a technical improvement, or a composition of a new product ("composition of matter"). A granted patent has a term of 20 years from the first date of filing. Most importantly there are minimum legal agreements that cover much of the world as set out by the "Trade Related Aspects of IP Rights Standards" by the World Trade Organization (WTO). It should be noted that not all countries are signatories of these WTO minimal standards. There are also multinational patent systems: the European Patent Office covers over 33 states and the Gulf Cooperation Council and African Regional Intellectual Property Organization cover several countries in Middle East and Africa, respectively. This is the most relevant means of protecting IP for genome-based diagnostics and biomarkers.

The Patent Cooperation Treaty (PCT) provides an internationally recognized process for filing patents. It allows a single PCT application to serve as a priority application for later filing in other member countries after 30 months. For non-PCT members, the Paris Convention allows for patent application outside the initial state within a year.

Once a patent has been considered (even before application), it is important to protect it for which the inventors should consult an IP expert. In principle, the timing of what is published should be reviewed carefully (but not stopped) and the use of material transfer agreements, non-disclosure agreements, and invention disclosure forms should become routine. Universities have technology transfer offices that can help with these formalities. An even greater role for TTOs, with regard to IP, is consideration of the non-inconsequential costs of patenting.

The initial role of the scientist is the **technical disclosure** of the invention and an **invention summary** that a patent agent will need. This results in the "detailed description" which summarizes the background data and the claims that they support. This will be an **enabling disclosure** that allows others to make or use the invention. The patent agent will then do a search for the closest **prior art**. This results in a finalization of claims.

The initial impetus for patenting DNA technology came from the work of Boyer and Cohen that allowed scientists to isolate specific segments of DNA and transfer them into the DNA of other organisms. In 1972, Ananda Chakrabarty applied to the US Patent and Trademark Office (USPTO) for a number of patents on a genetically engineered bacterium including patenting the process and the organism itself. The USPTO granted the former but not the latter. The latter was refused on the basis that the organism itself was (1) a product of nature and (2) living. Although subsequently granted as it was argued that a genetically engineered organism does not occur in nature, this ruling had consequences for genome-based diagnostics. The *Association for Molecular Pathology v. Myriad Genetics*, 569 US 12-398 (2013) was a case challenging the validity of gene patents in the United States, specifically challenging certain claims in issued patents owned or controlled by Myriad Genetics that cover isolated DNA sequences, methods to diagnose propensity to cancer by looking for mutated DNA sequences, and methods to identify drugs using isolated DNA sequences. Prior to the case, the US Patent Office accepted patents on isolated DNA sequences as a composition of matter. Diagnostic claims were already under question through other cases such as *Mayo v. Prometheus* 566 US_2012. The conclusion of the Supreme Court was "A naturally occurring DNA segment is a product of nature and not patent eligible merely because it has been isolated, but cDNA is patent eligible because it is not naturally occurring." There will likely be further legal challenges but for most researchers who want to see the results of their endeavor having a smoother path to market, the lesson is not to base their tests on a natural gene.

The salient legislation in the United States for commercialization—which has the lion's share of patents in genome diagnostics—is the 1980 Bayh—Doyle Act which entitles small business, universities, research institutes, and non-profit organization to keep the IP even if the original research was government funded. It should be noted that it still entitles the government to use the invention for internal use. This has spurned academic institutions, in particular, to develop novel genome-based technologies.

The lack of legally binding IP is not, however, a complete barrier to commercialization as keeping "commercial know-how" secret can also allow others from copying it. In the diagnostics arena this is a difficult one to pull off, however, as users will want to have validation that they are using the right tests. And there is no legal protection.

6. PRODUCT DEVELOPMENT: DEVELOPING A BIOMARKER OR COMPANION DIAGNOSTIC

Once a potential biomarker is identified and a decision is made to commercialize it, proving the relevance of a molecular biomarker is usually more difficult than most researchers envisage. The Centers for Disease Control in the United States has developed the 44-point **ACCE model** process for developing genetic tests. They have identified four main criteria for evaluating such a test:

1. **Clinical validity** describes the accuracy with which a test predicts a particular clinical outcome. As a diagnostic, clinical validity measures the degree of association of the test with the disorder. Prognostically, it measures the probability that a positive test predicts the appearance of a disorder in a stated timeframe.
2. **Clinical utility** is the likelihood that using the test result will influence health outcome. False positivity and false negativity are considerations to be taken into account under this item.
3. **Analytical validity** describes how accurately and reliably the test measures the genotype of interest.
4. *Ethical, legal, and social implications (ELSI)* refer to the context in which it is used. Although not referred to specifically in the ACCE guidelines, cultural considerations for marketing tests in different geographies come under this category.

The European Medicines Agency provides a similar model under the ENCePP register of studies.

Product development is more than just the basic diagnostic test. Consideration should be given to scalability, particularly if it is being developed as part of a company. Most successful diagnostic companies have also built platforms for their test systems. So the format of the test needs to be considered to maximize robustness and reproducibility, sensitivity and specificity, and minimize false negatives and positives. Quality issues can often take as long as the initial development of the test!

In addition to regulatory approval, the manufacture of medical devices (including IVDs) is covered by International Organization for Standards (ISO) quality management regulations. Over and above the generic quality management standard 9001/2000, medical devices for commercial use usually have to comply with ISO 13485:2003 and ISO 14971:2000. The former covers all aspects of quality management systems that demonstrate design, manufacture, sales, and monitoring of a medical device. The latter is a standard for risk assessment.

In Europe, Conformite Europeene (CE) Mark certification is required for most IVD devices sold in Europe. CE Marking indicates that an IVD device complies with the European In Vitro Diagnostics Directive (98/79/EC), and that the device may be legally commercialized and distributed in the European Union (EU). As part of the CE Marking approval process, IVD manufacturers must compile a technical file showing compliance with 98/79/EC. The IVD CE technical file must include information such as device design, intended use, risk assessment, and route to conformity with 98/79/EC requirements. Once completed, it must be made available to European Competent Authorities upon request.

CE marking on a product:

- is a manufacturer's declaration that the product complies with the essential requirements of the relevant European health, safety, and environmental protection legislation, enshrined in many of the so-called Product Directives;
- indicates to governmental officials that the product may be legally placed on the market in their country;
- ensures the free movement of the product within the European Free Trade Association and EU single market (a total of 28 countries);
- permits the withdrawal of the nonconforming products by customs and enforcement/vigilance authorities.

7. REGULATORY CONSIDERATIONS

Regulatory processes are designed to ensure the safety, performance, and consistency of a product. Diagnostic tests, including companion diagnostics, are generally classified as IVDs. As already mentioned, the regulatory guidelines for genomic-based tests are in a state of evolution and discussion is likely to be needed with regulators. Which regulator, depends on several factors: country of origin, the market potential at entry, and the level of advice required. In general, the US FDA and EMA have the most extensive advice, network, and most lucrative markets for new technologies.

The US FDA has a wide remit for safety and effectiveness for medical devices through its Center for Device and Radiological Health. Their remit is to cover medical devices, where a medical device is defined as "an instrument, apparatus, implement, machine, contrivance, implant, in vitro reagent, or other similar or related article, including a component part, or accessory which is:

- recognized in the official National Formulary, or the United States Pharmacopoeia, or any supplement to them;
- intended for use in the diagnosis of disease or other conditions, or in the cure, mitigation, treatment, or prevention of disease, in man or other animals;

• intended to affect the structure or any function of the body of man or other animals, and which does not achieve any of its primary intended purposes through chemical action within or on the body of man or other animals and which is not dependent upon being metabolized for the achievement of any of its primary intended purposes."

US Federal law (Federal Food, Drug, and Cosmetic Act, section 513), established the risk-based device classification system for medical devices. Each device is assigned to one of three regulatory classes: Class I, Class II, or Class III, based on the level of control necessary to provide reasonable assurance of its safety and effectiveness. The classes are:

• Class I (low-to-moderate risk): general controls
• Class II (moderate-to-high risk): general controls and special controls
• Class III (high risk): general controls and premarket approval (PMA)

There are several pathways to introduce diagnostic devices to the market. The two main paths are a PMA, which can lead to approval of a diagnostic device, and premarket notification, which can lead to clearance of a device. The latter is widely known as a 510(k), named after the relevant section in the Federal Food, Drug and Cosmetic Act.

In the United States, Clinical Laboratory Improvement Amendments (CLIA) program dictates federal regulatory standards that apply to all clinical laboratory testing performed on humans in the United States, except clinical trials and basic research. Three federal agencies are responsible for CLIA: The FDA, Center for Medicaid Services (CMS), and the Center for Disease Control (CDC). Each agency has a unique role in assuring quality of laboratory tests. CMS has overall responsibility for the program. CDC covers numerous aspects including training and quality standards.

Outline schema for the regulatory process with the FDA in shown in Figure 2.

In Europe, medical devices regulations are in the process of change. The EU has published a new IVD Regulation which will replace the current Directive 98/79/EC on in-vitro diagnostic medical devices (IVDD) around 2016. The original IVDD regulation has been in place since 1998 and the new regulation is designed to take into account advances in the industry. The current IVDD Directive 98/79/EC includes a series of requirements that manufacturers and authorized representatives need to comply with to ensure the efficacy, quality and safety of their products. A third party (notified body) assesses this compliance in a

(a)

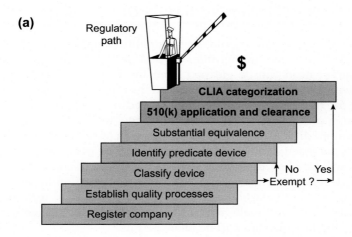

Regulatory
path

$

CLIA categorization

510(k) application and clearance

Substantial equivalence

Identify predicate device

Classify device   No   Yes
→ Exempt ? →

Establish quality processes

Register company

(b) FDA regulatory clearance class I/II diagnostic device (6–9 months)

- Register company as medical device manufacturer with FDA
- Establish quality process – design, packaging, labeling, and manufacturing
- Classify device-Class I exempt, Class I, or Class II for some tests. If exempt, apply directly for "CLIA categorization only"
- Identify predicate devices for application
- Establish substantial equivalence with approved tests
- Pre-market notification (510(k) submission); CLIA categorization request
- Post-marketing reporting

FIGURE 2   Outline schema for the regulatory process with the FDA.

process known as the "conformity assessment procedure." Manufacturers and authorized representatives can demonstrate compliance with the essential requirements through harmonized standards which are mandated by the European Commission to support the directive. Once the manufacturer confirms that their IVD meets the essential requirements, they can place a CE mark of conformity on the products. The new regulations will include a new risk classification system and increase both standards and harmonization across EU. Interestingly, "companion" diagnostics are now referenced in the new directive while they were not in the 1998 version.

A regulatory plan for a product needs to be aligned with product development and marketing strategy from the outset. Errors in the regulatory process have major consequences for the other aspects of product development as well as the support from those financing a project.

8. MARKETING AND SALES DEVELOPMENT

Product development should have identified unmet clinical need. This is not the same as unmet market need. Initial "ballpark" value of unmet market need can be gleaned from the many marketing reports that are available (mainly as paid publications but some information can be found for free). These estimates vary based on assumption and only give a rough estimate of the potential for a new test. The "addressable" market is more difficult to reckon as factors such as which current test(s) will be superseded, clinician acceptance, and reaction from current suppliers will be often unknowns. Just as important to consider is what else is in development that might be entering the market at or around the same time—particularly if any of these are "disruptive" technologies or products which will make what is being developed obsolete rapidly. Therefore, the key questions to be answered are:

- How does the test being developed stack up against the competition? If not better, then it will be difficult to gain market share.
- Who are the key competitors? If the competition is from major companies willing to defend market share then it is more difficult even with a better product.
- What is the total market potential? And how does that break down by market share?
- Are there key distributors? Are they likely to be interested in working with a small start-up?
- Who holds the competing or similar IP? Is it restrictive of competition?
- What other research is being done? Particularly "left of field" research.

Reimbursement strategy should be part of the marketing one. Even after reimbursement is agreed, procurement processes of major health care systems such as the United States and NHS in the UK can be complex and thought about early.

For start up companies, manufacturing and distribution are considerations to take into account. For most, these will involve partnerships with established contract manufacturers and partners who already have distribution networks in the territories sought.

9. CONCLUDING DISCUSSION

Genomic-based tests and biomarkers are being developed at an ever increasing pace as precision medicine becomes more routine. This will act as a catalyst for the entire diagnostics industry and probably for the pharmaceutical industry as many will also decide to develop their own companion diagnostics. Creating a successful commercial product is only minimally dependent on identifying the right technology. The business of science is multifactorial and success depends on identifying and getting these factors right. This chapter gives a broad overview but there are further details, particularly with respect to creating a successful start up that need to be addressed. A scientist wishing to do this would be strongly advised to seek support.

References

[1] Batelle Technology Research Partnership. Impact of genomics on the US economy. June 2013.
[2] Voora D, Ginsburg GS. Clinical applications of cardiovascular pharmacogenetics. J Am Coll Cardiol 2012;60(1):9–19.

Index

Note: Page numbers followed by "f" or "t" indicates figures and tables respectively.

Printed in the United States
By Bookmasters